T0262230

Climate Change: Human and Social Concerns

Climate Change: Human and Social Concerns

Edited by **Andrew Hyman**

New York

Published by Callisto Reference,
106 Park Avenue, Suite 200,
New York, NY 10016, USA
www.callistoreference.com

Climate Change: Human and Social Concerns
Edited by Andrew Hyman

© 2015 Callisto Reference

International Standard Book Number: 978-1-63239-114-8 (Hardback)

This book contains information obtained from authentic and highly regarded sources. Copyright for all individual chapters remain with the respective authors as indicated. A wide variety of references are listed. Permission and sources are indicated; for detailed attributions, please refer to the permissions page. Reasonable efforts have been made to publish reliable data and information, but the authors, editors and publisher cannot assume any responsibility for the validity of all materials or the consequences of their use.

The publisher's policy is to use permanent paper from mills that operate a sustainable forestry policy. Furthermore, the publisher ensures that the text paper and cover boards used have met acceptable environmental accreditation standards.

Trademark Notice: Registered trademark of products or corporate names are used only for explanation and identification without intent to infringe.

Printed in the United States of America.

Contents

Preface

Anomalous climatic effects like higher temperatures, intense rainfall and floods, constant and severe droughts have now reached a whole new level. Without proper adaptation methods, climate change is expected to worsen the vulnerability of society, pose serious risk to food security and human health, threaten the lives of growing urban population and obstruct in the aim of achieving sustainable development. The human and social dimensions of climate change, comprising of climate policy, play crucial roles in our response to several challenges arising from climate change. By emphasizing on a vast range of subjects and involving a number of reputed scholars, this book discusses the human and social dimensions of climate change and the topics usually overlooked or poorly understood by scholars and policymakers.

After months of intensive research and writing, this book is the end result of all who devoted their time and efforts in the initiation and progress of this book. It will surely be a source of reference in enhancing the required knowledge of the new developments in the area. During the course of developing this book, certain measures such as accuracy, authenticity and research focused analytical studies were given preference in order to produce a comprehensive book in the area of study.

This book would not have been possible without the efforts of the authors and the publisher. I extend my sincere thanks to them. Secondly, I express my gratitude to my family and well-wishers. And most importantly, I thank my students for constantly expressing their willingness and curiosity in enhancing their knowledge in the field, which encourages me to take up further research projects for the advancement of the area.

Editor

Climate Adaptation as Relational and Institutional Challenge

Community-Based Climate Change Adaptation for Building Local Resilience in the Himalayas

Pashupati Chaudhary, Keshab Thapa,
Krishna Lamsal, Puspa Raj Tiwari and Netra Chhetri

Additional information is available at the end of the chapter

1. Introduction

The Himalaya region has been experiencing the multitude of undesired change that cut across both biophysical and social realms. Observed biophysical changes include unpredictability in the timing and magnitude of rainfall, frequent occurrence of extreme heat during the summer season, glacial retreat and melting snow (Sharma et al. 2009; Gurung et al. 2010; MoE, 2010; Chaudhary et al. 2011). Temperature is rising over the past 100 years (Yao et al. 2006; Chapagain et al. 2009). These changes have already been posing serious threats on water, biodiversity, human health, agriculture, and consequently on food security throughout the region and downstream (Chaudhary and Bawa 2011). Vulnerable social and economic conditions pose further threat to the region. Recent social changes include rapid exodus of able-body manpower from the country, frequent economic crises, social and political unrest, and shrinking human capital. Since the region is the water tower of Asia and the lifeline for nearly one-fifth of world population, the current trend of climate change in the region will continue presenting an immense threat to humanity (Immerzeel et al. 2010). While any one of these factors will likely pose significant challenges on livelihoods of the people of the Himalaya region, the threat posed by changing climate and uncertainty associated with it cannot be ignored.

This paper (a) reviews an observed and perceived climate change in Nepal, (b) discusses impacts of climate change on ecosystems, biodiversity, agriculture and human wellbeing, and (c) offers, by drawing upon experiences of an NGO, Local Initiatives for Biodiversity, Research and Development (LI-BIRD), some community-based mitigation and adaptation techniques for curbing challenges resulting from climate change. The local level examples shared in this paper can be useful for other countries that share similar threats and socio-cultural and political challenges. People in other developing world can adopt the local technologies promoted by LI-BIRD in Nepal.

In the next section, we will provide background of Nepal, a Himalayan country that is sandwiched between two giants, India and China. Following that section, we present evidence of climate change, both observed and perceived, which is followed by impacts caused by climate change. Then we introduce an integrated climate adaptation model and some case studies drawn from LI-BIRD experience. Finally, we discuss the findings and provide some suggestions for the future.

2. Background

Nepal is a landlocked Himalayan country situated in between two economically booming giant countries, India and China. The country is characterized by geographical, ecological, and social diversity that give rise to wealth of biological diversity throughout the country. Not only eight of ten tallest mountains of the world are in Nepal, but there are more than 240 mountain peaks over 6000masl in the country. The country is facing tremendous pressure in its ecology and biodiversity resulting from social, economic and climate change, which is eventually affecting local livelihoods and wellbeing of people. People living in remote hills and mountains have limited ability to cope with changes and thus are under great threat.

Nepal has three physiographic zones (mountain, hill, and Terai) and five climatic zones (tropical, sub-tropical, temperate, sub-temperate, sub-arctic and arctic) across 800 meters length east-west and 200 meters width north-south and across elevation range of below 100 to 8848masl, the highest pick in the world. The mountains are experiencing rapid snow melting while the Terai is facing prolonged drought. Agriculture and other activities are adversely affected by these changes.

Both observed data and local perception reveals that climate change is no longer a future reality–it is already happening. As already mentioned, any departure from the expected "*normal*" climate poses serious threats to the livelihoods of the people of the Himalaya region (Chaudhary and Aryal 2009; Gurung et al. 2010; Sharma et al. 2009). In the last decade, the possible threat of changing climate change has received an overwhelming public attention (Maplecroft, 2011). However, we are hitherto short in understanding of climate change occurring in the nation and its impacts on ecosystems, biodiversity, agriculture and livelihoods, which have been impeding policy advocacy and real action on ground. There is a dearth of literature that describes the challenges and solutions of climate change (Chaudhary and Bawa 2011). Literature describing the roles of institutions in mitigating and adapting to climate change is also scanty (Chhetri et al. 2012). As a consequence, rhetoric has yet to be translated into action.

Even at global level, while information on observed and predicted change is amply proliferated and exchanged, adaptation techniques are poorly studied and exchanged at a wider level. Local people have been coping with changes using their own knowledge, resources, and skills (Chapagain et al. 2009) and such knowledge and skills can be taped to develop efficient adaptation plans. Adaptation techniques developed by Local Initiatives for Biodiversity, Research and Development (LI-BIRD), a national NGO in Nepal, using such local capitals and implemented with local participation have greater chance to become sustainable, low cost, and socially feasible (Thapa et al. 2012). LI-BIRD has been working with local communities to develop locally feasible adaptation measures with the participation of local knowledge, resources and skills.

The local level examples shared in this paper can be useful for other countries that share similar threats and socio-cultural and political challenges. People in other developing world can adopt the local technologies promoted by LI-BIRD in Nepal.

3. Evidence of climate change in Nepal

Nepal has been constantly experiencing changes in weather and climate throughout the country. Temperature data analyzed for the past clearly show the increasing trend of temperature over time, although the progression is not linear (Shrestha et al. 1999; Agrawala and Berg 2002; Shrestha and Devkota 2010), nor is the change uniformly distributed across the geographic regions (Shrestha and Bawa 2012). In Nepal temperature has increased at the rate of 0.03-0.06⁰C per annum between 1977 and 1994, with the higher altitude regions experiencing more rapid increase than the lower altitude regions (Shrestha et al. 1999; Shrestha and Bawa 2012). Another study showed a rate of 0.4⁰C per decade between 1981 and 1998 for the country (GoN 2004). In a district in Nepal (Gorkha), an average temperature for 2001-2006 was found 1.5⁰C greater than 1978-1982 (Lamichhane and Awasthi 2009). Other researchers have also found similar trend (Shrestha and Devkota 2010; Sharma et al. 2009). An analysis of data from Darjeeling hills of India also suggests that adjoining eastern hilly regions of Nepal are experiencing increase in average temperature, mainly during winter season (Chaudhary et al. 2011). Warming trend in the Tibet (Agrawala and Berg 2002) is also an indication of temperature rise in the mountains that are contiguous to Tibet. Not only average temperature, but several extreme temperatures (both high and low) have also been altering over time. For instance, in the past decade, both extreme temperatures were recorded the same day, which was not very common in the earlier decades (Rajbahak 2006).

Without any doubt, regional and seasonal variation of precipitation is also occurring (Gurung and Bhandari 2009). Shrestha (2000) observed an increase in average rainfall and precipitation over time. According to MoE (2010), pre-monsoon rainfall is decreasing and post-monsoon is increasing in some pockets of Western Nepal. Similarly, early monsoon and delayed withdrawal of monsoon has become normal (GoN 2004). However, it is paradoxical that precipitation has been increasing in the high rainfall regions and decreasing in low rainfall or drier regions (GoN 2004). For instance, precipitation has increased by 774mm in Lumle (high rainfall region) and decreased by 36mm in Mustang (low rainfall region) over the past four decades (DoHM 2007). Nepalgunj, one of driest cities, also sustained a wettest monsoon season in the last 123 years in 2006 (Sharma 2006).

Extreme events are also increasingly replacing normal monsoon seasons (Baidya et al. 2008); short-duration heavy downpours are more frequent than ever. In Laprak Gorkha, a total rainfall was recorded 341.8mm between 4pm to 7am (15 hours) in July 3, 1999. In Nepane

Kaski, 128mm rainfall occurred between 11pm to 12:30am (1.3hours) in July 15, 2006. Similarly, in Nepalgunj, Banke, 336.9mm rainfall was recorded in 24 hours in Aug 27, 2006 (SOHAM 2006; Gurung 2006).

Similarly, increase in average temperature of Rasuwa, Dhading, Banke and Bardiya districts was also noted between 1977 and 2007. The degree of increase was higher in Rasuwa (0.03⁰C), a high mountain area, followed by Dhading (0.02⁰C), a middle hill area, and Bake and Bardiya (0.01⁰C), the Terai area (CARE/LI-BIRD, 2009). The study also found out a gradual increase in maximum and minimum temperature in the Terai region and an abrupt change and erratic precipitation trends were observed in the high mountain areas.

Alteration of climate change is Corroborating with the observed changes is the perception of the people of the Himalaya region (Chaudhary and Bawa 2011). People have perceived change in their climate and are based on their day-to-day observations of weather change patterns and impacts resulting from it (Chaudhary and Aryal 2009). They also associate various effects occurring at local level with the change in weather and climate based on their experiences in planning agricultural activities according to weather patterns. From local perceptions, it is obvious that temperature is rising, rainfall patterns have become erratic and unpredictable, and snow is melting faster than before in the Himalayas (Sharma et al. 2009; Chaudhary and Bawa 2011; Chaudhary et al. 2011). A study done in the Siddhi village of Central Nepal also reported the farmers' perception in the increase of temperature as well as decrease in rainfall (Shrestha and Shrestha, 2010). Another perception study on the trend of climatic hazards done among 486 shifting cultivators of eastern and central Nepal found that 45% and 55% households have experienced heavy but short-duration rainfall and prolonged drought, respectively (Thapa et al. 2012). Studies done in the western Nepal also report similar trend of temperature and rainfall (Gurung et al. 2010).

People also experience hotter summer, shorter and less intense winter, intense but short downpours, less cold days, and reduction of frost and fog (Chapagain et al. 2009; Gurung et al. 2010; Chaudhary and Bawa 2011); the proportion of people observing these changes was more for high hills than the low hills (Chaudhary and Bawa 2011). Thapa (2012) suggests that there is a delay in onset of monsoon season (shift from June-August to July-September), which is in contrast to Chaudhary and Bawa (2011) findings from Darjeeling (India) and Ilam (Nepal) hills. As already mentioned, a report by the Government of Nepal suggests an early onset and late withdrawal of monsoon. The differences in trends reported in various literatures might be the function of spatial variation as sites considered by different studies differ. People also associate several local level impacts with weather and climate change. Some of impacts are discussed below.

4. Impacts of climate change

Above-mentioned changes have ensued the following impacts: drying up of water table, early flowering and budburst in some species; adaptation of natural vegetation, cultivated crops, weeds, crop pests, and mosquitoes to higher altitude regions; and early crop maturity leading to yield loss (Chaudhary and Bawa 2011; Chaudhary et al. 2011). The impacts are severe in several regions of the country.

As many as 70% people in a part of Nepal and Indian Himalaya believe water sources are drying up, which could have direct impacts on agriculture, wilder biodiversity, ecosystem health, and daily water use in already water-scarce hills and mountains. Erratic and intense but short-duration rainfall has been increasing the threats of landslide (MoE 2010), because short duration rain doesn't recharge or saturate soil as effectively as prolonged rain does. Impacts are also seen in watershed; for instance, lake ecosystems are greatly affected by increased siltation, decreased agricultural workforce, and declined productivity due to prolonged dry spells (Thapa et al. 2012). People may experience yield loss in major staple crops. Change in floral composition is also inevitable due to change in water regime on ground. On the other hand, several life-threatening GLOF are not far from reality (Ives 1986; Bajracharya et al. 2007). Scientists have alerted that about 25 glacial lakes in Nepal are prone to outburst, which may cause massive damage on forests, agricultural lands, physical properties, and human lives as history shows. Scientists have linked this catastrophe with increased snow melting and glacial retreat resulting from increased temperature (Bajracharya et al. 2007).

Using a remote sensing tool, Shrestha and Bawa (forthcoming) have observed a shift in flowering time in the Himalayan flora, which corroborates to local perceptions. Nearly 50% people among 250 the Darjeeling (India) and Ilam (Nepal) Himalayas have experienced an early flowering in selected species, which include rhododendron, magnolia, peach, pear, and marigold. The study found out food shortage, crop failure, loss of livestock, and water scarcity as the major shocks to the shifting cultivators, which were highly influenced by the climatic hazards and their variability (Thapa et al, 2012).

Temperature rise has also benefited the high altitude regions. An increase in temperature has resulted in the shifting of climatic suitability of crops in the Mustang (a mountainous) district of Nepal. In the case of apple, there is a clear notice of unsuitability of apple cultivation in lower elevation unlike in past years but apple cultivation is expanded in the higher altitudes (Pradhan et al, 2012). People now can also grow cabbage, cauliflower, tomato, chilly, mango and other tropical species that were not possible about a decade ago due to cold, frost, snow, and intense cold.

In Nepal's high mountain region, change in rainfall and temperature resulted in early flowering and vegetation shift. Similarly, frequent flooding in the mid and far western Terai washed away productive lands, reduction in crop yield and damage to infrastructure, livestock and human settlement; whereas the prolonged drought, drying of water sources and outbreak of pests have been threatening crop production in the western and central Nepal (CARE/LI-BIRD, 2009).

5. Local solutions: Experiences from LI-BIRD

Solutions that are locally developed by using indigenous knowledge, local resources, and participatory methods are socially acceptable and viable in long run. LI-BIRD has been promoting several such local and participatory initiatives in rural areas of Nepal. Framework of action LI-BIRD follows and some successful case studies are presented below.

5.1. Climate integration framework

Addressing the challenges of climate change needs interventions both in the forms of mitigation and adaptation. As mitigation is a long-term process and costly in many cases, adaptation is a better choice to respond to ongoing and immediate threats. To make adaptation effective, LI-BIRD employs the following four-tier approach: (a) vulnerability assessment, (b) understanding of local knowledge in responding to climate change risk, (c) blending local and scientific knowledge, if necessary, and (d) implementing or tailoring new knowledge. For implementing new knowledge, LI-BIRD develops an appropriate policies and implements them both at national and local levels; builds local and national capacities to implement policy; generates internal and external support to provide appropriate support to local communities; and strengthens networks and partnership to improve public awareness, information sharing, community participation, lobbying, and policy advocacy. The process discussed above is depicted in figure 1.

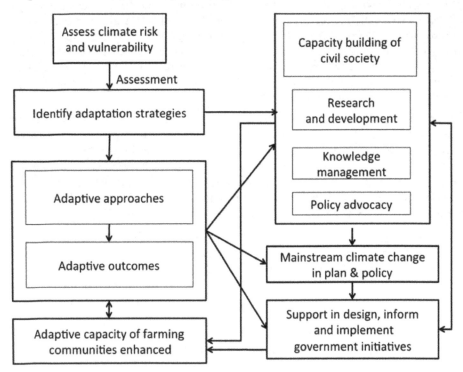

Figure 1. Climate change mainstreaming framework of LI-BIRD

LI-BIRD factors in or integrates climate change into its work. As such LI-BIRD has made its working approaches climate sensitive or has embraced climate adaptive approaches. By implementing these climate adaptive approaches in participation with local stakeholders, we have been able to generate climate adaptive outcomes, which contribute to improving the adaptive capacity of farming communities. Furthermore, LI-BIRD is using its experiences in adaptive approaches and outcomes to build institutional capacity of the Civil Society, who also need to integrate climate change. It is done through the following steps:

1. Assess climate risk and vulnerability through participatory tools and methodologies.
2. Identify potential adaptation strategies and approaches (based on community responses, LI-BIRD's own experiences of research and development, and scientific recommendations) that address the climate risks and vulnerabilities, documenting local knowledge and practices with focus on climate change adaptation.
3. Implement the adaptive approaches and strategies in the communities.
4. Build capacity of grass root organizations working in agriculture, food security, livelihood improvement, natural resources management and environment.
5. Carry out participatory research and development in collaboration with the grass root organizations and research organization to refine and develop adaptive technologies and options that are suitable to the local context.
6. Document and disseminate the field realities of climate change and climate change related knowledge and technologies generated to a wider stakeholders (communities, local government, national government, policy makers, donors, scientists and academicians, NGOs).
7. Carry out policy advocacy to mainstream climate change responses in the national development plans and policies.
8. Support national and local government and non-government organizations to design and implement climate change responsive policies, programs, and plans.

5.2. Case studies

There are several case studies that can be drawn from LI-BIRD projects to demonstrate how local initiatives and innovations can be effective to mitigate and adapt to climate change. Most noteworthy out of those case studies are presented below.

5.2.1. Community based biodiversity management (CBM)

Community based biodiversity management is a participatory approach of managing plant and animal genetic resources including the wild through conservation, utilization and value addition of the genetic resources. This approach is implemented in the community through documentation and assessment of biodiversity of the community through diversity fairs and community biodiversity register (CBR). During this process, genetic resources for conservation and utilization are identified which are then conserved and promoted through a number of tools and methodologies[1], and is helpful in assessing and identifying the

---

[1] http://www.cbmsouthasia.net/components-of-the-cbm-approach/

threatened crop species as well as the crops and their varieties having climate stress tolerating traits. With such tools and methodologies, CBM approach has enriched local communities with increased access to market and financial institutions, technology(ies) and practices to respond drought and flood, strengthened social institutions and cohesion thereby enhancing integrity among their livelihood resources and building resiliency of farmers in rural communities.

CBM approach has contributed to conserving and utilizing the genetic resources and associated knowledge, increasing access of genetic resources through exchange and diversifying income sources. The approach has promoted community seed banks, registration of agro-biodiversity, value addition of farm products through breeding (participatory plant breeding) and non-breeding techniques (value addition and market linkage), identification of crops/varieties that thrive best in harsh climate, establishment of group fund, and strengthening of local social institutions. Such methodologies to conserving and utilizing genetic resources have contributed to increasing adaptive capacity of poor and marginalized farmers by:

i.   Increased farm income through value addition, marketing of local products (entrepreneurship development), and reduction of external inputs
ii.   Increased agriculture production through integrated soil and nutrient management
iii.   Provided varietal options to the farmers through conservation and utilization of diverse genetic resources
iv.   Enhanced poor and marginalized people and women's access to healthy and a variety of foods, saving and credit facility (CBM fund), social institutions such as cooperatives
v.   Conservation and utilization of local genetic resources and the associated traditional knowledge that have potential to contribute to climate change adaptation
vi.   Empowering poor and disadvantaged women for value addition and conservation of genetic resources and their participation in farmers' institutions
vii.   Developed coherence and social harmony in the communities through gender and social inclusion of disadvantaged groups in farmers' institutions and networks

---

Case Study I

Pratigya Cooperative in the Rupa Lake of Kaski Nepal was established in 1997 with 43 share members. In 2010, it has 78 members with 38 female members. The cooperative is instrumental in conserving local drought tolerant and underutilized crops, maintaining a diversity of crops in their field, and supporting the poor and marginal farmers to sustain their livelihoods have enhanced their capacity to deal with those climate hazards.

Value addition of local crops has enhanced their access to market and information. Their major focus is on local landrace of sticky rice, taro, finger millet, and ginger marketing. The cultivar of taro conserved and promoted as well as practices to manage marginal lands along with crop diversity in their field and seed bank are also enhancing communities capacity to respond to drought there by reducing its risk. The cooperative members at current have promoted marketing of sticky rice, local landrace of sticky rice, after the members realized the market potential for its medicinal, cultural, and traditional value; and marketing of taro products through value addition. They have been selling products from all parts of the plant (root, pseudo stem and leaves) of taro and sticky rice through cooperatives. They sold taro products of USD 200 in 1999, whereas they earned USD 450 by selling the products in 2009. In addition, the members of the cooperative are maintaining field gene banks of different taro species and conserving the medicinal and aromatic plants.

Inclusion of indigenous *dalit* community members in the cooperative and supporting them with income generating activities are also enhancing their capacity through increased income and well-being among the community that is very important to deal with climate change uncertainties. Seed banks are established as an effective means of crop insurance at the community level. Individual member of the cooperative are assigned to maintain the seed production and conservation of those threatened crop varieties to maintain the important seeds.

5.2.2. Home garden diversification

Diversification of home gardens and its efficient management for women, disadvantaged groups including indigenous communities (*janajatis*) and *dalits* is adaptive approach at household level to address and respond to impacts like drought and floods. This adaptive approach has significantly contributed to improve nutritional status and reduce economic vulnerability of poor and disadvantaged communities in Nepal. The various components, and their efficient management, of home gardens such as kitchen waste water management, cultivation of neglected and underutilized crop species, integration of small farm animals and fish, cultivation of vegetables, integration of fruits and fodder, and market linkage have contributed to increase their adaptive capacity and make home gardens resilient to climate stress by

i.   Diversified family nutrition and increased nutrition sources
ii.   Increased household income and income generating options (such fruits, fish, vegetables and livestock)
iii.   Increased homestead biodiversity
iv.   Optimized utilization of available resources through integrated management
v.   Sustainably managed social seed system l
vi.   Increased institutional capacity for decision making, planning, resource utilization and benefit sharing
vii.   Increased access to service provider organizations
viii. Establishment of group funds and access of resource poor and disadvantaged groups to group fund mobilization

---

Case Study II

Poor and disadvantaged communities of some parts of Kanchanpur, Kailali, Bardiya, Rupandehi, Gulmi, Ramechhap, Dolakha, Sindhupalchowk, Jhapa and Ilam have got benefits from home garden diversification. The farmers of Dudharakshya village of Rupandehi district have improved their nutritional status through vegetable consumption and selling in the nearby market. Livestock integration has become an important livelihood asset in the home gardens, which can be an important source of income for most of the poor and disadvantaged farmers. Some of the specific cases are (Pudasaini, 2009):

- The intervention has increased species diversity in most of the home gardens (n=690). After the intervention, households growing 26-50 species in their home gardens increased from 32% to 72%.
- Adoption of better management practices in home gardens has increased the product marketing of the households (from 15% to around 40% households). Almost all the people (n=690) who used to buy vegetables from market have increased their self-sufficiency and reduced the amount of money going out of the family for daily used vegetables.
- Number of families with 12 month vegetable sufficiency from the family home garden increased from 13% to 56%.

5.2.3. Development and promotion of stress tolerant crop varieties through PPB and PVS

Participatory variety selection (PVS), participatory plant breeding (PPB) and client oriented breeding (COB) are participatory research and development activities that help in development and promotion of stress resistant crop varieties. Participatory approaches of variety selection and breeding adopted by LI-BIRD are need based and demand driven approaches. These approaches focus on traits preferred by farmers in which farmers' preferences are determined by socioeconomic situation, land type, availability of varietal options and environmental conditions (climatic factors-niche specificity like rainfall pattern, temperature and humidity). By utilizing these adaptive research and development approaches, LI-BIRD has been developing and testing cereal crop and legume varieties that tolerate environmental extremities such as drought and flood. The outcome of these approaches is increased productivity of agro-ecosystem with changed cropping pattern, increased crop diversity and improved soil health. This approach has also contributed to enhance the adaptive capacity of farming communities to the impacts of climate change by:

i. Increasing use of stress tolerant and high yield crop varieties to improve farm income by developing high yielding and farmers' preferred crop varieties

ii. Promoting local business through local variety and employment opportunities at local level through community based seed production (CBSP) groups

iii. Providing technological option (varieties) and management option (agronomic interventions such as legume integration in farming system) to deal with climate stress, improve soil health and system productivity

iv. Increasing the access of communities to seeds (technology) and financial institutions (through CBSP groups) and market (through community based seed production groups)

v. Establishing strong social seed networking (seed and knowledge exchange) and strong social institutions (CBSP)

vi. Building capacity of communities for seed production and variety selection including flood and drought tolerant species

vii. Contributing to policy and regulation for enhancing access of poor and marginal farmers to technology and lobbying for equal benefit sharing mechanism

---

Case Study III

Mansara is a landrace grown in hills especially Western Mid hills of Nepal and it is highly adapted to poor soil and low-input production system. It has very low productivity, poor eating and cooking quality due to which it fetches low market price but it is adapted to marginal environments it is grown by resource poor rice farmers who do not possess varietal options. Thus, using participatory plant breeding (PPB) approach the quality and yield of the *Mansara* landrace has been improved by crossing it with a quality modern variety *Khumal 4*. Through PPB the better yield potential and good eating and cooking quality traits from modern variety has been transferred without losing adaptive traits of *Mansara* landrace. Participation of farmers to select the type of *Mansara* variety they wanted was highly valued during the breeding process. Now improved *Mansara* is superior to the original *Mansara* in terms of eating and cooking quality along with better productivity yet still adapted to the marginal growing environment, thus providing options to farmers in marginal rice growing environments.

Sugandha 1 is an aromatic rice variety with medium maturity developed by LI-BIRD using client oriented breeding (COB). The variety has a unique combination of high yield with aroma. It is highly suitable for very marginal, low input, rain-fed rice growing environments. Most of the aromatic rice varieties are prevalent in irrigated medium to lowland conditions. But *Sugadha 1* is unique as it thrives best in stress prone environments. This again demonstrates how PPB provides varietal options to farmers in marginal environments.

Judi 582, Barkhe 1027, and *Barkhe 1036* are other rice varieties developed using COB and are suitable for drought prone and rain-fed environments. Development of these short duration varieties has supported the farmers for growing winter cereals, legumes and vegetables making their farming system more resilient. Similarly, *Barkhe 3004* and *Barkhe 3019* are rice varieties developed using COB suitable for lowland rice growing environments and they have some tolerance to flooded conditions.

5.2.4. *Payment of ecosystem services*

It is a right based and ecosystem based mechanism for the sustainable management of ecosystems and natural resources, ultimately contributing in building resilient ecosystem. It is a market driven approach to natural resource management by involving the buyers and sellers in the utilization of ecosystem services. LI-BIRD in partnership with IUCN has established a reciprocal benefit sharing mechanism in the management of natural resources in the watershed between the upstream and downstream communities to demonstrate payment of ecosystem services in Rupa Lake Watershed, Kaski (Regmi et al, 2009; Pradhan et al, 2010). The outcome of this approach is rehabilitation of degraded watershed and management of associated biodiversity. Such practice offers adaptation led mitigation opportunity from a climate change perspective and enhances the adaptive capacity of watershed communities and resilience of watershed at landscape level by

i.   Providing option for reducing siltation and conserving natural resources
ii.   Enhancing ecological integrity of the catchment
iii. Increasing income of sellers through diversified livelihood options (fishery, value addition and marketing, ecotourism)
iv.   Conserving biodiversity (white lotus, wild rice, water birds, *Sahar* fish and NTFPs)
v.   Promoting traditional knowledge on rehabilitation of degraded catchments
vi.   Building capacity of watershed communities for watershed management
vii. Increase in social cohesion and harmony

Case Study IV

A reciprocal benefit sharing mechanism between upstream and downstream communities is established effectively for ensuring stakeholders substantial rights to environment in Rupa Lake watershed in Nepal. The voluntary payment mechanism in the management of Rupa Watershed by downstream communities through Rupa Lake Rehabilitation and Fishery Cooperative to the upstream communities is the only documented case of payment of watershed services in Nepal.

Every year the cooperative shares some of its income in these watershed management practices to the upstream community members through their institutions. It has also mobilized mother groups to conserve the biodiversity of lake such as wild rice, birds, indigenous fishes, and white lotus. The cooperative annually pays 25% of the total income of the cooperative from fish selling to the upstream communities (19 schools, 52 students and 17 community forest user groups).

In addition, the inclusion of disadvantaged community members in the cooperatives and supporting socially marginalized community members by income generating activities like goat raring, etc. has enhanced the socioeconomic status of these community members. More importantly, it has enabled those members to send their children to the school. Social cohesion, market promotion of lake products, diversification of livelihood options, inclusion of socially marginalized community members in the cooperatives, coordination with upstream community members to reduce siltation in the lake and conservation of biodiversity in the watershed are the outcomes of this payment of ecosystem services mechanism.

These approaches helped to identify how multiple (social, economic and cultural) co-benefits can be taken into account for establishing well functional payment for ecosystem services and promoting ecosystem based adaptive mechanism on watershed management. Hence, the role of ecosystem services in reducing the climate vulnerability through multi-sectoral and multi-level approaches has been effective in increasing access to the poor and marginalized communities in their livelihood resources.

5.2.5. Capacity building through public awareness and network

LI-BIRD has established a mechanism of raising public awareness on climate change through publication of research findings, field cases and information on climate change through mass media. Capacity building of mass media and determining the role that the media can play in order to improve information sharing, both from science to local level and from local level to policy makers, are urgent matters (Lamsal, 2011). In the country like Nepal, poor dissemination of information to the marginalized and climate vulnerable communities and their access to such information is very limited, which has also limited their capacity to adapt to the impact of climate change. In this context, the approach of utilizing mass media especially local FM radio network for climate change communication is one of the strategies to increase access of marginalized, poor and climate vulnerable communities to information and then the adaptive technologies and options.

The NGO Network on Climate Change (NGONCC) network established and facilitated by LI-BIRD is another adaptive mechanism to build the capacity of civil society organizations of Nepal. The ultimate goal of this network is to enhance the adaptive capacity of those poor and climate vulnerable communities through implementing adaptation strategies and policy advocacy at local, district and national level. Through this network, more than 120 NGOs get information on climate change issues and are regularly getting updated on the various adaptation and mitigation issues related to climate change.

Case Study V

Information sharing and capacity building is an important part of enhancing adaptive capacity in response to climate change. The members of NGO Network on Climate Change in Nepal have enhanced their capacity through a regular sharing of information related to climate change and through involving in various capacity building programmes related to climate change. The network in Nepal was initiated in 2007 by LI-BIRD. Initially, there were 12 NGOs from Kathmandu, Pokhara, Dhading and

---

Chitwan. Realizing the importance of human capital for dealing with climate change issues, the network has now expanded to all development regions of Nepal. There are 120 NGOs from all development regions of Nepal covering 34 districts in the NGONCC.

Some of the NGOs in the network have mainstreamed climate change issues in their programs and projects to enhance community awareness and support adaptation to climate change. LI-BIRD has been promoting awareness raising, capacity building, action researches, and community based adaptation interventions in Nepal. Youth Acting for Change (YAC), Dhangadi, Resource Identification and Management Society (RIMS), Namsaling Community Development Center (NCDC), Ilam, Dalit Welfare Organization (DWO), Banke, Environment Camps for Conservation Awareness (ECCA) and LI-BIRD host climate change information and resource center at ground level for effective information transformation and dissemination.

More importantly, the network has contributed to the preparation of National Adaptation Program of Actions (NAPA) and piloting of Local Adaptation Plans of Action (LAPA) document by providing ground realities on climate change through transect appraisal exercises and technical inputs.

6. Conclusions

Climate change research is nascent in the developing countries that bear the major portion of consequences resulting from climate vagaries. Hence, more research effort is needed to aptly understand real problems, and driving forces of such problems, and devise efficient adaptation and mitigation measures. This should be achieved by employing participatory approach as developing countries lack infrastructure to record quantitative data to measure change and make a precise prediction. While bottom up approach to identifying locally feasible adaptation tools will be a first step to generate knowledge, investment should be made to introduce advanced technologies in order to generate quantitative information so that we can easily discern change, make prediction for future, and accordingly devise national strategies with more confidence. Longitudinal research is necessary if a country can afford.

Many local adaptation techniques might be used as such while several others can be combined with scientific knowledge generated using modern tools and techniques, to make is more robust, reliable, replicable, relevant, remedial, resilient and resource conserving in nature. Several of such knowledge thus can be adopted at cross-country levels but with possible modifications to tailor to local needs and capacities.

It is also important to raise public awareness at all levels as national and local level planners, policy makers, implementers and victims of climate change have little knowledge about change process, driving factors and remedies. Even capacities of scientists and climate advocates need to be strengthened through latest scientific inventions as they lack access to information due to costly journal fees and unreliable access of internet. Exchange of knowledge is also not efficiently done among stakeholders within country. It is important to share national, regional, and global policies, treaties, legislations and strategies with all national and local partners through their proper networks and help them tailor priorities and allocate resources accordingly. This will help them draw more resources from international and global community. More specifically, the global financial commitment on adaption in the least developed countries should be increased and committed funds (e.g. Least Development Country fund, special climate change fund, adaptation fund, climate investment fund, and green climate fund) transferred timely and appropriately through proper channels. Furthermore, the climate vulnerable countries must prioritize implementation of their adaptation programs and plans (e.g. NAPAs) to build community resilience to climate change, making sure the funds disbursed are properly distributed among target beneficiaries.

To achieve aforementioned goals, we not only require research and development fund — both public and private—but also existence of several organizations like LI-BIRD to generate, translate, and disseminate knowledge, build community capacity in adaptation, and strengthen NGO networks to expand collaboration, scale out good practices, and foster policy advocacy. There is a lot to learn from LI-RBID approach and practices, which are viable, robust, need-based, and thus responsive to real challenges. It is also important to properly integrate LI-BIRD framework with NAPA and LAPA, where the countries have privilege of mainstreaming these strategies into climate change and development plans of the country.

Author details

Pashupati Chaudhary, Keshab Thapa, Krishna Lamsal and Puspa Raj Tiwari
Local Initiatives for Biodiversity, Research and Development (LI-BIRD), Pokhara, Kaski, Nepal

Netra Chhetri
School of Geographical Sciences and Urban Planning the Consortium for Science, Policy and Outcomes, Arizona State University, USA

7. References

Agrawala S. and M. Berg 2002. Development and climate change project: concept paper on scope and criteria for case study section, COM/ENV/EPOC/DCD/DAC/Final, Paris OECD.

Baidya S. K., M. L. Shrestha and M. M. Sheikh 2008. Trends in daily climatic extremes of temperature and precipitation in Nepal.

Bajracharya, B., Shrestha, A.B. & Rajbhandari, L. 2007. Glacial Lake Outburst Floods in the Sagarmatha Region. Hazard Assessment Using GIS and Hydrodynamic Modeling. *Mountain Research and Development*, 27:336–344.

CARE Nepal/LI-BIRD. 2009. Climate Change Impacts on Livelihoods of Poor and Vulnerable Communities and Biodiversity Conservation : A Case Study in Banke, Bardia, Dhading and Rasuwa Districts of Nepal. CARE Nepal, Kathmandu, Nepal.

Chapagain B. K. R. Subedi, and N. S. Paudel 2009. Exploring local knowledge of climate change: some reflections. *Journal of Forest and Livelihood*, 8(1):106-110.

Chaudhary P. and K. Aryal 2009. Global warming in Nepal: challenges and policy imperatives. *Journal of Forest and Livelihood*, 8: 4-13.

Chaudhary P. and K. S. Bawa 2011. Local perceptions of climate change validated by scientific evidence in the Himalayas. *Biol. Lett.*, published online, 27 April 2011; doi: 10.1098/rsbl.2011.0269.

Chaudhary P., S. Rai, S. Wangdi, A. Mao, N. Rehman, S. Chettri and K. S. Bawa 2011. Consistency of local perceptions of climate change in the Kangchenjunga Himalaya landscape. *Current Science*, 101: 504-513.

Chhetri N., Pa. Chaudhary, P. R. Tiwari and Ram Baran Yadaw 2012. Institutional and technological innovation: Understanding agricultural adaptation to climate change in Nepal. Applied Geography 33: 142-150.

DoHM 2007. Meteorological Data of Pokhara Valley. Pokhara: Department of Hydrology and Meteorology.

Government of Nepal (GoN) 2004. Initial National Communication to the COP of UNFCCC. Kathmandu: Ministry of Population and Environment, Government of Nepal.

Gurung N. 2006. Cause of Laprak landslide in Gorkha district and Nepane landslide in Kaski district of Nepal and their remedial measures. In Proceedings of Geo-disaster, infrastructure management and protection of World Heritage Sites. Novermber 2006. Nepal Engineering College, Changunarayan, Bhaktapur, Nepal (pp:25-26).

Gurung G. B. and D. Bhandari 2009. Integrated approach to climate change adaptation. *Journal of Forest and Livelihood*, 8(1):90-98.

Gurung G. B., D. Pradhananga, D., Karmacharya, J., Subedi, A. K. Gurung, and S. Shrestha 2010. Impact of climate change – voices of people: Based on field observations, information and interactions with the communities in Nepal. Practical Action, Kathmandu, Nepal, 2010.

Government of Nepal (GoN) 2004. Initial national communication to the conference of the parties of the United Nations Framework Convention on Climate Change. July 2004. Ministry of Population and Environment. Government of Nepal.

Immerzeel W.W., van Beek, L. P. H. & Bierkens, M. F. P. 2010 Climate change will affect the Asian Towers. Science, 328, 1382–1385.

IPCC. 2007 Summary for Policymakers. In Climate Change 2007: Climate Change Impacts, Adaptation and Vulnerability. Working Group II Contribution to the Intergovernmental Panel on Climate Change Fourth Assessment Report. See http://www.ipcc.ch (accessed 12 April 2007).

Ives J. D. 1986. Glacial Lake Outburst Floods and Risk Engineering in the Himalaya, ICIMOD, Kathmandu, p. 42.

Lamichhane B. R. and K. D. Awasthi 2009. Changing climate in a mountain sub-watershed in Nepal. *Journal of Forest and Livelihood*, 8(1):99-105.

Lamsal, K. (2011). A perspective on communicating climate change, NGO Network on Climate Change Bulletin, Mainstreaming for sustainable livelihoods, 4th issue, Local Initiatives for Biodiversity, Research and Development, Pokhara.

Maplecroft 2011. Climate change vulnerability index 2011. http://www.washingtonpost.com/wp-srv/nation/green/pdfs/ClimateChangeVulnerabilityIndex2011.pdf [retrieved on 31 March 2012]

MOE. 2010. National Adaptation Programme of Action (NAPA) to Climate Change. Ministry of Environment, Government of Nepal. http://www.napanepal.gov.np/pdf_reports/NAPA_Report.pdf

Mool, P.K., Bajracharya, S.R., & Joshi, S.P. 2001. Inventory of Glaciers, Glacial Lakes and Glacial Lake Outburst Flood Monitoring and Early Warning System in the Hindu Kush-Himalayan Region, ICIMOD: Nepal. [pp 364-365].

Pradhan, NS; Khadgi, VR; Schipper, L; Kaur, N; Geoghegan, T (2012) Role of Policy and Institutions in Local Adaptation to Climate Change – Case studies on responses to too much and too little water in the Hindu Kush Himalayas. International Center for Integrated Mountain Development.

Pudasaini, R. 2009. Wrap up survey report of enhancing family nutrition and income for improved livelihoods of resource poor and disadvantaged groups through integrated home gardens in Nepal: home garden project, phase II (2006-2008). Local Initiatives for Biodiversity, Research and Development (LI-BIRD), Pokhara, Kaski, Nepal.

Rajbahak M. 2006. Weak rainfall activity in 2005 over Nepal. Disaster Review 2005. Series XIII, Department of Water Induced Prevention, GoN. (pp:26-28).

Regmi, B.R., G. Kafle, A. Adhikari, A. Subedi, R. Suwal, and I. Poudel. 2009. Towards an innovative approach to integrated wetland management in Rupa Lake Area of Nepal. Journal of Geography and Regional Planning Vol. 2(4), pp. 080-085, April, 2009. http://www.academicjournals.org/jgrp/PDF/PDF%202009/Apr/Regmi%20et%20al.pdf [retrieved on 24 Jan 2011].

Sharma K. 2006. Hydrologic extremities of South-West Nepal in 2006. In Proceedings of International Symposium on Geo-disaster, infrastructure management and protection of World Heritage Sites. Novermber 2006. Nepal Engineering College, Changunarayan, Bhaktapur, Nepal (pp: 224-230).

Sharma E., N. Chettri, K. Tse-ring, A. B, Shrestha, F. Jing, P. Mool and M. Eriksson 2009. Climate change impacts and vulnerability in the Eastern Himalayas, ICIMOD, Kathmandu.

Shrestha A.B., Wake, C.P., Mayewski, P.A. & Dibb, J.E. 1999. Maximum Temperature Trends in the Himalaya and its Vicinity: An Analysis Based on Temperature Records from Nepal for the Period 1971–94. Journal of Climate, 12: 2775–2786.

Shrestha, A.B., Wake, C.P. & Dibb, J.E. 2000. Precipitation Fluctuations in the Himalaya and its Vicinity: An Analysis Based on Temperature Records from Nepal. International Journal of Climate, 20: 317–327.

Shrestha, S. and A. Shrestha. 2010. Gender perspective: integrating energy resource use into climate change adaptation. A research report submitted to National Adaptation

Program of Action (NAPA) Project office, Ministry of Environment, Singha Durbar, Kathmandu, Nepal

Shrestha, A. B. and L. P. Devkota 2010. Climate change in the Eastern Himalayas: observed trends and model projections. In Climate Change Impact and Vulnerability in the Eastern Himalayas– Technical Report 1, ICIMOD, Kathmandu.

Shrestha, U. B. and Bawa, K., Widespread climate change in the Himalayas and associated changes in local ecosystems Forthcoming.

SOHAM 2006. Newsletter. SOHAM-Nepal (Society of Hydrologists and Meteorologists-Nepal).6(2).

Thapa, K., G.B. Sharma, B.B. Tamang, P. Limbu, B. Ranabhat, R.C. Khanal, B. Joshi, and A. Shrestha. 2012. Regional Project on Shifting Cultivation (RPSC): Promoting Innovative Policy and Development Options for Improving Shifting Cultivation in the Eastern Himalayas-Land Use Option and Extension Approaches in Shifting Cultivation System of Nepal. Local Initiatives for Biodiversity, Research and Development.

Yao T. D., X. J. Guo, T. Lonnie, K. Q. Duan, N. L. Wang, J. C. Pu, B. Q. Xu, X. X. Yang, and W. Z. Sun 2006. Record and temperature change over the past 100 years in ice cores on the Tibetan plateau. *Science in China: Series D Earth Science,* 49(1):1-9.

Chapter 2

Technological Solutions for Climate Change Adaptation in the Peruvian Highlands

Ann Marie Raymondi, Sabrina Delgado Arias and Renée C. Elder

Additional information is available at the end of the chapter

1. Introduction

Climate change is one of the most pressing and complex problems facing humanity. Climate change was first presented as a biophysical phenomenon, one that would manifest itself through sea level rise, melting polar ice caps, and warming temperatures. It has since become an issue of international significance and the subject of ongoing dialogue in both political and academic settings. As the risks of climate change deepen, so too does the contention and confusion surrounding how best to proceed. How we address climate change is continuously redefined and expanded in academia and policy. As the focus has gradually shifted from mitigation to adaptation, the dialogue has shifted to include the social dimensions of climate change, resulting in a sharp rise in research on the human dimensions of climate change. No longer just a biophysical phenomena, climate change has become a lens through which we analyze poverty, inequality, vulnerability, and development. It has created intense debate about who is responsible, who is most susceptible, and how best to prepare for the consequences of a changing climate. To address these questions, climate change scholarship has developed the concepts of vulnerability and adaptation.

The rise in vulnerability and adaptation research has been accompanied by an increase in methodological frameworks and theoretical approaches to understand and assess vulnerability and design and implement appropriate adaptation strategies. Through this work, it has become increasingly apparent that vulnerability to the various threats posed by climate change is not just a function of the threats themselves, but is shaped by the cultural, institutional, and socioeconomic contexts in which these risks occur. Similarly, the capacity to adapt to climate change is influenced by the very same factors. Much of the scholarship on vulnerability and adaptation has focused on those most vulnerable to the impacts of climate change, particularly resource-dependent, agricultural societies. Food production is of fundamental importance to the global community and many agricultural societies will undeniably need to adapt to new biophysical conditions in order to persist. The livelihoods of resource-dependent farmers are inextricably linked to their biophysical environments and there is a great need to understand these relationships to better inform adaptation options.

Within the context of resource-dependent agriculture, institutions and technology have been identified as two factors that shape both vulnerability and the capacity to adapt. To understand how innovation in both institutions and technology can be used to address climate change, the South American country of Peru will serve as a case study and the backdrop to a discussion on the dynamics between vulnerability, adaptation, institutions, and technology. This chapter will focus on the Peruvian highlands, an expansive steppe characterized by vast mountain ranges, tropical glaciers, and indigenous, resource-dependent agricultural communities. Peru's highlands cover a little over 30% of the country and yet contain over 70% of the world's tropical glaciers. Runoff from these glaciers forms the life support system for highland agriculture and for urban consumption in swelling coastal cities such as Lima. Warming temperatures and more variable precipitation have led to accelerated melting, with far reaching implications for agriculture, industry, and urban users. The region is also undergoing considerable socioeconomic changes, as population growth has increased the need for agriculture, mining, and recreation, changing the nature of traditionally rural, subsistence landscapes.

In the last several decades, the role of environmental and humanitarian NGOs in climate change adaptation has become far more pronounced in the region. Often, these NGOs provide resources, assistance, and in some cases, technologies to farmers. They are a part of an emerging network of climate change practitioners focused on preserving the agricultural and cultural traditions of highland communities. In these communities, traditional, pre-hispanic agricultural technologies are being revived and in some cases combined with modern technologies to create innovative adaptation efforts. This chapter will illustrate the dynamics between institutions, technology, and local-scale adaptation efforts. It will present a background on the past, current, and projected changes in the region's glaciers and climate, it will discuss how these changes will impact local communities, and will explore the various methods communities are using in response to these impacts. It will review important theoretical concepts such as vulnerability, adaptive capacity, and adaptation. Andean agricultural communities offer an excellent case to examine the role of technology and institutions in shaping adaptation pathways in agricultural communities. In many regions of the world, climate change impacts are already a reality, underscoring the need to understand how communities are impacted and how they are adapting.

1.1. Why glaciers matter

Tropical glaciers respond quickly to changes in climate and therefore serve as excellent indicators of climate change. Since glaciers are distributed worldwide, they serve as an exemplar of global climatic conditions. Many of the world's glaciers are in close proximity to settlements, causing the effects of climate change to be readily observed by communities [1]. Additionally, alpine glaciers are oftentimes connected to environmental, social, and economic dimensions of communities in surrounding lowlands; consequently, any changes to mountain glaciers have immediate and significant impacts on populations [2]. This is especially true in Peru, where glaciers and the water they provide are intimately associated with indigenous culture, customs, and religion.

In regions where glaciers provide an important source of freshwater, there is growing concern regarding how climate change may alter the relationship between alpine glaciers and freshwater availability. There has been rapid and accelerating recession of mountain glaciers worldwide during the last 50 years, with increasing concern for potential environmental, economic, social, and political implications should these trends continue [3]. Andean glaciers have displayed continued recession, with the most pronounced retreats beginning in the early 1980s. Many of these glaciers have been in existence for centuries, yet they exhibit some of the most rapid mass and surface area losses in the world [4]. Scientists project serious implications in the future if atmospheric warming continues.

1.2. The Cordillera Blanca

The Cordillera Blanca is part of the Peruvian Andes and is the country's largest and most northerly mountain range, covering over 130 km between 8° and 10°S latitude [5]. More than 99% of all the world's tropical glaciers are located in the South American Andes, and roughly 70% of these are found in Peru alone [6]. The Cordillera Blanca is Peru's most extensively glaciated mountain range, with a quarter of the world's tropical glaciers in glacial valleys with elevations ranging from 3000 to 6800m [4,6-7]. The range contains approximately 722 glaciers covering more than 723 km² [9]. Similar to the global trends, considerable glacier retreat has been documented in the Cordillera Blanca over the 20th century [6]. Studies indicate that the Cordillera Blanca has lost over 25% of its coverage since 1970, during which, the average temperature has increased 0.35°C - 0.39°C per decade [8]. Persistent and accelerated glacial shrinkage in the Cordillera Blanca is anticipated for the future, with severe consequences for the region's water supply and those who depend on it [6].

The Cordillera Blanca's glaciers provide the fundamental water source for irrigated agriculture, domestic livestock, human consumption, and hydropower generation in the region [7]. The region's dependence on this resource will invite serious implications in decades to come as scientists expect atmospheric warming to continue to accelerate. Additionally, increased melting causes the formation of glacier lakes at the glacier terminus; these lakes become dammed by unsorted till deposited by the retreating glaciers, leaving nearby developments at risk for glacier lake outburst floods (GLOFs). Consequently, "accelerated melting in the region is of great concern as it poses a threat to the local water resource and increases in the risk of GLOFs associated with moraine-dammed lakes" [7].

1.3. Climate regimes

In the tropics and extra-tropics, moisture availability is governed by the oscillation of the Inter-Tropical Convergence Zone, which creates variations in the amount of moisture received [6]. The climate of the Cordillera Blanca region features low thermal seasonality with an average annual temperature range from 0°C to 9°C, while diurnal variations in temperature are typically greater. In alpine areas, the steep topography also creates a steep temperature gradient. Monthly precipitation displays significant seasonality with the most precipitation falling during the wet season occurring between October and April [5,9]. More than 80% of precipitation falls during the wet season, resulting in higher stream discharge during those months [9]. Glacier mass accumulation occurs during the wet season while ablation, mass loss through evaporation and melting, can occur year round [5,7]. Most melting occurs during the dry season, which is important since this coincides with decreased precipitation. As a result, water is made available from glaciers when it is most in demand [4]. The Central Andes display fairly moist conditions at higher altitudes, while the coastal lowlands experience extremely arid conditions. Thus, the further one travels down Peru's mountain slopes, the more important glacier runoff becomes.

1.4. Climatic changes

Although the impacts from climate change are still uncertain for many parts of the world, alpine regions in tropics and subtropics are already experiencing the ramifications of climate change. Studies conducted on tropical glaciers reveal that they are particularly sensitive to climatic changes [6]. Climate in the tropical Andes has undergone measurable changes over the past 50 years, with an observed temperature increase of approximately 0.1 °C per decade with a general temperature increase of 0.68 °C from 1939 to 2008. A limited number of modeling studies conducted in the region indicate that warming will continue throughout the 21st century, suggesting further warming of approximately 2-4°C [10]. Following the IPCC's Special Report on Emissions Scenarios A2, scientists project that the tropical Andes may even experience extreme warming, on the order of 4.5–5 °C [11]. Warming of this magnitude suggests that an additional 30-50% of existing mountain glacier mass could disappear by 2100 [10].

In addition to the increase in temperature, changes in precipitation are of equal concern, especially in the Peruvian Andes where rain-fed agriculture supports a large part of the population. There are several limiting factors when it comes to assessing precipitation in the Andes, however. Precipitation records are low in quality and have not been collected systematically, hindering detailed assessments of long term trends. Researchers have been able to identify a small increase in precipitation during the latter half of the 20th century. Perhaps of more importance though, are the projected increases in seasonal variability, including the timing and amount of precipitation in both the wet and dry seasons [11].

1.5. Glacier response to climatic forcing

Variations in climate are commonly linked to glacier advances and retreats [12]. Despite some transient advances, glaciers in the Cordillera Blanca have shown continued recession since the mid-19th century [11]. Every glacier has an accumulation zone, where it gains mass, predominantly from snow, and an ablation zone where it loses mass to evaporation, sublimation and melting [4]. The annual equilibrium-line altitude (ELA) delimits these two zones, above which accumulation exceeds ablation [12, 4]. ELAs can be understood through changes in temperature and precipitation; increased temperature and decreased precipitation causes the ELA to ascend, and decreased temperature and increased precipitation causes the ELA to descend [4, 12]. Persistent rising of the ELA of glaciers in the tropical Andes has been documented in recent decades, by as much as 300 meters in some cases [4]. Glaciers of the Cordillera Blanca lost 22.4% of their area from 1970 to 2003. Due to higher temperature increases at lower elevations, much of this loss has been in small, low-lying glaciers [7].

In addition to implications of precipitation and temperature, atmospheric humidity also influences melting and sublimation [11]. A decrease in atmospheric humidity is thought to be one of the major reasons for glacier retreat throughout the tropics at the end of the Little Ice Age. Studies suggest that in the 1930s and 1940s, temperature accounted for one third of glacier retreat in the Cordillera Blanca while factors promoting decreased humidity accounted for the remainder. Alternately, accelerated recession since the 1980s is attributed to increased air temperature and increased air humidity [13]. Observations show that humidity increases during the wet season are particularly responsible for elevated melt rates. In general, however, quality atmospheric humidity records are absent for the Peruvian Andes and most studies suggest that temperature has greater relevance to glacier mass balance [11].

Glacier mass balance is clearly sensitive to changes in both temperature and moisture, and these are also tied to climatic anomalies such as the El Niño-Southern Oscillation (ENSO) [14]. ENSO is identified as the largest influence on interannual variability of weather patterns and climate fluctuations on the global climate regime [15-16]. As a result of ENSO on interannual climate variability in the Andes, mass balance is greatly influenced by Pacific sea surface temperature (SST) anomalies and their impact on precipitation. Due to warmer SSTs in the Pacific, the El Niño period is typically warmer as well as drier. ENSO events have been linked to significant negatives in mass balance, which is attributed to low precipitation, low albedo, and as a result, increased radiation exposure [17]. La Niña periods are characterized by cold temperatures and more abundant snowfall, during which glaciers usually experience balanced or positive mass balances [17-19].

Pacific SST anomalies in the tropics induce large scale forcing on interannual time scales, resulting in negative mass balance during El Niño years and above average results during La Niña years. Changes in the upper-tropospheric flow aloft associated with ENSO conditions determine snowfall magnitude during the wet season, thus, impacting mass balance. "This teleconnection mechanism is spatially unstable and oscillates latitudinally along the subtropical Andes and affects the Cordillera Blanca in most, but not all years" [11]. Therefore, the connection between ENSO and glacier mass balance in the Cordillera Blanca features inconsistencies which have occurred more frequently since the 1970s, when El Niño and La Niña events featured above average mass balances and negative mass balances, illustrating that glaciers in the region are altered through the ENSO events [11]. In addition, climate model research regarding the impacts on climate change on ENSO occurrences suggest climate change will promote an increased frequency of ENSO events along with increased intensity which is expected to be a further detriment to glacier mass balance and seasonal water availability in the future [15-16, 20].

1.6. The future of water in the Cordillera Blanca

The effects of climate change are underway in the Peruvian Andes and many of the Cordillera Blanca's smaller glaciers are projected to disappear within a few decades [11]. Current and projected warming trends combined with the observed changes in glaciers thus far suggest that the retreat of the Cordillera Blanca's glaciers will continue unabated. It is believed that regardless of any mitigation measures the international community takes now to reduce the impacts of climate change, low altitude glaciers will not be able to recover [4]. This continued retreat will alter runoff patterns, the timing of discharge, and will decrease water availability during the dry season [21]. Thus, declines in glaciers will create significant uncertainty and vulnerability for the region's water supplies and the tens of millions of highland communities and lowland urban dwellers who depend on them [21]. Such drastic changes in the Andean highlands are fundamentally altering the relationship between subsistence communities and their surroundings. The impacts are often noticeable, as retreating glaciers leave once white mountains exposed; the black slopes a stark indicator of fundamental change to alpine environments. These biophysical changes are at the heart of a time of drastic shifts in the ecological and social fabric of life in the Andean highlands. Of great importance is an understanding of the mechanisms and factors that shape vulnerability in these communities.

2. Vulnerability: Theoretical development

Vulnerability has its origins in geography and natural hazards research, but it is now used a concept in climate change studies [22]. The biophysical components of vulnerability have been widely studied in both the hazards and risks literatures, where much of the focus is on the biophysical threat itself. The social dimensions are not as well studied, primarily because it is often difficult to quantify these dimensions, but a growing body of work has emerged to address the social aspects of vulnerability. In 2001, the IPCC defined vulnerability as:

The degree to which a system is susceptible to, or unable to cope with, adverse effects of climate change, including climate variability and extremes. Vulnerability is a function of the character, magnitude, and rate of climate variation to which a system is exposed, its sensitivity, and its adaptive capacity [23].

This definition, as the IPCC recognized in its fourth assessment report in 2007, has been challenged and expanded to include social vulnerability [24]. Brooks provided a critical clarification of vulnerability by distinguishing between biophysical vulnerability and social vulnerability, where biophysical vulnerability refers to impacts of a hazard event, while social vulnerability refers to "an inherent property of a system arising from its internal characteristics" [25]. Accordingly, climate change hazards, such as water scarcity or extreme weather events are the physical manifestations of changing climates.

A variety of disciplines engaged in climate change research are proposing methodological approaches to quantify and assess vulnerability. Assessment of vulnerability includes place-based studies, the development of indicators and, increasingly, scenario-building methodologies and stakeholder-driven processes [26]. A recent attempt to measure vulnerability in the Cordillera Blanca is indicative of the types of variables that are typically included in vulnerability assessments. This method divides social vulnerability into three factors: preparedness, prevention, and response, and divides physical vulnerability into the number of people who may be impacted by the magnitude and trajectory of a hazard event. Each factor is assigned numerical values to allow for measurability and to reduce complexity. These indicators are selected based on academic literature, local conditions, and the availability of data. To maximize mathematical and conceptual transparency, the researchers assigned three classes of scores to each indicator, low, medium, and high. The indicators were then combined to calculate social and physical vulnerability, and combined in total for an integrated vulnerability measure [27].

When the method was applied to the town of Huaraz, it revealed high physical vulnerability and middle to high social vulnerability. Huaraz—an important tourist center—is located in what is known as the Callejón de Huaylas or Alley of Huaylas, which sits in between the Cordillera Blanca and the Cordillera Negra mountain ranges. High physical vulnerability is centered on the Quilcay River and is exacerbated by the outburst potential of Lake Palcachoca and to the high population density of the area. Social vulnerability was attributed to the lack of organized preparedness and prevention practices at a state level and a household level. In addition, preparedness at the individual level was found to vary depending on age, socioeconomic status, and awareness. Integrated vulnerability was found to be highest where exposure, high population density and poverty coincided [27]. The researchers conclude that in order to reduce vulnerability, prioritization of intervention strategies should be based on the underlying causes of high vulnerability in Huaraz: high exposure coinciding with high population density. This method is designed for the scale and conditions of the Cordillera Blanca, but it could be used at other scales if adapted to specific local characteristics and data availability. Studies such as these underscore that while vulnerability is experienced locally, the factors that shape vulnerability occur at different spatial, social, and even temporal scales [28].

There has been significant progress in developing sets of vulnerability indicators to analyze vulnerability at different scales. These indicators can then be used to prioritize adaptation efforts [29]. Oftentimes, these indicators are measures of socioeconomic and political variables [30]. For instance, at a national scale, relevant indicators include literacy rates, life expectancy, governmental flexibility and responsiveness [30]. At a community scale, indicators that measure poverty, inequality, access and representation can be used to quantify vulnerability [29-30]. At a household level, age distribution, labor availability, livelihoods, assets, and resource dependence can shape vulnerability [31]. Thus, vulnerability studies vary immensely on the spatial and temporal scale of analysis and on contextual factors. More and more, the emphasis in vulnerability studies is on the complexities and interconnectedness between social and ecological variables in coupled human-environment systems [32].

2.1. Operationalizing vulnerability: The national scale

Beginning at a large-scale, the first vulnerability for Peru and other nations is the failure to reach binding agreements on emission cuts for carbon emissions and other greenhouse gases (GHG) by the international community. The lack of consensus on emission cuts, witnessed at the 2009 United Nations Climate Change Conference in Copenhagen, is cited for imperiling glaciers worldwide [33]. For many, this ensures that glaciers will continue to retreat rapidly, exposing and exacerbating vulnerabilities in the process. Glacier retreat is a hazard that is associated with a multitude of risks such as loss in minimum ecological stream-flows and reductions in freshwater. These risks affect a variety of sectors, including both urban and rural populations, hydropower production, and the equilibrium of mountain ecosystems [33].

In Peru, reductions in freshwater will greatly affect the livelihoods of people living in the arid coastal plain between the Andes Mountain and the Pacific Ocean [34]. Approximately half of the population, now an estimated 29.2 million, lives along the coastal plain, while one-third live in the mountainous region of the Andean Cordillera, also referred to as the Andean highlands [35]. The largest cities are located in the coastal region and have the greatest economic activity, including Peru's capital, Lima, with 8.7 million inhabitants. For the people living in this region, the primary source of water is from the many rivers that flow down the western slopes of the Andean Cordillera [35]. A key issue, however, is that water resource distribution is highly heterogeneous in Peru: only 1.8% of the average annual freshwater availability flows to the Pacific Ocean, compared to 97.7% to the Atlantic Ocean and 0.5% to Lake Titicaca [36].

Migration to coastal cities is increasing and it is expected that more people will move into the coastal region and place more pressure on water resources as both demand and industry expands [34]. In 1940, 24% of the population lived in coastal regions compared to 65% in the Andean highlands; today, half live along the coasts and 36% live in the highland regions [35]. In 2010, Peru resumed the growth that it had been experiencing before the world recession and ended the year with over 8% economic growth, driven primarily by private investment and high government spending. Delayed spread of growth to non-coastal areas due to poor infrastructure will also aggravate pressure on the coastal region [37]. Ultimately, economic growth, the majority of which occurs along the coast, will exert pressure on the national economy as more investment and action will be needed to address water scarcity. Thus it becomes clear that vulnerability exists along the entire elevational gradient from the Pacific coast to the Andean highlands, but both are linked through national discourse and policy on climate change.

In 2007, then president Alan García promoted water exports to take advantage of the water abundance from melting glaciers. García suggested that excess water could be sold to Brazil. When asked what Peru would do if these water reserves ran dry, he proposed desalination of ocean water [38]. According to Garcia, "We should not pay attention to those who consider melting and disappearing glaciers a threat....the water will never stop trickling down" [38]. This national dialogue reveals conflicts in how experts and decision makers view the consequences of glacial retreat, which may obscure the issue for Andean mountain communities [38]. Andean communities have a lack of understanding regarding the potential consequences that glacial retreat can have on their livelihoods [33]. This vulnerability could be magnified by social inequities arising from differences in how water is governed [39].

The recent increase in flow levels from enhanced glacial retreat has also created vulnerabilities at the local scale. In the short term, glacier retreat has provided a temporary increase in runoff, and populations have quickly responded by taking advantage of these surpluses [11]. Increased water availability has promoted increased agricultural production and development, which has led to sustainability concerns [21]. These surpluses will be a short-lived phenomenon and the eventual decline in water will create adjustment problems for populations that have become dependent on temporarily higher flows [40]. Uncertainty regarding the temporal duration of increased water supplies necessitates consideration for future action if and when the glaciers finally vanish. Thus, priority must be given to studying potential impacts of climate change on water resource management in the region and identifying sustainable adaptation strategies.

2.2 Vulnerability in the Andean Highlands- the Cordillera Blanca

In the Cordillera Blanca, glaciers provide the water necessary to sustain river flows during the dry-season and during extended drought. Inhabitants of the region are dependent on the availability of this water resource for social and economic activities [9]. Water usage can be classified by sector, including consumptive use for agriculture, industry, and households and, non-consumptive use for energy production and mining activities [36]. According to this classification, out of a hypothetical 100 liters supplied by Peru's ecosystems, 51 are used for agriculture, 36 for hydroelectric power generation, 12 for human settlements and industry, and 1 for mining respectively [36].

Climate change will create vulnerabilites along elevation gradients impact many different sectors along these elevation gradients. Rain-fed subsistence agriculture traditionally dominated agriculture in the Andean highlands, but increasingly, farmers are adopting commercial crop production, which often requires extensive irrigation [9]. Irrigated agriculture is highly inefficient in these regions; for every 100 liters used in agriculture, an average of 60-65 liters end up in the Pacific Ocean [36]. During the rainy season, glaciers accumulate mass and store water that is then released during the dry season [41]. Glacial retreat is accelerating this process, leading to the formation of pools and lakes at the terminus of glaciers. These glacial lakes develop in loose sediment formations and can rupture after tectonic activity or from the impacts of falling ice, which presents another major risk in the form of devastating floods and avalanches [41]. The National Aeronautics and Space Administration (NASA) for example, identified a crack in a glacier overlooking Lake Palachoca, located in the Cordillera Blanca, which led to a warning to cities around the lake of the potential for a devastating flood [42]. The predictions turned out to be accurate; on April 13, 2010 a large chunk of ice fell into a lake in the Cordillera Blanca near the town of Carhuaz [43, 44]. The impact led to the formation of a 23 meter high tsunami that flooded nearby communities with sediment and water and destroyed a water processing plant that supplied water to 60,000 people [43]. Scientists attributed the event to climate change, heavy rains, and possibly an earthquake that occurred one month earlier in the region [43, 44].

The Peruvian government had taken some measures to prevent floods like the one near Carhuaz. A drainage tunnel was constructed for the lake. However, since the volume of the ice chunk doubled in size as it descended, its impact created a splash that breached the lake's 60 foot containment banks. Engineers with Peru's National Water Agency, who oversaw the construction of the drainage tunnel, estimated that the drainage tunnel only protects Carhuaz against 80 percent of icefalls [45]. Local experts have also found that retreat below a lake's surface causes the lake reservoirs to become deeper and longer, leading to an increased potential for disaster despite the drainage [41]. Going back to the NASA findings, it is important to note that some scientists, specifically those at the Peruvian National Institute for Natural Resources (INRENA), felt that more scientific study was needed in order to prove the dangers posed by glacial retreat [42]. The following was the stance taken by the head of the Climate Change Unit of Peru's National Council for the Environment (CONAM), Patricia Iturregui:

We need to make an important effort to plan disaster management and prevention of risks in the future. The most important measures to be taken are to organize local communities and to organize an institutional framework able to respond to these adverse effects. We are in the process of desertification. The retreat of the glaciers is definitely going to mean a shortfall in the water supply in years to come [42].

As this example highlights, how a potential risk is perceived could either hinder or advance action towards its management and prevention. It could also lead to confusion as to what should be considered an imminent danger. Both CONAM and INRENA are important environmental institutions in Peru. CONAM is the central environmental agency charged with all creating and implementing national environmental policy [46]. INRENA serves as the principal agency for the management of natural resources and is responsible for, among other areas, managing soil and water resources [46]. INRENA is also concerned with the impacts of climate change on water resources [35]. One of the most important limitations for CONAM results from its sector-based management arrangement, where sectoral ministries are responsible for both promoting and ensuring compliance of management activities. This arrangement represents a conflict of interest for stakeholders as the efficiency and neutrality of ministries is oftentimes mistrusted [46]. Another big challenge for CONAM is achieving inter-institutional coordination to prevent the ambiguity created from jurisdictional overlap [46]. Lack of resources, staff and administrative capacity are other limitations for both CONAM and INRENA.

Any sector that uses glacial water is vulnerable to hydrological changes, but whether or not these sectors are acutely vulnerable depends on factors not readily apparent. These hydrological changes and the risks they pose will be translated into different outcomes via the particular socioeconomic contexts in which they occur [28]. These outcomes will likely depend on the social inequities present within and beyond communities, and especially, the policies that create them [39]. Access to water is a critical issue in many regions in Peru, especially in highland agricultural communities. One of the issues with water access and allocation stems from that fact that Peruvian water policy defines water as an economic good [39]. Commoditizing water has the potential to create scarcity by excluding particular groups and increasing demand in certain sectors, which ultimately increases competition and deepens these inequalities.

We identified some of the important vulnerabilities, risks, and hazards facing the Cordillera Blanca of Peru. Regional and local vulnerabilities will ultimately have to be assessed based on the local socioeconomic contexts, institutional power-plays, and the value placed on water resources. Though vulnerabilities are communicated, perceived, and measured at different scales, it is clear that vulnerabilities are interrelated. Key stakeholders (including the public) will have to recognize the interconnectedness of these vulnerabilities and risks in order to create suitable adaptation strategies.

3. Adaptation: Theoretical development

Vulnerability is closely tied to another important concept in climate change scholarship: adaptation. Both adaptation and vulnerability have become indispensable to the study of the human dimensions of climate change. When analyzing a particular system's vulnerability to a climate hazard, or set of hazards, the ability of the system to respond, to cope, or to adapt is a complementary consideration. To relate the two, adaptation can be viewed as an attempt to reduce the vulnerability associated with climate change [25]. The relationship between adaptation and vulnerability is not usually linear however [47]. For example, some adaptive measures might reduce vulnerability in the short term while creating vulnerability to a different hazard in the long-term [47].

Adaptation did not make up a significant portion of the early years of international climate change discourse because much of this discourse focused on the mechanisms of biophysical impacts or on strategies to mitigate carbon emissions [48]. In the fourth assessment report of the IPCC, adaptation received more attention. This report defined adaptation as "actual adjustments or changes in decision environments, which might ultimately enhance resilience or reduce vulnerability to observed or expected changes in climate" [48]. Adaptation measures can consist of a range of practices carried out at different scales by many different actors. They include, but are certainly not limited to, policy changes and investment in new technologies or infrastructure at national levels and behavioral adjustments and changes in economic decisions at the household level. There are also temporal dimensions to adaptation. Adaptation can be targeted for current risks or in anticipation of future risks. In addition, adaptation can be further defined by sector, such as practices specific to agriculture, water resources, health, etc. [48]. As in vulnerability, adaptation is also influenced by the hazard in question. Specific hazards may require specific adaptation strategies, and not all systems may have the ability to adapt to all hazards [25]. Thus, adaptation can be highly differential in scope and in purpose.

Another closely related concept to adaptation is adaptive capacity, which is the ability of a system to prepare for and respond to climate change stresses [49]. Adaptive capacity can be thought of as a system's potential for adaptation [25]. The capacity for individuals or communities to adapt depends on factors that are not necessarily related to climate, but to socioeconomic factors such as education levels, access to information, resources, and social networks, and even the ability to experiment and innovate [50]. Adaptive capacity can include tangible assets such as financial and technological resources and less tangible assets such as information sharing [51]. Ethnographic studies of pastoral communities in the Andean highlands have shown that livestock ownership is related to the ability of households to adapt to changing and uncertain environmental conditions [52]. Livestock are an important economic asset in these communities, serving as financial capital, a food base, and a means to produce exchangeable goods such as meat, fur, and milk; livestock owners also hold greater positions of power and have more influence in community decision-making, which has resulted in social differentiation based on livestock ownership [52, 53]. During periods of variable weather and/or socioeconomic conditions, households with more livestock are better able to cope by selling or trading livestock or their goods, making livestock a key aspect in the household's adaptive capacity. Households lacking livestock and the associated flexibility lack this capacity [52]. As this case reveals, adaptive capacity depends upon available assets such as livestock and the social, political, and institutional context through which adaptation decisions take place [51, 54]. This context is further characterized by the absence or presence of social networks, institutions, and political influence that may operate at different scales [54]. It is important to recognize that local vulnerabilities and adaptive capacity may be influenced by broader, and in some cases global, forces [55]. For example, international free trade agreements mediated at the national level may remove price supports for local crops, exposing local producers to global market forces and exacerbating local scale vulnerabilities.

Adaptive capacity can be influenced by a variety of factors such as governance, institutions, economic resources, technology, information and skills, infrastructure, and access to resources and services [49]. Poverty is commonly associated with lower levels of adaptive capacity because impoverished communities are typically resource poor and have limited power and representation [56]. Other work has shown that analyzing household characteristics such as education levels and farm vs. non-farm income may indicate why households adopt adaptation technologies and practices [57]. In an in-depth analysis of two rural agricultural communities in Burkina Faso, researchers found that contrasting worldviews within two cultural groups influenced their adaptive capacity through their willingness to adjust behaviorally to changing climates [58]. Studies such as these reveal that deep-seated socioeconomic, cultural, and political forces within communities often shape adaptive capacity. Through in-depth studies of local communities, it is becoming clear that many variables interacting at different scales influence adaptive capacity. Furthermore, adaptive capacity is uneven across and within societies and there may exist barriers and even limits to adaptation. It is important to understand what barriers, limits, or opportunities exist within given societies for adaptation so policies and interventions have a better chance at success [48].

3.1. Pro-poor adaptation as a means for sustainable development

Adaptation may provide broader benefits, beyond its role in reducing vulnerability to climate change. Climate change adaptation is also viewed as a potential pathway in sustainable development (47). There has been a refocusing in vulnerability assessments to shed light on the causal factors that shape household vulnerability, such as income, employment, representation, and access [28]. These approaches focus on impoverished communities and lead to recommendations for "pro-poor" adaptation strategies. To some climate change researchers, herein lies the opportunity to combine adaptation strategies and development efforts. Regardless of whether the goals of adaptation and sustainable development align, development pathways will influence vulnerability and adaptive capacity and vice versa. The manner in which societies choose to develop may lessen or create vulnerabilities to future climate changes, leading to the idea that adaptation and development goals should be combined when possible [49]. Some degree of caution is needed when pursuing "pro-poor" adaptation practices for sustainable development. Since both are promoted in complex social-ecological systems where the drivers that create and maintain poverty and contribute to vulnerability are interrelated and often hard to identify, interventions must be carefully devised so as to avoid creating future vulnerability [47]. This underscores the routinely requested need for systematic and in-depth understanding of the factors that both contribute to vulnerability and facilitate adaptation.

3.2. Resource-dependent, agricultural communities

Agriculture is one of the most fundamental livelihood activities for human societies and climate change is anticipated to heavily impact agricultural systems. Changing climates will not only challenge agricultural systems in the Peruvian highlands, but the more fundamental relationship between humans and these systems will also be tested [21, 59]. Agricultural systems are sensitive to temperature, water availability, crop disease, and extreme weather events, all of which are influenced by climate. Adaptation in agricultural systems aims to promote practices that build long-term resilience to climatic changes. They may include attempts to increase productivity, water delivery and storage, and soil conservation. They may also involve attempts to enhance relations and networks between farmers to create better access to tools, resources, and financial instruments for investment capital [40]. Since climate change is projected to cause further retreats in alpine glaciers and more variability in the timing and amount of precipitation, from the broadest view, adaptation in Peru centers on preparing for the implications of reduced water availability. Rural farmers are one of the most vulnerable groups to changing climates, especially to changes in water availability, temperature, and the timing of precipitation events [21]. Glacial melting and retreat combined with rising temperatures and more variability will shift where crops can be grown, reduce water availability, and will make predictions more difficult [52]. The situation is complex however, because in the short term, glacial melting has led to increased runoff and has opened new terrain to farming and grazing. As the ice continues to disappear however, water availability will eventually decline and with it the buffer that glaciers provide during extended droughts and seasons where the rains are delayed [52].

It is often stated "that all adaptation is local." Since impacts occur in particular places along defined temporal scales, then responding to climate change indeed will also occur through a locally bound set of actions. Even if a community carries out adaptation in a specific sector, it can be supported, coordinated, or mediated through a network of international funding, national initiatives, and regional collaboration between NGOs and communities. Recent work on climate change adaptation emphasizes that good governance structures can advance adaptation. The social, financial, and political framework within which adaptation occurs can either enhance or undermine adaptation practices. Thus, there are multiple and interacting scales to adaptation, which necessitates careful deliberation and planning in any adaptation endeavor. Adaptation and especially adaptive capacity are influenced by the unique cultural, institutional, and socioeconomic contexts in question, highlighting the complex nature of climate change adaptation.

Though agricultural systems are equally complex, there are several dimensions that are especially relevant to climate change adaptation: local ecological knowledge, technology, and institutions. Indigenous and rural agricultural systems are characterized by local knowledge and practices that co-evolved over time, resulting in systems that are finely tuned to the prevailing biophysical environment. Much of the attention on climate change adaptation in agricultural systems supports technology as a strategy to help farmers increase productivity and conserve resources. However, recent scholarship has recognized that appropriately scaled institutions are a critical factor in technological research, development, transfer, and adoption. In the Peruvian highlands, efforts are underway to develop integrated adaptation that embraces local actions and institutional support. Local knowledge can be used as a foundation to enhance adaptation efforts, warranting further discussion of the relationships between technology, institutions, and agricultural adaptation.

3.3. Technology in adaptation for farmers

Technological research and development are central features of agricultural growth and development [60]. Building of the precedent set in the Green Revolution, technology has been one of the more advocated strategies for climate change adaptation in agricultural systems [61]. Technology can provide an important mechanism for farmers to adapt to changing conditions and has many applications including crop development (new varieties with pest and disease resistance or suitability to temperature and moisture conditions), weather and climate forecasting systems, and management innovations (irrigation, conservation tillage) [62, 63]. Furthermore, technologies are considered farm-level resources and are associated with adaptive capacity [63]. One of the larger challenges in implementing technology in agricultural systems lies in its transfer and adoption, highlighting that technology is not just a strategy, but is also a part of a larger process of research and development, farm-level adoption, and feedback for further innovation [61].

3.4. Institutions in climate change adaptation

Adaptation is the result of decisions made by individuals, by groups within communities, or by organizations, governments, and even international groups on behalf of communities [65]. Thus, adaptation is influenced by the rules and norms that structure decision-making and social interactions, which is to say that adaptation processes are influenced by institutions [66]. In this context, institutions are the formal and informal channels that shape and mediate human behavior [67]. The role of institutions in climate change adaptation has been the focus of recent attention in climate change research, especially those that are involved directly in adaptation practices adaptation practices [68]. Institutions play a critical role in shaping adaptive capacity by supporting and mediating adaptation options and especially by determining how the resources necessary for adaptation flow to different communities [66, 68].

Institutions can operate at a variety of scales from local, to regional, national, or international and they can also emerge within local communities or can be initiated by donor agencies external to the communities that ultimately receive support [68]. Institutions at the national level can promote investments in physical assets such as transportation, water, and energy infrastructure [47]. Local level institutions may influence how resources and information flow to and within communities [66]. It has been shown that institutional factors such as farmer to farmer networks can increase access to information on production techniques and enhance the adoption of various technologies for adaptation [57]. Social networks enable individuals to act collectively, something which climate change practitioners view as necessary to climate change adaptation [65]. Institutions shape the social dynamics of adaptive capacity and adaptation processes are interdependent on these dynamics [65]. The power relationships within communities for example, are oftentimes analyzed in adaptation research because these relationships influence which community actors may or may not have access to representation and resources. Thus, power relationships may create heterogeneous vulnerability even within a community.

Nongovermental organizations are at the forefront of local scale development, conservation, and climate change adaptation efforts. Oftentimes, non-governmental organizations (NGOs) focus on building local capacity and representation, which often entails working within communities and enhancing local institutional capacity [66]. Institutions need to be effectively designed to work within the power and relational dynamics that characterize local groups and some studies show that institutions that fail to take these complex factors into consideration produce weaker outcomes [66, 68]. It is through appropriately scaled institutions that better interactions between farmers, local experts, scientists, and individuals from NGOs can take place. Institutions such as NGOs can be thought of as new configurations that facilitate co-production, the preservation of local knowledge, the integration of expert knowledge, and the sharing of information and technology [69].

It is relevant to consider institutions in the context of technological approaches to climate change adaptation because institutions oftentimes mediate the availability, transfer, and adoption of agricultural technologies. In-depth studies of climate change adaptation on rural farms in Argentina and Mexico have found that financial, technical, or social support is requisite to the adoption of technologies. For example, adoption of irrigation improvement required a combination of tax incentives, technical assistance, and support from the greater community [63]. At the farm level, availability of technology, information, and other resources influence the performance of farmers in terms of productivity and social outcomes [63]. These findings suggest that off-farm institutions play a critical role in agricultural practices in general and are an important consideration in climate change adaptation [66, 70]. Thus, institutions can be thought of as leverage points that determine how technological resources flow to communities [66].

Adaptation is a process through which decisions are made over extended periods of time in response to multiple stimuli that may include climatic and non-climatic dimensions [62]. In agricultural systems, technology and institutions have been identified as two important pathways for building adaptive capacity and facilitating climate change adaptation, especially those institutions that transfer or enhance the use of technology [69]. The broader institutional environment can impose both opportunities and constraints for climate change adaptation, highlighting the need to closely examine how institutions influence adaptive capacity in agricultural systems [63]. The relationship between farmers, technology, agricultural research, and climate change is complex and the following section will use the Andean highlands as a case study to explore these dimensions and to illustrate the role of technology and the institutions in climate change adaptation.

3.5. Local ecological knowledge and technologies

Indigenous populations in Andean regions have always lived in environments characterized by variable weather. To cope with climate uncertainty and risk, farmers developed a suite of strategies tried against the test of time to cope, organize, and adapt to meet these challenges [71]. These strategies exist within long-standing belief systems that are referred to in academic settings as local ecological knowledge (LEK). Local ecological knowledge, also referred to as traditional knowledge or local knowledge, is "gathered over generations by observers whose lives depended on this information and its use" [72]. Local ecological knowledge includes not only the practices, but the social mechanisms behind these practices such as the rituals and institutions that facilitate the sharing and internalization of knowledge [72]. A stark contrast between local ecological knowledge and more modern belief systems lies in the fact that practices built upon local knowledge tends to work within and adapt to environmental limits while modern practices aim to control variability and increase production within agro-ecosystems through the use of high energy inputs and other practices [70, 72]. In the Andean highlands, local ecological knowledge developed from years of observations of and experiences in extreme environments [73]. These experiences led to the development of indigenous crop varieties, irrigation systems, forecasting methods and the cultural and social processes that reinforced these practices. Peruvian farmers developed intimate relationships with their biophysical environments and engage in a number of strategies to cope with variable temperature and precipitation regimes [74]. Since local ecological knowledge in this context contributes to adaptive capacity, it has received attention as an opportunity for climate change adaptation. The following section will introduce aspects of local ecological knowledge in highland communities.

In the Andean highlands, local ecological knowledge is built upon responding to and developing the means to predict climatic variability and this knowledge forms the basis of many of the farming practices of rural inhabitants [21, 70]. Instead of relying on instrumentation to predict weather events, many farmers rely on direct observations of various indicators to determine planting and harvest times. Some groups of Andean farmers forecast climatic conditions by observing constellations. In June, months before the rainy season commences, farmers assess the brightness of the Pleiades constellation to predict the timing of its arrival. When the constellation appears bright, the rains are expected on time and farmers plant accordingly. A dim constellation is thought to mean that the rains will be delayed, and farmers postpone planting to compensate. This annual observation is a part of traditional ritual, but anthropologists who studied the communities were skeptical that it could accurately predict the arrival of the rains. Researchers who have studied these practices however found that there is a fair degree of legitimacy to the practice [75].

During El Niño years, high velocity winds originating off the Pacific Ocean disrupt the flow of moisture-laden tropical air from the Amazon, which delays the onset of the rains and leads to drier conditions. The El Niño winds allow clouds to accumulate over the Andes, which partially obstruct the night sky and cause constellations such as Pleiades to appear dimmer. During normal years, Amazonian air currents deliver the rains as expected and prevent the formation of clouds, causing the constellations to appear bright. Thus, local populations are able to observe these signals from their environment to forecast climatic conditions and plan accordingly. This work is significant because it highlights how worldviews and spiritual beliefs are interconnected with climate and agricultural practices, and all will impacted by climate change. Though some may question the importance of worldviews and subjective beliefs such as these in a discussion of adaptation, others believe that understanding the cultural dimensions of climate change is vital to promoting effective adaptation strategies [76].

The elevation and climate in the Peruvian Andes create a spectrum of microclimates that test the flexibility and adaptability of Andean farmers. To cope with these environmental conditions, farmers have developed an equally broad range of crop varieties for different elevations and microclimates. Potatoes are an important staple in Peruvian highlands and have been cultivated by farmers since the Incan Empire. Potatoes are also culturally significant in the region and some communities pay homage to deities associated with potatoes [77]. There are roughly 2,700 varieties of this important staple crop that are suited to different microclimates and varying water and nutrient requirements. Despite this impressive variety, potatoes are susceptible to extreme cold. Potatoes are typically planted at the start of the rainy season and once planted, soil moisture conditions must fall within a certain range to support growth. If soil moisture falls below a certain level, potatoes may fail to produce strong enough shoots. Potatoes are also susceptible to frosts and if the ground freezes, plants may be damaged or killed. The timing of planting is thus crucial, which necessitates accurate forecasting [75]. Some varieties are more resistant to certain pests [78]. Other tubers such as maca and olluca and the grain-like quinoa were bred to be hardy against frosts and low temperatures and to withstand high altitude environments and the accompanying extreme fluctuations in temperature and precipitation [59]. These crops not only have high resistance to climate related impacts, but are also high in nutritional value [79]. Following Spanish colonization, European crops such as barley replaced indigenous crops [59]. The locally developed techniques and customary practices associated with native crops were lost in some cases as well [59].

Many indigenous farmers engage in a practice known as parcel-zoning diversification. One household will cultivate two or more parcels at different elevations or even in different ecological zones to spread the risk in the event of extreme climate events. If a climatic event negatively impacts a parcel in one zone and not the other, the household will still have yields at the end of the season. However, there are constraints to this practice such as the lack of access to adequate irrigation and limits to parcel size [59]. Many households also engage in livelihood diversification so that they do not depend heavily on any one resource [52]. Households will engage in livestock husbandry, tourism, or construction to supplement incomes [52]. The previous strategies are integral components of farming livelihoods, but none is so important perhaps as securing and maintaining adequate water resources.

Andean communities developed irrigation systems to effectively manage and conserve water, a highly fluctuating resource [80]. These systems were developed by pre-Hispanic cultures, most notably, the Incas. These water storage and delivery systems consisted of terraces, ditches, canals, reservoirs, and raised field agriculture known as Waru Waru. Irrigation in the Andean highlands allows farmers to extend their growing season by providing crops an early start in the dry season [81]. Once the rainy season commences, farmers switch from irrigated to rain-fed agriculture. Ceremonious and collective action formed the foundation of these traditional water systems [82]. Maintenance depended on a combination of labor, worship, and pilgrimages to pay homage to mountain springs and gods [82]. These ceremonious and social processes are essential to regulating this extremely scarce resource and ensuring its equitable and efficient distribution [80].

Raised field agriculture, known as Waru Waru, is a farming technique that was used extensively in the Andean highlands prior to Spanish arrival. The system consists of a series of elevated soil platforms embedded in canals. These canals provide moisture in the event of drought and by absorbing sunlight during the day, protect crops against the lethal frosts and low nighttime temperatures that are common at high altitudes. In addition, raised fields also increase cultivated area and soil depth and the nutrient-rich silt that accumulates in canal bottoms can be recycled into the raised beds to augment soil fertility and increase productivity. Because collective labor is often required to construct the canals and beds, the system has also been found to improve social relations by strengthening local capacity and uniting farmers [83]. These attributes make Waru Waru an effective agricultural strategy for variable weather conditions. Despite multiple benefits and being well suited for the Andean region, this type of agriculture was abandoned in many regions after the Spanish conquest.

Indigenous farmers in the Peruvian highlands engage in a variety of practices that are well suited to extreme environments, but climate change is not the only stressor in these communities. Highland communities are also being exposed to socioeconomic changes that challenge the ability of local ecological knowledge and traditional practices to provide adequate livelihoods [21, 52]. Socioeconomic and political factors such as population pressure, global market forces, and national policies that promote commercial agricultural production are rapidly changing rural economies. Migration and the corresponding loss in traditions and native languages further compound these changes. Despite the erosion of such knowledge, local ecological knowledge is regarded as an extremely valuable means for climate change adaptation and many local efforts seek to preserve and in some instances revive it. "Local populations have been coping with climate and social changes for centuries using local knowledge; this local knowledge could be used to supplement technical knowledge to generate a more comprehensive understanding of the problems resulting from climate change" [52].

These past and current practices and the associated beliefs make up a body of knowledge that ties highland communities to their environments. Local knowledge has been fundamental in enabling rural farmers to adapt in the past, and because of this, the strategies, technologies, and belief systems reflected in local knowledge are viewed as a means to enable communities to adapt to the challenges posed by climate change. The following section will provide three cases of institutional and technological innovation that have incorporated local knowledge in adaptation. Each case provides important lessons and together they provide a review of efforts currently underway.

3.6. Technology and institutions in adaptation

Starting in the 1980's, NGOs and governments recognized the potential agricultural and social benefits of Waru Waru and the role it could play in sustainable development and in climate change adaptation. In 1996, through the United Nations Educational, Scientific, and Cultural Organization (UNESCO), 2.5 million U.S. dollars were donated to Peru to implement Waru Waru in Andean communities. The funds were mediated through the non-profit CARE and covered training sessions, capacity building, and construction. Despite the positive publicity and large amounts of external funding, progress reports revealing abandonment of the newly created fields started to appear, which began to raise questions of the appropriateness of the practice. [83] found that some of the NGO's promoting these raised fields offered incentives, such as food, pay, seed, and tools to encourage farmers to participate. Once the incentives were removed however, farmers abandoned projects. Despite good intentions, some fields were built in inappropriate locations or constructed at the wrong time of the year. In some instances, crops that were ill suited to local climatic conditions were imposed on communities by the NGOs and in other instances, agencies provided misleading information to farmers about the expected harvests and risks. Collectively, these factors worked against the widespread adoption of the practice. Erickson noted that fields that were constructed by individual farmers without the incentives from NGOs had more staying power. According to [83], "why they worked or did not is a complex matter and has more to do with social, cultural, and economic factors than with labor or technology issues".

Despite the challenges experienced in the 1980's and 1990's, in 2007, the United Nations Framework Convention on Climate Change (UNFCCC) identified Waru Waru as a potential local coping strategy to climate change because when implemented correctly, it reduces the risks from droughts and frost events and can provide farmers with greater harvest security. Waru Waru was also viewed favorably because of its cultural importance. According to the UNFCCC, local coping strategies are important elements of climate change adaptation. Local coping strategies originate within communities and provide "efficient, appropriate, and time-tested ways of advising and enabling adaptation to climate change in communities who are feeling the effects of climate changes due to global warming" [84]. Though these strategies take place on the local scale, the UN recommends that they be used in synergy with governmental interventions. As Erickson's work revealed however, the successful adoption of Waru Waru may depend on appropriate incentives and empowerment at the local level to ensure that the practices are continued. An effective agricultural practice is but one piece in an adaptation strategy. Other important factors include adequate standards of living, ownership, and especially in agricultural settings, access to markets [83]. This example highlights the issues of scale, goals, and governance in adaptation decision-making [48]. It is important to understand who is making the decision to adapt, what may be influencing this decision including goals or values, and at what scale the adaptation will occur. In other words, institutions that recognize subtleties such as power differences and community perceptions may have greater success than those that fail to do so. The ability of groups to adapt depends on factors that may not be related to climate, but rather to the social context within which climatic changes occur. Factors such as technology, education, information, creativity, and innovation become increasingly important in shaping adaptive capacity and enabling adaptation responses [50]. The Waru Waru system provides a structural adaptation measure, but non-structural components such as the social dynamics that support its functioning are just as important.

As discussed previously, the preservation of crop diversity is viewed as crucial for climate change adaptation. Agricultural diversity, or agrobiodiversity as it is sometimes referred to, has an important role to play in climate change adaptation [64]. Maintaining crop diversity and variety may help buffer the risk of climate variability and extreme weather events. Some varieties may be able to withstand different temperature conditions and rainfall patterns and it provides farmers with more options in the event of climatic uncertainty. Agricultural diversity can contribute to more diverse, adaptable, and ultimately resilient agro-ecosystems. Seedbanks that store crop varieties and the agricultural science extension programs where they are tested under experimental conditions are two important agricultural strategies for climate change adaptation. The presence and availability of technologies certainly is of importance, but the deeper questions include what technologies farmers choose to use, what factors, either external or internal, induce farmers to adopt certain technologies, and what other factors will influence the ultimate success of these technologies against changing climates. With the assistance of NGOs, farmers in Cuzco province are encouraged to test potato varieties to identify which varieties are best adapted for local conditions. In addition, farmers diversify along elevation gradients to identify which varieties work best at certain elevations [78]. These efforts have become an important mechanism for risk management and increasingly, in climate change adaptation [21]. They also provide an important lesson on the relationship between poverty and adaptation. Typically, resource-reliant communities are viewed both as acutely vulnerable and lacking in adaptive capacity, but Andean farmers have confronted climatic variability for centuries through the careful selection and cultivation of crop varieties [85].

Much of the more recent technological research for climate change adaptation surrounds one of the Andes most well-known indigenous crops: the potato. The Potato Park (El Parque de la Papa) in the southeastern portion of Cusco province in southcentral Peru has become an internationally recognized exemplar of collective action in agricultural research and preservation. The park spans an area of 29,000 acres and is home to over 6,500 members of six indigenous communities. Formed in 2000 by the Cusco based NGO, the Association for Nature and Sustainable Development (ANDES), the park's mission is to preserve Andean agricultural and cultural diversity to promote local rights and livelihoods. Traditional knowledge, in the form of agricultural practices and the customs and beliefs associated with these practices, is viewed as essential to the survival of indigenous communities and culture. The park has provided a shared mechanism for agricultural research and information sharing founded on co-production. Local communities elect individuals to network with other communities to share and transfer new information as it develops. Research within the park is participatory and focuses on identifying distinct varieties, breeding, and crop improvement. In more recent years, the park has widened its scope to work with other NGOs, the Peruvian government, and with the international agricultural community. Dozens of community seed banks and indigenous conservation associations similar to the Potato Park are forming, and they too are organized by NGOs that operate at local, regional, and national levels. The potato park represents both social and technological innovation and is an example of how indigenous communities are complementing local knowledge with social innovation and modern tools to protect indigenous social-ecological systems [64].

A third case of institutional and technological innovation is exemplified by the work of the Asociación Andina Cusichaca (AAC), an NGO founded in 2003 in the southcentral Andean highlands. The organization is best known for its work on rehabilitating prehispanic agricultural terraces and canal systems using locally sourced materials. These terraces prevent surface runoff and soil erosion, create a favorable microclimate for growth, and enhance productivity. Thus, they provide an effective way to buffer the risk of extreme frosts and prolonged droughts, which makes them suitable as an adaptation strategy. The ACC has devoted time and energy into cultivating local involvement and leadership. It provides information sessions, classes, and specialized knowledge exchange sessions on a regular basis. The AAC also offers internships to local youth, creating links to younger generations, which will ensure that knowledge persists within communities.

The AAC created a training program to instruct local residents on how to restore and maintain the systems. This training is ongoing and participants have traveled to other districts and have been involved in regional and even national seminars on rehabilitation. The revival of the canals and terraces has also renewed traditional water festivals and canal cleaning ceremonies that had disappeared in some districts. Now farmers in some districts are beginning to rehabilitate terraces on their own, an indication that communities are reinstituting pride and belief in Andean traditions. This is yet another example of the relationship between institutions and technology; when strengthened, the ties between community members can enhance the success and sustainability of technological solutions.

These three examples make the case that technological change that is grounded in local knowledge and mediated through local networks can increase the adaptability of agricultural systems faced with climate change. They also support the notion that adaptation requires approaches where traditional knowledge systems and science are integrated within institutions that create avenues to transfer information while empowering local communities. The role of institutions in the adoption of new technology may be as important as the technology itself. In some instances, changes in agricultural practices simply cannot be made at the individual level and require community and outside support [64]. Future research on the institutional dimensions of climate change adaptation is needed to assess the social and ecological outcomes of these institutional arrangements on both short and long term horizons. Practitioners and researchers alike may wish to identify what features are best able to facilitate technological transfer and adoption and ultimately, what features are best able to enhance the adaptive capacity of local systems.

4. Conclusion

The future of Peru's tropical glaciers appears bleak, as glacial retreat is projected to accelerate. Increase in atmospheric temperatures is projected to continue, accompanied by decreases in precipitation, raising serious concerns for water availability. There is a need for continued and expanded research throughout the region on glacial mass balance studies, climate change impacts, ENSO occurrences, and glacier behavior. This research will allow for better knowledge of climate forcing on glacier mass balance, which can inform better management of water resources. These biophysical changes present formidable challenges to the social-ecological systems in highland communities. Though these communities have always existed on the edge of climatic extremes, climate change combined with socioeconomic changes may push communities beyond their range of adaptability [21]. Migration between areas and to major cities has led to the loss of local ecological knowledge [86]. In addition, the pressures of globalization and the rise of export agriculture have caused local knowledge to erode as traditional crop varieties are replaced with fast-growing, imported varieties for export markets [87]. Despite these complex and dynamic challenges, indigenous knowledge and technologies may contain great potential for some communities, which will benefit from continued and renewed attention on how to create appropriately scaled institutions to support their full integration.

Technology in and of itself is no silver bullet. However, institutions that are sensitive to local contexts and receptive to changing conditions can provide an avenue for the revival of old technologies and the integration of new ones in agricultural systems. The relationship between technological innovation and institutional change within farming communities, and especially, the processes and channels by which it occurs presents an important direction for future analysis. Climate change adaptation in the Peruvian highlands displays a continuous process of innovation made possible by enhancing, and in some instances reviving, traditional knowledge and technology. These intersections represent an area of rich study and future insights into the capacity of communities to adapt to climate change.

Vulnerability and adaptation in the Peruvian Highlands are interrelated and dynamic. Both the biophysical and social environments in this region are rapidly changing. Agriculture in the Andean highlands will need to undergo measurable shifts to less water intensive crops and more efficient irrigation practices, and in many instances, this shift will continue to be mediated through institutions and technology. A fusion of local ecological knowledge, technological innovation, and renewal of old technologies complemented by institutional support may offer the best mix of strategies to help communities to adapt. The retreat of glaciers has significant implications for Peruvian society as a whole. Though the most vulnerable are often considered ill equipped to adapt to climate change, inhabitants of the Andean highlands have adjusted to harsh climatic conditions for millennia. Despite a bleak picture, there are opportunities for experimentation with adaptation strategies in the Andes, which can offer important lessons and insights on the ability of human societies to meet and overcome the many challenges of climate change.

Author details

Ann Marie Raymondi
School of Sustainability, Arizona State University, Tempe, USA

Sabrina Delgado Arias
Consortium for Science Policy and Outcomes, Arizona State University, Washington DC, USA

Renée C. Elder
School of Geographical Sciences and Urban Planning, Arizona State University, Tempe, USA

5. References

[1] Beniston M (2003) Climate Change in Mountain Regions: A Review of Possible Impacts. Clim. change. 59: 5-31.

[2] Kaser G (2006) Mountain Glaciers. In: Knight PG editor. Glacier Science and Environmental Change. Malden: Blackwell Publishing. pp. 268-271.

[3]   Stern N (2007) The Economics of Climate Change: The Stern Review. Cambridge: Cambridge University Press. 700 p.

[4]   Painter J (2007). Deglaciation in the Andean region. New York: United Nations Development Programme. 21 p.

[5]   Mark BG, Seltzer GO (2005) Evaluation of Recent Glacier Recession in the Cordillera Blanca, Peru (AD 1962-1999): Spatial Distribution of Mass Loss and Climatic Forcing. Quat. scie. rev. 24: 2265-2280.

[6]   Hofer M, Molg T, Marzeion B, Kaser C (2010) Empirical-Statistical Downscaling of Reanalysis Data to High-Resolution Air Temperature and Specific Humidity Above a Glacier Surface (Cordillera Blanca Peru) J. geophys. res., 115:D12.

[7]   Racoviteanu AE, et al. (2008) Decadal Changes in Glacier Parameters in the Cordillera Blanca, Peru, Derived From Remote Sensing. J. Glaciol. 54: 499-510.

[8]   Bury J, et al. (2007) Glacier Recession and Human Vulnerability in the Yanamarey Watershed of the Cordillera Blanca, Peru. Glob. & plan. change. 59:189-202.

[9]   Bury J, et al. (2011) Glacier Recession and Human Vulnerability in the Yanamarey Watershed of the Cordillera Blanca, Peru. Clim. change. 105:179-206.

[10] Beniston M (2003) Climate Change in Mountain Regions: A Review of Possible Impacts. Clim. change. 59: 5-31.

[11] Vuille M, et al (2008) Climate Change and Tropical Andean Glaciers: Past, Present and Future. Earth scie. rev. 89: 79-96.

[12] Condom T, et al. (2007) Computation of the Space and Time Evolution of Equilibrium-Line Altitudes on Andean Glaciers (10°N–55°S). Glob. and plan. change. 59:189-202.

[13] Kaser G (1999) A Review of Modern Fluctuations of Tropical Glaciers. Glob. and plan. change. 22: 93 – 103.

[14] Coudrain A, et al. (2005) Glacier Shrinkage in the Andes and Consequences for Water Resources. Hydrological Sciences – Jounals des Sciences Hydroliques. 50: 9225-932.

[15] Collins M (2000) Understanding Uncertainties in the Response of ENSO to Greenhouse Warming. Geophy. res. lett. (21): 3509 -3512.

[16] Tsonis A, et al. (2002) On the Relation Between ENSO and Global Climate Change. Meteor. and atmos. phys. 000:1-14.

[17] Smith J, et al. (2008) The Timing and Magnitude of Mountain Glaciation in the Tropical Andes J. quat. scie. 23;609-634.

[18] Francou B, et al. (2004) New Evidence for an ENSO Impact on Low-Latitude Glaciers: Antizana 15, Andes of Ecuador, 0°28′S. J. of geophy. res. 109:1-17.

[19] Jomelli V, et al. (2009) Fluctuations in Glaciers in Tropical Andes over the Last Millennium and Paleoclimatic Implications: A Review. Palaeogeo. paleoclima. paleoecol. 281:269-282.

[20] Timmermann A, et al (1999) Increased El Nino frequency in a Climate Model Forced by Future Greenhouse Warming. Lett. to nat. 398:694-697.

[21] Perez C, et al (2010) Small-scale Farmers. Climate Change in the High Andes: Implications and Adaptation Strategies for Small-scale Farmers. Int. j. env. cult. econ. soc. sust. 6:1-16.

[22] Füssel HM (2005) Vulnerability to Climate Change: A Comprehensive Conceptual Framework. In: Breslaur Symposium Paper 6. Berkeley:University of California International and Area Studies. 34 p.

[23] IPCC (2001) Climate Change: Impacts, Adaptation, and Vulnerability. In: McCarthy JJ, Canziani OF, Leary NA, Dokken DJ, White KS, editors. Cambridge: Cambridge University Press. 1032 p.

[24] Parry ML, et al (eds) (2007) Chapter 2: New Assessment Methods and the Characterisation of Future Conditions. In: Climate Change 2007: Impacts, Adaptation and Vulnerability. Cambridge: Cambridge University Press. 976 p.

[25] Brooks N (2003) Vulnerability, Risk and Adaptation: A Conceptual Framework. Tyndall for climate change research working paper. 38: 1–16.

[26] Malone EL, Engle NL (2011) Evaluating Regional Vulnerability to Climate Change: Purposes and Methods. WIREs clim. chan. 2: 462-474.

[27] Hegglin E, Huggel C (2008) An Integrated Assessment of Vulnerability to Glacial Hazards: A Case Study in the Cordillera Blanca, Peru. Mount. res. and dev. 28.299-309.

[28] Ribot J (2010) Vulnerability Does Not Fall From the Sky: Toward Multiscale, Pro-Poor Climate Policy. In: Means R, Norton A, editors. Social Dimensions of Climate Change: Equity and Vulnerability in a Warming World. Washington DC: The World Bank. pp. 47-74.

[29] Adger WN (1999) Social Vulnerability to Climate Change and Extremes in Coastal Vietnam. World dev. 27:249–269.

[30] Brooks N, Adger NW (2005) Assessing and Enhancing Adaptive Capacity. In Lim B, Spanger-Siegfried E, editors. Adaptation Policy Frameworks for Climate Change: Developing Strategies, Policies and Measures. Cambridge: Cambridge University Press/ UNDP-GEF. pp. 165-181.

[31] Eakin H (2005) Institutional Change, Climate Risk, and Rural Vulnerability: Cases From Central Mexico. World dev. 33:1923-1938.

[32] Turner BL, Kasperson RE, Matson PA, McCarthy JJ, Corell RW, Christensen L, et al (2003) Framework for Vulnerability Analysis in Sustainability Science. Proc. natl. acad. scien. 100: 8074-8079.

[33] Hoffman D (2010) Andean Glaciers Vanish, Add Socio-Economic Strains. FOCALPoint. 9: 13-15. Available: http://www.focal.ca/pdf/focalpoint_february2010.pdf . Accessed 2011 Sept 18.

[34] Hennessy H (2005) Peru's Glaciers in Retreat. BBC News. Available: http://news.bbc.co.uk/2/hi/americas/4720621.stm. Accessed 2011 Sept 26.

[35] Chevellier P, Pouyard B, Suarez W, Condom T (2011) Climate Change Threats to Environment in the Tropical Andes: Glaciers and Water Resources. Reg. envi. chan. 11: 179-87.

[36] Elgegren C (2009) Biodiversity and Ecosystems: Why these are Important for Sustained Growth and Equity in Latin America and the Caribbean- Peru Country Case Study. Lima, Peru: UNDP.

[37] (2011) The World Factbook: Peru. Central Intelligence Agency. Available: https://www.cia.gov/library/publications/the-world-factbook/geos/pe.html. Accessed 2012 Mar 25.

[38] Kozloff, N (2010) Peru No Rain in the Amazon: How South America's Climate Change Affects the Entire Planet. Palgrave Macmillan.

[39] Deutsch Lynch, B (2005) Equity, Vulnerability and Water Governance: Responding to Climate Change in the Peruvian Andes. Ceará Convention Center Fortaleza. Available: http://www.icid18.org/files/articles/566/1277944530.pdf. Accessed 2011 Nov 15.

[40] (2009) Peru: Country Notes on Climate Change Aspects in Agriculture. World Bank. Available: http://siteresources.worldbank.org/. Accessed 2012 Mar 25.

[41] Spang E (2006) Alpine Lakes and Glaciers in Peru: Managing Sources of Water and Destruction. Tufts University Institute of Environment. Unpublished.

[42] (2003) Melting Glaciers Threaten Peru. BBC NEWS. Available: http://news.bbc.co.uk/2/hi/americas/3172572.stm. Accessed 2011 Sept 26.

[43] Josephs J (2010) BBC News: Melting Glacier Floods Towns in Peruvian Andes. YouTube. Available: http://www.youtube.com/watch?v=9FiQBugs1-E. Accessed 2011 Sept 26.

[44] (2010) Deadly Tsunami in Andes Lake, Peru. Fire Earth. Available: http://feww.wordpress.com/2010/04/13/deadly-tsunami-in-andes-lake-in-peru/. Accessed 2011 Nov 15.

[45] Grossman D (2011) Peru: Global Warming Makes a Splash. Pulitzer Center. Available: http://pulitzercenter.org/articles/carhuaz-peru-climate-change-global-warming-glacier-flood. Accessed 2011 Nov 15.

[46] Giugale MM, Newman JL, Fretes-Cibils V (2007) An Opportunity for a Different Peru: Prosperous, Equitable, and Governable. Washington, D.C.: World Bank.

[47] Mertz O, Halsnaes K, Olesen JE, Rasmussen K (2009) Adaptation to Climate Change in Developing Countries. Environ manage. 43:743-752.

[48] Adger WN, et al. (2007) Assessment of Adaptation Practices, Options, Constraints and Capacity. In: Parry ML, Canziani OF, Palutikof JP, van der Linden PJ, Hanson CE, editors. Climate Change 2007: Impacts, Adaptation and Vulnerability. Contribution of Working Group II to the Fourth Assessment Report of the Intergovernmental Panel on Climate Change. Cambridge: Cambridge University Press. pp. 717-743.

[49] Engle NL (2011) Adaptive Capacity and its Assessment. Glob. env. change. 21:647-656.

[50] Martínez A, et al. (2006) Vulnerability and Adaptation to Climate Change in the Peruvian Central Andes: Results Of A Pilot Study. In: Proceedings of the 8th Annual ICSHMO. Brazil: National Institute for Space Research. pp. 297-305.

[51] Chapin FS, et al. (2006). Building Resilience and Adaptation to Manage Artic Change. Ambio. 35:198-202.

[52] Postigo JC, Young KR, Crews KA (2008) Change and Continuity in a Pastoralist Community in the High Peruvian Andes. Hum. ecol. 36:535-551.

[53] Browman DL (1974) Pastoral Nomadism in the Andes. Curr. Anth. 15: 2188–196.

[54] Smit B, Wandel J (2006) Adaptation, Adaptive Capacity and Vulnerability. Glob. env. change. 16:282-292.

[55] Adger NW, Eakin H, Winkels A (2008) Nested and Teleconnected Vulnerabilities to Environmental Change. Front. ecol. environ. DOI:101.1890/070148.

[56] Dow K, Kasperson RE, Bohn M (2006) Exploring the Social Justice Implications of Adaptation and Vulnerability. In: Adger WN, Paavola J, Huq S, Mace MJ, editors. Fairness in Adaptation to Climate Change 79. pp. Cambridge: Massachusetts Institution of Technology. pp. 79-96.

[57] Deressa TT, Hassan RM, Ringler C, Alemu T, Yesuf M (2009) Determinants of Farmer's Choice of Adaptation Methods to Climate Change in the Nile Basin of Ethiopia. Glob. Env. change. 19:248-255.

[58] Nielsen JØ, Reenberg A (2010) Cultural Barriers to Climate Change Adaptation: A Case Study from Northern Burkina Faso. Glob. env. change. 20:142-152.

[59] Rubio ET (2007) Climate Change Impacts and Adaptation in Peru: The Case of Puno and Piura. New York: United Nations Development Programme. 15 pp.

[60] Crosson P (1983) A Schematic View of Resources, Technology and Environment in Agricultural Development. Agr. ecosy. env. 9:339–357.

[61] Smithers J, Blay-Palmer A (2001). Technology Innovation as a Strategy for Climate Adaptation in Agriculture. Appl. geo. 21:175-197.

[62] Smit B, Skinner MW (2002) Adaptation Options in Agriculture to Climate Change: a Typology. Mit. adap. strat. glob. chang. 7:85-114.

[63] Wehbe M, et al. (2006) Local perspectives on adaptation to climate change: Lessons from Mexico and Argentina. AIACC Working Paper. Available: http://www.aiaccproject.org/working_papers/Working%20Papers/AIACC_WP39_Wehb e.pdf. Accessed 2012 Mar 25.

[64] Mijatovic D, et al. (2010) The Use of Agrobiodiversity by Indigenous and Agricultural Communities in: Adapting to Climate Change. Rome:Platform for Agrobiodiversity Research. 31 p.

[65] Adger N (2003) Social Capital, Collective Action, and Adaptation to Climate Change. Econ. geog. 79: 387-404.

[66] Agrawal, A (2008) The Role Of Local Institutions In Adaptation To Climate Change. Social Dimensions of Climate Change Conference Proceedings. Washington DC: World Bank. 65 p.

[67] Ostrom E (1990) Governing the Commons: the Evolution of Institutions for Collective Action. Cambridge: Cambridge University Press. 281 p.

[68] Upton C (2012) Adaptive Capacity and Institutional Evolution in Contemporary Pastoral Societies. App. geo. 33:135-141.

[69] Chhetri NB, Easterling WE (2010) Adapting to Climate Change: Retrospective Analysis of Climate Technology Interaction in the Rice-Based Farming System of Nepal. Ann. assoc. ameri. geog. 100: 1156-1176.

[70] Bebbington A (1990) Farmer Knowledge, Institutional Resources, and Sustainable Agricultural Strategies: A Case Study from the Eastern Slopes of the Peruvian Andes. Bull. latin am. res. 9:203-228.

[71] Dillehay TD, Kolata AL (2004) Long-Term Human Responses to Uncertain Environmental Conditions in the Andes. Proceed. natl. acad. scien. 12: 4325-4330.

[72] Berkes F, Colding J, Folke C (2000) Rediscovery of Traditional Ecological Knowledge as Adaptive Management. Ecol. appl. 10:1251-1262.

[73] Gilles J, Valdivia C (2009) Local forecast Communication in the Altiplano. Bull. amer. meteor. soc. 90:85–91.

[74] Young KR, Lipton JK (2006) Adaptive Governance and Climate Change in the Tropical Highlands of Western South America. Climatic change. 78:63-102.

[75] Orlove B, Chiang J, Cane M (2002) Ethnoclimatology in the Andes: A Cross-Disciplinary Study Uncovers a Scientific Basis for the Scheme Andean Potato Farmers Traditionally use to Predict the Coming Rains. Amer. scien. 90:428–435.

[76] O'Brien K, Hochachka G (2010) Integral Adaptation to Climate Change. J. int. theor. pract. 5: 89–102.

[77] Silberner J (2008) In Highlands Peru, a Culture Confronts Blight. All Things Considered. Available: https://www.npr.org/templates/story/story.php?storyId=87811933. Accessed 2010 Mar 25.

[78] Conger L (2008) A Quest for the Perfect Potato. Newsweek mag. Available: http://www.thedailybeast.com/newsweek/2008/07/25/a-quest-for-the-perfect-potato.html. Accessed 2010 Mar 25.

[79] Jacobsen SE, Mujica A, Ortiz R (2003) The Global Potential for Quinoa and Other Andean Crops. Food. rev. intl. 19:139-148.

[80] Trawick PB (2001) Successfully Governing the Commons: Principles of Social Organization in an Andean Irrigation System. Hum. ecol. 29:1-25.

[81] Mitchell WP (1976) Irrigation and Community in the Central Peruvian Highlands. Am. anthro. 78: 25-44.).

[82] Delgado JV, Zwarteveen M (2008) Modernity, Exclusion, and Resistance: Water and Indigenous Struggles in Peru. Dev. 51:114-120.

[83] Erickson CL (2003) Agricultural Landscapes As World Heritage: Raised Field Agriculture In Bolivia And Peru. In: Managing Change: Sustainable Approaches to the conservation of the Built Environment. Oxford: Oxford University Press. pp. 181–204.

[84] United Nations Framework Convention on Climate Change (2007) Impacts Vulnerability and Adaptation in Developing Countries. Bonn: UNFCCC. 64 p.

[85] McSweeney K, Coomes OT (2011) Climate-Related Disaster Opens a Window of Opportunity for Rural Poor in Northeastern Honduras. Proc. natl. acad. scie. 108:5203-5208.

[86] Valdivia C, et al. (2010). Landscapes, Capitals, and Perceptions Shaping Rural Livelihood Strategies and Linking Knowledge Systems. Ann. assoc. amer. geog. 100:818-834.

[87] Kirkland E (2011) Peru Looks to Traditional Adaptation Practices to Cope with Climate Destruction. Earth island j. Available: http://www.earthisland.org/journal/index.php/elist/eListRead/peru_looks_to_traditiona l_practices_to_cope_with_climate_disruption/. Accessed 2012 Mar 25.

Chapter 3

Communicating the Needs
of Climate Change Policy Makers to Scientists

Molly E. Brown, Vanessa M. Escobar and Heather Lovell

Additional information is available at the end of the chapter

1. Introduction

In the confusion of the national conversation on climate change issues, a clear and explicit narrative can help cut through the chatter. Science can provide information to improve societal outcomes by focusing debate and guiding policy in ways that are transformative. The science that is done to support climate change policy, however, must be focused and relevant. The purpose of this chapter is to suggest ways that policy and decision-maker needs can be communicated to scientists working to improve the understanding of processes, relationships and products in climate change science. A partnership between science and policy must be forged at multiple levels and at many time scales in order to be effective. Many organizations are developing programs that seek to increase the relevance of its science and data products to decision makers grappling with science, influencing not only the scientific questions that are asked, but also the format, resolution and scale of the data output. It is only through two-way communication and relationship building that effective partnerships can be built which will help policy makers have the scientific foundations they need.

This chapter will describe the challenges that earth scientists face in developing science data products relevant to decision maker and policy needs, and will describe strategies that can improve the two-way communication between the scientist and the policy maker. Climate change policy and decision making happens at a variety of scales – from local government implementing solar homes policies to international negotiations through the United Nations Framework Convention on Climate Change. Scientists can work to provide data at these different scales, but if they are not aware of the needs of decision makers or understand what challenges the policy maker is facing, they are likely to be less successful in influencing policy makers as they originally intended. This is because the science questions they are addressing may be compelling, but not relevant to the challenges that are at the forefront of policy concerns.

In this chapter we examine case studies of science-policy partnerships, and the strategies each partnership uses to engage the scientist at a variety of scales. We examine three case studies: the global Carbon Monitoring System pilot project developed by NASA, a forest biomass mapping effort for Silvacarbon project, and a forest canopy cover project being conducted for forest management in Maryland. In each of these case studies, good relationships between scientists and policy makers were critical for ensuring the focus of the science as well as the success of the decision-making.

1.1. Background

Meeting the needs of decision makers requires a transformational change in how environmental research is organized and incorporated into public policy in the United States (NRC 2009). Although there has been much discussion in the literature on the need for scientists to clearly and accurately discuss their results (Pettricrew et al. 2004), little attention has been paid to how to communicate the needs of the policy community to scientists. The information needs of decision makers, and how they use scientific information needs to be clearly presented and communicated to scientists so that they can do the necessary research and focus on the processes that are truly important to society.

Increasing the usage of scientific evidence in policy-making therefore requires that scientists increase their understanding and engagement with these organizations and individuals (Jones and Walsh 2008). By making explicit and testing assumptions underlying the way a policy is supposed to work, researchers can identify additional questions for which existing empirical evidence can be sought. In this way, sequences of evidence can be gathered and accumulated to provide a rounded and appropriate evidence base for decision-making (Davies 2005).

As scientists we need to publish our results in multiple venues, including those where policy makers can find and understand our results. A researcher can greatly increase the likelihood that their results will be used and will influence climate change policy by documenting their research findings in clear, detailed and uncomplicated writing. Policy makers and other users of research evidence are usually quite aware that the scientific issues surrounding policies are complex (Davies 2005). However, the transformation of technical language used in scientific reports into user-friendly terms is worthwhile, but often requires a two-way conversation between the policy maker and the scientist to ensure the relevance of the science.

It has also been argued that researchers would help policy makers use research evidence more effectively if they could identify, report and present the key findings with greater clarity. Involving policy makers and other research users throughout the research process, and identifying the implications for policy and practice, might also enhance the utilization of research evidence in policy making (Davies 2005). In the end, in order to be relevant to policy and decision makers, scientific conclusions need to be important to the known users and relevant throughout the development process. The outcomes not only need to have a societal impact but in order to be relevant they must also be financially feasible.

Scientific research, which is often not bound by time constraints, is difficult to integrate with the time sensitive demands of politicians who are compelled to work under tight deadlines to produce short- term, tangible policy results. However, policy-makers often struggle to stay apace of new scientific thinking, especially in terms of developing relevant policies and infrastructure to enable as well as regulate the implementation of scientific and technological advances (Alcock 2002). Fostering an ongoing, interactive relationship between the two communities, and clearly addressing each groups' sensitivity to implementation, quickly lessen these issues.

In addition to this valuable range of practical issues related to the climate change/science-policy interface, there are a number of academic studies that are useful to consider in terms of their insights into the relationship between science and policy (Jamison 2001; Jasanoff et al. 1995; Litfin 1994; Wynne and Irwin 1996), as well as the nature of the policy process (in particular how policy change takes place) (Kingdon 2003; Sabatier 1999; Smith 1997). One finding from these studies which is pertinent to the work of NASA and our specific case studies discussed below, is that the process of policy change, much like science, is uncertain and tends to be 'bumpy'; characterized by long periods of stability with little change or progress, interspersed with times of rapid innovation and upheaval of established ideas and ways of doing things. In literatures on policy change and science innovation this pattern of change has been termed 'punctuated equilibrium' (John 2003; Phillimore 2001; True et al. 1999). The relevance of this insight for the role of NASA (and science more generally) is in conceptualizing what NASA and other science agencies do as *providing the science base for policy*. In other words, the science findings from NASA studies might well not provoke rapid immediate change in policy (sometimes this does happen, but it is rare), but rather that these findings will be there and available to policy makers as a 'solution' as and when a particular policy problem arises that demands them.

The work of the US political scientist John Kingdon (2003) eloquently explains this matching of policy problems and solutions in his book 'Agendas, Alternatives and Public Policies'. In his discussion of 'the policy primeval soup' – his metaphor for describing the chaotic nature of policy in which a messy mix of policy problems, politics and solutions floats around US government chambers and policy circles - Kingdon explains how a policy problem is much more likely to rise on the government agenda if a solution is already there and worked out, as he explains (2003: 142):

"It is not enough that there is a problem, even quite a pressing problem. There is also generally a solution ready to go, already softened up, already worked out."

Thus the role of climate change science is to engage with government, to be part of the 'policy primeval soup', but also to work to provide science-based solutions to current policy problems as well as emerging future problems, which are as yet only hazily defined. It is with this in mind that we turn to consider our case studies: three different projects are examined that seek to bring together policy and decision makers with scientists working to do relevant science. In each project, the challenges scientists face are different, but the solution of increased interaction, product clarification and connection between the users of science and the producers, is the same.

2. Case study 1: NASA's Carbon Monitoring System

In 2007 the US National Research Council released the first earth science decadal survey report recommending "a suite of satellite missions and complementary activities that serve both scientific and applications objectives for the nation" (NRC 2007). The report presented a vision for developing new satellite data products that have specific user communities' needs and requirements at the forefront of the mission development. Meeting this objective will require a transformation of the way that NASA traditionally does business. The NASA Carbon Monitoring Systems initiative is meeting this objective by re-evaluating priorities and integrating the local needs of society into the development of carbon science products. Two of the NRC report's priorities over the next decade are (1) to develop the science base and infrastructure to support a new generation of coupled Earth system models to improve attribution and prediction of high-impact regional weather and climate; and (2) to strengthen research on adaptation, mitigation and vulnerability. The Carbon Monitoring System (CMS project) addresses both of these issues with a consortium of end users and policy decision makers.

2.1. NASA's Carbon Monitoring System pilot project

The Carbon Monitoring System (CMS) is a NASA initiative designed to make significant contributions in characterizing, quantifying, understanding, and predicting the evolution of global carbon sources and sinks. The study uses satellite observations and model outputs to calculate human produced carbon dioxide (CO^2) while discussing effective delivery mechanisms with policy bodies such as the Environmental Protection Agency (EPA), the US State Department and others.

NASA CMS conducts pilot studies to provide information across a range of spatial scales that seeks to improve measures of the atmospheric distribution of carbon dioxide. NASA has initiated this work by building on its global measurement capability for carbon. Other agencies and organizations have ongoing activities that are related to CMS, that support national carbon policy objectives and resource management; most notably the National Oceanic and Atmospheric Administration (NOAA)'s Carbon Tracker program, the US Geological Survey's carbon sequestration efforts, and National Institute for Standards and Technology's Greenhouse Gas Measurements and Climate Research Program. Thus coordination across these and other climate programs is critical to ensure long-term utility.

Emissions from vegetation disturbance and land-use and land-cover change are the most uncertain component of the global carbon cycle (Prentice et al. 2000). The CMS pilot project is designed to address the urgent need for geospatially explicit, observed (not modeled) carbon and biomass inventory information to inform national and international policy-making. The project addresses two objectives: 1) to develop prototype data products of national and global biomass (carbon storage and emissions) that can be assessed with respect to how they meet the nation's needs for Monitoring, Reporting, and Verification

(MRV) of carbon inventories; and 2) to demonstrate readiness to produce a consistent global biomass/carbon stock distribution using the existing in situ and satellite observations to meet the MRV requirements (Pawson and Gunson 2010).

The CMS flux pilot involved multiple institutions (four NASA centers, as well as several universities) and over 20 scientists in its development. This pilot study strives to use complimentary models to transform satellite-derived observations into quantities that are both meaningful and useful for carbon cycle science and policy. The CMS pilot will generate CO_2 flux maps for two years, using observational constraints in NASA's models. Bottom-up estimates (the movement of carbon dioxide from the land surface to the atmosphere) of the CO_2 flux will be computed using data-constrained land and ocean models; comparison of the different techniques will provide some knowledge of uncertainty in these estimates. Ensembles of atmospheric carbon distributions will be computed using an atmospheric general circulation model (GEOS-5), with perturbations to the surface fluxes and to transport. Top-down flux estimates (absorption of carbon dioxide by plants on the land from the atmosphere) will be computed from observed atmospheric CO_2 distributions and model retrievals alongside the forward-model fields, in conjunction with an inverse approach based on the CO_2 model (Figure 1). The forward model ensembles will be used to build understanding of relationships among surface flux perturbations, transport uncertainty and atmospheric carbon concentration. This will help construct uncertainty estimates and information on the true spatial resolution of the top-down flux calculations. The relationship between the top-down and bottom-up flux distributions will be documented (Pawson and Gunson 2010).

Because the goal of NASA CMS is to be policy relevant, the scientists involved in CO2 flux modeling pilot need to understand and be focused on the needs of the climate policy community. How should the data be presented? What analysis of the data would be most useful for policy makers? What is the time scale of the information needed by decision makers (daily fluxes, annual, 5-year)? What is the optimal spatial resolution of these products? What is the needed accuracy of the information? If the answers to these questions are communicated to scientists working on the pilot study, it is more likely that the project will be relevant and produce the answers that are needed by policy and society.

2.2. Policy and NASA's CMS System

Because of its ambitious goal to produce products relevant to policy, NASA has organized meetings between policy makers, decision makers and CMS scientists to ensure that the data products being developed are relevant and responsive to the needs of policy makers. In September 2011, a meeting in Washington D.C between NASA CMS flux scientists and local DC policy decision makers provided an overview of the status of the NASA CMS flux pilot and data products under development, and provided a forum to discuss how to better characterize uncertainty in CO2 measurements. The focus of the meeting was to ensure that the data are able to meet the needs of other agencies and organizations engaged in flux measurement and monitoring. Early product development conversations such as this will enable NASA to generate better overall products in support of agency needs. Much of the discussion during the CMS flux meeting focused on how the CMS pilot products could contribute to US carbon policy and decision making.

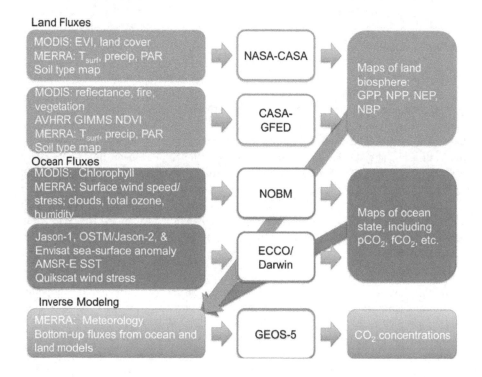

Figure 1. NASA Carbon Monitoring System Flux project data inputs, outputs and connections.

The CMS flux products are based on satellite observations of land, ocean, and atmosphere, as well as CO2 concentrations. The CO2 flux estimation that can be attributed to a specific location on the ground and could complement Global Climate Models and direct CO2 atmospheric observations. Were a mitigation policy be put into place, decision makers would need a mechanism to know if the policy was making an impact.. The CMS effort will be able to provide information on the underlying emissions, irrespective of whether a policy intervention requires voluntary or mandatory actions. NASA can work to ensure that CO2 models are used with observations from satellite observatories to provide information on the success of mitigation efforts.

In order to make a difference with climate policies, we need to know CO_2 trends through time. Sustained observational monitoring is necessary for carbon management. NASA is well positioned to do this task and no one else in the federal government has this responsibility in the federal government. There is a significant need for scientific infrastructure to determine if regulations and policies put in place (on the local, state and federal levels) are making a difference. This need for scientific infrastructure is usually forgotten. It is also difficult to fund because it is perceived as unimportant and requires continuous support, despite it being at the center of effective programs and policies. However NASA's engagement between scientists and end users is designed to remind society of the relevance of scientific structure. NASA CMS will provide a key to better understand what such a system will look like. CMS will enable us to estimate the impacts of our policies through the use of satellite observations. We need to ensure that the resolution, time step and uncertainty of the CMS CO_2 flux products are adequate for these needs - keeping an open line of communication with the scientist will be necessary for a developing a successful product.

Through briefings and presentations at meetings, scientists involved with CMS have learned about policy maker needs. This knowledge will affect how the CMS project moves forward. Questions regarding next steps in the project, such as working to improve the spatial resolution or to improve the fidelity of ocean models, for example, can be decided with policy objectives in mind. This is important, as the group working on CMS flux models is large, interdisciplinary, and is fundamentally interested in producing an output relevant to policy makers.

3. Case study 2: Mapping the forests for REDD

In 2010, the United Nations climate negotiations launched the Reducing Emissions from Deforestation and Forest Degradation (REDD) program. REDD is an effort to create financial value, as an incentive for the carbon stored in forests, offering developing countries environmental and financial benefits for reducing emissions from forested lands and to invest in low-carbon paths to sustainable development. The REDD program goes beyond deforestation and forest degradation. It includes the role of conservation, sustainable management of forests and enhancement of forest carbon stocks. Silva Carbon is the United States Government's contribution to the REDD methods through the GEO Forest Carbon Tracking task, a component of the Global Earth Observation System of Systems (GEOSS), which provides data and information about a variety of Earth observations to users around the world. The program is designed to strengthen global capacity to understand changes in land cover as well as monitor and manage forest and terrestrial carbon.

The United Nations is setting up systems of Measurement, Reporting and Verification (MRV) of forests in order for countries to benefit from the United Nations treaty. Thus countries will need to develop cost-effective, robust and compatible national monitoring systems. The REDD agreement defines MRV as:

- Measurement – The process of data collection over time, providing basic datasets, including associated accuracy and precision, for the range of relevant variables. Possible data sources are field measurements, field observations, detection through remote sensing and interviews with stakeholders.
- Reporting – The process of formal reporting of assessment results to the United Nations Framework Convention on Climate Change (UNFCCC), according to predetermined formats and according to established standards.
- Verification – The process of formal verification of reports, for example, the established approach to verify national communications and national inventory reports to the UNFCCC.

Understanding of how ground information can be used in conjunction with aerial measurements of forest height and canopy, together with satellite remote sensing data, is central to REDD and will influence the research that scientists are doing. It is no longer enough to be developing new models or to do novel, publishable research. REDD set a standard of being 'cost effective, robust and compatible'. Knowing this, how do scientists working on methodological approaches to map biomass engage with the REDD countries to ensure that they can meet this standard? How do they simultaneously engage with REDD, progress in their own careers and publish the work that they do?

3.1. Biomass mapping and REDD

Accurate and precise quantification of the amount of biomass in forests has become a key issue for policy makers as it is a key requirement of REDD for climate mitigation strategy. Active aerial instruments measuring the height and structure of vegetation (using lidar and radar observations) will quantify carbon stock and changes, improve our knowledge of the geographic distribution of carbon sources and sinks, and help us understand where carbon is being sequestered in the landscape. The distribution of biomass and carbon storage produced from the existing remote sensing and in situ measurements will provide sub-optimum, but necessary information to develop national and international scale REDD policies and MRV frameworks (Goetz et al. 2009).

The NASA contribution to Silvacarbon and REDD is a biomass mapping project designed to address the urgent need for geospatially explicit, consistent carbon and biomass inventory information to inform national and international policy. The project will address two objectives: 1) To develop prototype data products of national and global biomass (and carbon storage/emissions) that can be assessed with respect to how they meet the nation's needs for MRV of carbon inventories; and 2) to demonstrate our readiness to produce a consistent global biomass/carbon stock distribution using the existing in situ and satellite observations to meet the REDD monitoring, reporting and verification requirements (USAID 2011).

Biomass mapping can be the basis of a tool that could be used by investors to target REDD projects. Land cover and carbon density maps can be used together with information on agriculture and opportunity costs of land. This is especially relevant in addressing the needs of developing countries who have tropical forests and would like to have an MRV capacity, thus capturing REDD funding. This has resulted in the US government's development of the SilvaCarbon program. This program focuses on enhancing the scientific capacity of countries worldwide to map and monitor biomass in forests. SilvaCarbon will draw on the scientific expertise of the U.S. scientific and technical community including experts from government, academia, non-governmental organizations, and industry (USAID 2011). Working with developing countries and international institutions, SilvaCarbon works to enhance scientific capacity by identifying, testing, and disseminating good practices and cost-effective, accurate technologies for monitoring and managing forest and terrestrial carbon.

3.2. Communication challenges between scientists and decision makers

Organizations and government agencies are actively working to adjust to conservation in the context of REDD standards, which may take five to ten years to implement. Bilateral and multilateral agreements are now in place and are currently giving developing countries money to be part of REDD. The question now is how to make biomass-mapping part of the policy discussion here in the United States. Research, communication and relationships must be forged in a way that provides a metric for producing affordable, repeatable measurements that are spatially explicit.

We need large-scale datasets that have some defensibility, with clear estimation of the uncertainty of the data both in space and in time. For Silvacarbon, the social and economic factors that affect the success of a REDD program are uncertain, so improved ways of calculating biomass as well as better data acquisition methods are important. Each country will need to be able to implement the methodology for biomass monitoring at the scale.

In order to connect policy makers to scientists, the U. S. Geological Survey (USGS) and REDD hosted international scientists at a SilvaCarbon Workshop in September 2011. Scientists received satellite data and training for the data, which applied to their areas of study, while policy makers had the opportunity to explain the challenges they face in implementing REDD globally. A big part of this challenge was the spatial uncertainty that is due to different land histories and species contribution. Many biomass mapping methodologies do a poor job of estimating uncertainty, which affects the broader policy and program implementation. Thus new science that is done, seeking to be REDD relevant must use older technologies that are inexpensive and develop models that are rigorously tested, but simple to implement. The Silva Carbon workshop provided improved communication on the technological and scientific needs, and ensure that they were relevant to the MRV requirements of countries involved in REDD. Linking satellite observations to measurements taken from the ground and from independent instruments on airplanes is another strategy that can lead to new, inexpensive but highly accurate estimates of forest biomass that meet the needs both of scientists and of the community.

The SilvaCarbon Workshops are designed to coordinate with project partners in distribution of products to organizations in need and to help address issues of deforestation and carbon reduction. Each workshop has participants sharing discourse on projects and accomplishments in their regions, accessing and downloading datasets pertinent to their studies, and meeting with leading scientists working on biomass. Two additional workshops are planned in 2012. As the science of using satellite remote sensing to estimate biomass evolves, understanding the challenges of local, regional and international actors working to implement REDD will affect the way this science is focused.

4. Case study 3: Developing forest canopy change maps for forest managers

The Baltimore Washington Partners for Forest Stewardship (BWPFS) was formed in 2006 and is a coalition of federal landowners who have joined with leaders from the Maryland Department of Natural Resources and the Center for Chesapeake Communities to promote collaborative strategies for the restoration, conservation and stewardship of shared forested ecosystems and managed lands in the Baltimore Washington corridor. Current BWPFS partner agencies include the U.S. Department of Agriculture Beltsville Agricultural Research Center, U.S. Fish and Wildlife Service Patuxent Research Refuge, NASA/Goddard Space Flight Center, U.S. Army Fort George G. Meade, Cities of Greenbelt and Bowie and Town of Cheverly, Maryland-National Capital Park and Planning Commission, University of Maryland, U.S. Secret Service, U.S. Forest Service, and the U.S. Geological Survey. The 2011 partners' semi-contiguous boundaries have an area totaling over 69,000 acres, 38.3% of which is forested. This region is critical for ensuring that the Baltimore-Washington's water resources, air quality and other basic ecosystem services are maintained (Costanza 1996).

One of the issues that BWPFS community forest managers are coping with is a significant new reporting requirement under a Chesapeake Bay federal mandate. In 2011, the Chesapeake Bay was placed on the *Federal Impaired Waters List for Nutrients and Sediment*. This was the result of a successful 2008 lawsuit against the EPA by Chesapeake Bay watermen in two states and the Chesapeake Bay Foundation. The resulting watershed implementation plan resulted in significant reporting requirements as well as strict new storm water runoff and regulatory requirements that apply to federal, state and local jurisdictions. Forest cover is a critical input to these requirements, as they serve as a buffer around streams and tributaries that filter storm water, reducing sediment and pollutants before they reach the Bay. Climate change policy is a second important input for these communities, with the State of Maryland implementing new programs relevant to climate change that implicate forest management. Thus the BWPFS partners use satellite and other environmental data, but have needs that are not met by the current suite of products, particularly those that describe change through time at a sufficiently high resolution for community-based forest management. These needs include repeatable, quantitative and high-resolution tree canopy cover percentages and change through time, maps of impervious surfaces, and integration of forest information into storm water hydrological models for estimation of pollutants and sediment contribution to the Chesapeake Bay.

4.1. Science and decision makers working together

The BWPFS aims to promote forest stewardship through best management practices for contiguous forest in the Baltimore/Washington corridor. To encourage communication between scientists and decision makers, a meeting was held in September 2011 that focused on identifying areas where NASA data and applications can be utilized by partners to improve forest management or to promote forest stewardship. In turn, the forest managers had the opportunity to identify needs that cannot be met by currently available data and systems. The meeting was attended by 25 people from 20 different entities.

During this meeting a consensus was reached that the science community needs to improve their ability to produce temporally comparable products that can be used by decision makers at multiple agencies and incorporated into policy. Current systems for valuing ecosystem services are insufficient in determining the value of a given plot of forest. For example, a 1-acre plot of forest that is between a shopping mall and a stream may have greater ecosystem value than a 1 acre plot in a rural setting. This is because the forest near a stream in an urban setting absorbs runoff water coming from parking lots and buildings, catches and retains sediment and absorbs nutrients and keeps them from entering the water system. Being able to map the location and health of these urban tree plots is a critical part of the forest management in the Baltimore-Washington region. Scientific, remotely sensed data, can contribute to the monitoring and evaluating of forest health, critical for environmental management in the region. Regional and national datasets can help bridge the needs for continuous and independent information for reporting to the Federal and State governments, though local information will be needed to supplement.

By improving relationships with these communities, it is possible to develop approaches that provide data and information needed to ensure that the products developed by scientists are both regulation compliant as well as scientifically robust and repeatable into the future. Although forest mapping is on the agenda for many agencies and individuals, few products provide the information needed by decision makers (to include forest canopy percentage) that can be repeatedly measured through time. By bringing out the needs of these local decision makers, scientists can report this secondary product from models used to estimate biomass to address local environmental challenges.

5. Conclusions

In order to produce scientific data that is readily useful, it is important for scientists and potential end-users to exchange information and ideas early (and often) in the science product development process. Scientists and policy makers need to work together as much as possible within the chaotic 'policy primeval soup' (after Kingdon, 2003) to use science to identify policy problems as well as provide solutions. Many research organizations, have as their goal, to make products useful to a wide community of scientists, managers and policy makers. The voice of the user (i.e. not only those working directly in government, but also decision makers from business, local communities, charities etc. is helpful not just to these scientific programs, but to the entire community working on related activities. As decisions are made throughout the research development process, scientists need the voice of the user to, for example, specify needs and site details, so that policy relevant science is delivered.

In this chapter, we have provided examples from three research programs where scientists and decision and policy makers have been brought together to increase communication and understanding of each group. Ensuring strong relationships and knowledge of the problems policy makers have in their efforts to address climate and environmental change at a variety of scales is critical to ensuring science relevance. Our need for policy relevant scientific data products will continue to grow, in particular with the demands of managing climate change impacts at local, regional and national levels. Only through improved relationships and effective communication forums will we ensure that these needs are met and delivered to society.

Author details

Molly E. Brown
NASA Goddard Space Flight Center, Greenbelt MD, USA

Vanessa M. Escobar
Sigma Space/NASA Goddard Space Flight Center, Greenbelt, MD, USA

Heather Lovell
School of GeoSciences, The University of Edinburgh, UK

6. References

Alcock, F. (2002). Mobilizing Science and Technology for Sustainable Development. In. Cambridge, MA: Forum on Science and Technology for Sustainability

Costanza, R. (1996). Ecological Economics: Reintegrating the Study of Humans and Nature. *Ecological Applications, 6*, 978-990

Davies, P. (2005). What is Needed From Research Synthesis From a Policy Making Perspective? . In J. Popay (Ed.), *Putting Effectiveness Into Context*. London: Prime Minister's Strategy Unit, Cabinet Office, United Kingdom

Goetz, S.J., Baccini, A., Laporte, N., Johns, T., Walker, W.S., Kellndorfer, J.M., Houghton, R.A., & Sun, M. (2009). Mapping & monitoring carbon stocks with satellite observations:a comparison of methods. *Carbon Balance and Management*, 4

Jamison, A. (2001). Science, technology and the Quest for Sustainable Development. *Technology Analysis and Strategic Management, 13*, 9-22

Jasanoff, S., Markle, G.E., Petersen, J.C., & Pinch, T.J. (Eds.) (1995). *Handbook of Science and Technology Studies*. London: Sage

John, P. (2003). Is There Life After Policy Streams, Advocacy Coalitions, and Punctuations: Using Evolutionary Theory to Explain Policy Change? *Policy Studies Journal, 31*, 481-498(418)

Jones, N., & Walsh, C. (2008). Policy Briefs as a communication tool for development research. In, *Background Note*. Overseas Development Institute

Kingdon, J.W. (2003). *Agendas, alternatives and public policies*. New York: Harper Collins College Publishers

Litfin, K.T. (1994). *Ozone discourses: science and politics in global environmental cooperation*. New York: Columbia University Press

NRC (2007). Earth Science and Applications from Space: National Priorities for the Next Decade and Beyond. In. Washington DC: National Research Council

NRC (2009). Restructuring Federal Climate Research to Meet the Challenges of Climate Change. In. Washington DC: National Research Council of the National Academy of Science

Pawson, S., & Gunson, M. (2010). NASA CMS 2010, Pilot Study: Surface Carbon Fluxes. In. Greenbelt, MD: NASA

Pettricrew, M., Whitehead, M., Macintyre, S., Graham, H., & Egan, M. (2004). Evidence for public health on inequalities: 1: The reality according to policymakers. *Journal of Epidemiology and Community Health, 58*, 811-816

Phillimore, J. (2001). Schumpeter, Schumacher and the Greening of Technology. *Technology Analysis and Strategic Management, 13*, 23-37

Prentice, I.C., Heimann, M., & Sitch, S. (2000). The carbon balance of the terrestrial biosphere: Ecosystem models and atmospheric observations. *Ecological Applications, 10*, 1553-1573

Sabatier, P.A. (1999). *Theories of the Policy Process: theoretical lenses on public policy*. Boulder: Westview Press

Smith, A. (1997). Policy networks and advocacy coalitions: explaining policy change and stability in United Kingdom industrial pollution policy? *Environment and Planning C, 18*, 95-114

True, J., Jones, B.D., & Baumgartner, F.R. (1999). Punctuated-Equilibrium Theory: Explaining Stability and Change in American Policy making. In P.A. Sabatier (Ed.), *Theories of the Policy Process* (pp. 97-115). Boulder: Westview Press

USAID (2011). USAID Silvacarbon Fact Sheet. In. Washington DC: US Agency for International Development

Wynne, B., & Irwin, A. (Eds.) (1996). *Misunderstanding science? The public reconstruction of science and technology*. Cambridge: Cambridge University Press

Understanding and Addressing the Challenges of Climate Change in Urban Areas

Climate Change: Innovative Approaches for Modeling and Simulation of Water Resources and Socioeconomic Dynamics

Attila Fur and Flora Ijjas

Additional information is available at the end of the chapter

1. Introduction

This chapter shows an innovative approach for handling the rising scarcity of water resources caused by climate change. It introduces a new way of the modeling and simulation of socioeconomic adaptation and mitigation to water scarcity. But the model can be used for all the resources that humans have to share on planet earth.

It tries to find an explanation for the reactions of different societies to water availability problems caused by climatic change. For that purpose a Knowledge Attributed Petri Net based discrete simulation model is used modeling these different reactions. The model is based on integrated psychosocial development theories.

The Knowledge Attributed Petri Net model is implemented in the CASSANDRA (Cognizant Adaptive Simulation System for Applications in Numerous Different Relevant Areas) system developed by the McLeod Institute of Simulation Sciences Hungarian Center at the Budapest University of Technology and Economics.

2. Climate change, water availability and socioeconomic problems

2.1. Climate change and water resources

Many climate scenarios, socioeconomic models and digitized river networks show that water stress is already high in many parts of the world.

Latest edition of the UN World Water Development Report, Managing Water under Uncertainty and Risk (WWAP, 2012) launched at the World Water Forum in Marseille on 12 March 2012 warns that unprecedented growth in demands for water are threatening all major development goals. The growing pressure on global water resources comes from rising food demand, urbanization and climate change. According to the Report climate change will alter rainfall patterns, soil humidity, glacier-melt and river-flow and also causes changes to underground water sources. Floods or droughts are already rising in frequency and intensity.

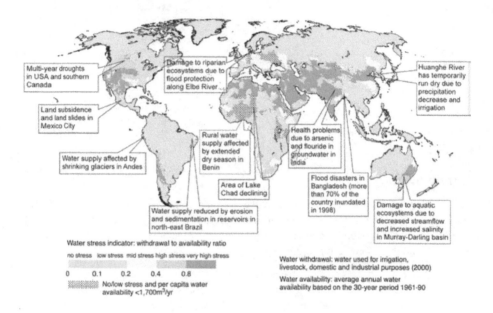

Figure 1. Examples of current vulnerabilities of freshwater resources and their management (Alcamo et al., 2003)

2.2. Climate change and water conflicts

The Report (WWAP, 2012) says that by the middle of the century 70 per cent more food will be needed. It means at least 19 per cent increase in the water required for agriculture. Many countries have already started to respond to water scarcity by acquiring fertile land outside their jurisdiction. Transnational land acquisition has risen from 15-20 million hectares in 2009 to more than 70 million hectares today. This is a problematic new tendency, a new kind of agro imperialism particularly happening in Africa by water scarce richer Arabian and western countries. Also as about 40 per cent of the world's population is living in river basins shared by many countries the risk rises that global climate change and the anticipated alteration of precipitation patterns will very likely lead to water-based conflicts in these regions.

2.3. Climate change, water and related socioeconomic challenges

According to the Report (WWAP, 2012) climate change will drastically affect food production in South Asia and Southern Africa between 2012 and 2030. By 2070, water-stress will also be felt in central and southern Europe, affecting up to 44 million people. These pressures will exacerbate economic disparities between countries, sectors or regions within countries. Better governance of water resources is required including investments in infrastructure from both private and public sectors (as for instance more than 80 per cent of the world's waste water is neither collected nor treated).

According to Intergovernmental Panel on Climate Change Technical Paper on Climate Change and Water (IPCC, 2008) the amount of water available for withdrawal is a function of runoff, groundwater recharge, aquifer conditions, water quality and water supply infrastructure. Safe access to drinking water depends more on the level of water supply infrastructure than on the quantity of runoff. However, the goal of improved safe access to drinking water will be harder to achieve in regions where runoff and/or groundwater recharge decreases as a result of climate change. In addition, climate change leads to additional costs for the water supply sector, e.g., due to changing water levels affecting water supply infrastructure, which might hamper the extension of water supply services to more people. This leads, in turn, to higher socio-economic impacts and follow-up costs (IPCC, 2008).

2.4. Environmental problems or social problems?

Researches show that current levels of human consumption, in combination with growing population are contributing to climate change (Dietz & Rosa, 1994; Myers & Kent, 2003; Stern, Dietz, Ruttan, Socolow, & Sweeney, 1997). Population growth will increase the global emissions anyway but a much larger increase would result if per capita emissions from energy consumption in developing countries increased to the developed countries level.

This makes it more imperative to understand how people make decisions that influence climate change through their behaviors (e.g. consumption) and to examine the values underlying that behavior. Consumption patterns represent classes of behaviors that explain the ways that human behavior contributes to climate change. In order to understand and address the links between consumption and climate change, it is useful to understand psychological, social, and cultural drivers of consumption and to understand what it is about consumption that influences climate change (APA, 2011).

Consumption is influenced strongly by social and cultural context. Cultural norms identify appropriate and desirable behaviors. It is important to see to which extent cultures value consumerism. Whereas environmental consumption is the use of resources and energy and economic consumption is spending money to acquire goods, consumerism is "a belief and value system in which consumption and acquisition rituals (e.g., shopping) are naturalized as sources of self-identity and meaning in life, goods are avidly desired for non-utilitarian reasons such as envy provocation and status seeking, and consuming replaces producing as a key determinant of social relations" (p. 231; Zhao & Belk, 2008).

Disciplines ranging from hydrology, politics, and international relations to law have in the recent past been tempted by the idea of „war over water". This is a common belief or a projection that if a country runs out of water it will start a war (Ijjas, Valkó, 2011). However this belief is just a belief and as such it is the product of a certain belief system (also called value system). Such beliefs change automatically when the value system changes. Value systems mostly change when life conditions change. Climate change will definitely indicate changing life conditions. As a result conflicting water scarce countries of shared river basins will escape water deficit by economically invisible and politically silent processes just as likely as they will start a war over water.

2.5. Climate challenge vs. psychosocial challenge

It seems that there is a strong cross-coupling between climate change, population growth, economic development, and social development. Therefore the relevance of these subsystems and links between them cannot be neglected.

The way how people manage their resources that their environment provides them with is strongly determinded by the thinking of these people. Both individual and social acts are triggered by the set of values and thinking modes that those individuals and social groups have. Let it be acts as consumption behavior (energy consumption, water consumption, food consumption or buying consumer goods) or developing strategies and programs by companies or by the government, decisions are always based on certain set of values and cognitive structures. By missing the qualities or levels of these subjective realms of decision making, resources management cannot fully be understood and finding solutions for environmental-social-economical problems will easily be dismissed. The question is how to link the subjective with the objective - how to understand why people are handling conflicts related to rising water, food or energy demand in a certain way.

In our case in order to have a more complete view of real water availability it is necessary to consider interactions among climate change, integrated water management and human systems including societal adaptations to water scarcity.

Grumbine (1994) states that management scenarios for climate change, water and economical, human systems should make the role of human values explicit because people base their commitments on values rather than on facts and on logic. People having different value systems want to fulfill different needs. This applies to scenario development concerning river management in developing, and particularly in developed areas, where the landscape is largely man-made. There are already cases where values systems have been taken into account by managing water resources. The psychosocial evolution model of river management is a good example for that. The model has been created within the cooperation of the Utrecht University and another research institution in order to have a guide for

selecting and positioning specific landscaping measures within a changing climatic, human
and economical surrounding (Straatsma *et al.*, 2009).

3. Psychosocial development

By social development we mean human activities organized at ever higher levels achieving
greater results. Psychosocial development further integrates patterns of psycho
development analogous to social development resulting in an integrated development
model of both individual and collective human structures. Psychosocial development takes
place when life conditions change in time. For example when water resources become scarce
- different societies react differently according to their value systems and they may organize
themselves at a higher level (regression to lower levels is also possible).

3.1. Models of psychosocial development

Water related problems are human related problems and for that they are not to be solved
without showing and understanding the human factor that is behind them. In order to
understand the role of value systems in forming different adaptation and mitigation
techniques - according to water scarcity issues caused by climate change - several
psychosocial developmental model can be used.

Psychosocial models relevant for our issue have been developed by Hamilton M. (Doctor of
Philosophy in Administration and Management, Columbia Pacific University) Graves C. W.
(Prof. Dr. Emeritus in Psychology, New York Union College) Cook-Greuter S. (doctorate for
Postautonomous Ego Development, Harvard University) Maslow A. (Prof. Dr. Head of
American Psychological Association) the psychologist Piaget J. or the developmental
psychologist Erick Erickson (Harvard Medical School).

We've found that these models are based on the same principles and they are following
similar dynamics. What we have done was synthetizing these models to get an integrated
model. The most focus was given to Graves' model as it proved to be the most applicable in
practical questions and most applicable for modeling and simulation.

The psychosocial development model of Graves (1970) is also called an emergent, cyclical,
double-helix model of adult psychosocial systems development. Graves identified (1974)
eight levels of existence that can be described by life conditions and the brain's coping
conditions with those certain life conditions. The eight levels are: A-N *Automatic*; B-O *Autistic*,
C-P *Egocentric*, D-Q *Absolutistic*, E-R *Multiplistic*, F-S *Relativistic*, G-T *Systemic*, and the H-U
Differential levels. 'A' stands for the neurological system in the brain upon which the
psychological system is based. 'N' stands for the existential problems that can be coped with
the 'A' neurological system. In the 'A-N' *Automatic* state 'N' problems of existence arise and
the 'A' neurological system is switched on in the brain. This is the first existential state (A-N
state) when the human being is living in conditions where it is only focusing on satisfying its
physiological needs. In B-O *Autistic* state man must assure the continuance of his first
established way of life, in C-P *Egocentric* state he must survive as an individual, in D-Q

Absolutistic state he must obtain lasting security in his existence, in E-R *Multiplistic* state he must assert his independence as a person, in F-S *Relativistic* state he must live in a non-competitive way together with other humans in community, in G-T *Systemic* state he must truly learn life is interdependent, and in H-U Differential state he must learn to fashion a life that honors and respects all the different levels of human being. The different states arise and come to stage center in man's mind as each successive set of human problems are resolved.

The model was later transformed by Wilber (1997); Beck and Cowen (1996) to be applicable in conflict resolution. (The result is called Spiral Dynamics Integral (SDI).) By further integrating the work of other researchers (e.g. Kohlberg, Armon, Mumford, Howe, Rawls, Piaget, Erikson, Maslow, Loevinger, Fromm) and inspired by the Book of Changes (Blofeld, 1965) we have formed a developmental model of ethical values in social systems which is useable to simulate optional adaptation strategies to the water challenges of our times.

3.2. Model of psychosocial complexity evolution

This chapter describes the eight value systems that are forming society's value systems and behaviors such as climate change mitigation and adaptation strategies according to shared and shrinking resources. The next chapter maps these behaviors into a Knowledge Attributed Petri Net model that simulates a case study taking river basins shared by many countries with different value-memes.

Following table shows these major value systems we have formed with the referring levels of other researchers such as Graves, Beck and Cowen and the main life conditions, with the mind/culture coping conditions and main needs of each level:

Fűr-Ijjas levels	Graves-Beck-Cowan levels	Main needs	Life conditions	Coping tools
Surviving individuals	A-N Beige	biophysical needs	N - State of nature and biological urges and drives: physical senses dictate the state of being	A - Instinctive: as natural instincts and reflexes direct; automatic existence
Superstitious clans/tribes	B-O Purple	temporary security within group	O - Threatening and full of mysterious powers and spirits that must be placated and appeased	B - Animistic: according to tradition and ritual ways of the group/tribe
Egocentric warriors	C-P Red	individual security	P - The world is a jungle where the strong prevail and the weak serve. Nature is to be conquered.	C - Egocentric Ego wants dominance, conquest and power; exploitive, aggressive

Conformist groups	D-Q Blue	long term safety within group	Q - Higher authority punishes bad behavior but rewards good work.	D- Absolutistic Obedient, conforming; conservative, hierarchic, driven by guilt
Creative hedonists	E-R Orange	long term individual safety, mental, behavioral independence	R - The environment is full of resources; the world is full of possibilities.	E - Multiplistic Pragmatic to achieve results; testing options, rational, modern, effective, selfish, arrogant, creative
Communities of human beings	F-S Green	long term individual safety within group, emotional freedom	S - Humanity is living in a habitat wherein people can find love and purpose through affiliation and sharing.	F - Relativistic Responding to human needs, affiliative, consensual, fluid, accepting, less efficient
System-thinking humans	G-T Yellow	finding and realizing Self-worth	T - The world is a chaotic organism where change is the norm and uncertainty is an acceptable state of being.	G - Systemic Functional, integrative, interdependent, existential, flexible, questioning, needs more time for complexity
Holistic communities of human beings	H-U Turquoise	finding and realizing self-worth within holistic system	U - Turquoise A delicately balanced system of interlocking forces in jeopardy at humanities hands	H - Holistic: experiential: transpersonal; collective consciousness; collaborative; interconnected

Table 1. Levels of psychosocial development 1

The defined levels represent also eight forms of behaviors that are related to the outer
manifestation of an entity (group of human beings) in the environmental space. This relation
is not easy to identify therefore oracles from different cultures have to be adopted and
composed in such way that the common representatives can be examined within scientific
frames. In order to understand how it can become possible to analyze a social action (e.g.
regarding environmental resources) based on its internal level of psychosocial development

linear independent types of attributes has to be found. These types of attributes should give the most elementary common description to each level regardless of their actions taken.

The elementary types of attributes that aim to the description of the connection between the level and its environment are following:

- set of needs (N)
- way of the actions taken to suffice needs by acquiring resources (A)
- set of resources that are handled by the entity (R)

Fig. 2. highlights the direction of materialism of the three main model elements from the inner (soft) aim to the outer (sharp) result.

Figure 2. The 3-tuple of the basic psychosocial-environmental relation

After determining the elements of the basic psychosocial-environmental relation the domain of analysis should be appointed. This question is of great importance and also of enormous complexity. The needs can be ordered to clusters in many different ways and can also own different relevance in space and time regarding the environment. The set of actions, interventions taken by a social entity can also be infinite consequently the exact description and fragmented clustering of these elements is not suggestible at that level of examination. In point of the resources we face the same situation therefore the complicated formal descriptions can be replaced by binary qualities.

These binary qualities give relevant information about the elements of the 3-tuple of the basic psychosocial-environmental relation:

- Needs can have two subsets:
  - Inner needs (e.g. need for express the existence, need for belonging to, need for freedom, need for "happiness" etc.)
  - Outer needs (e.g. need for nutriment, need for heat, need for space, etc.)
- Resources can have also two subsets:
  - Inner resources (e.g. religion, art, education, social institutes, etc.)
  - Outer resources (e.g. water, soil, flora, fauna, fossil fuels, etc.)
- Actions can be taken in two different ways:
  - Actively (the psychosocial entity makes effort to achieve the expected state in case of presence of need). This phenomenon is also called "need-driven" action.

- Passively (the psychosocial entity is awaiting the optimal circumstances for its needs to be sufficed independently from the presence of need). This way of acting is marked as "resource-driven" action.

Fig. 3. highlights the possible behaviors through 8 different interconnections between needs and resources.

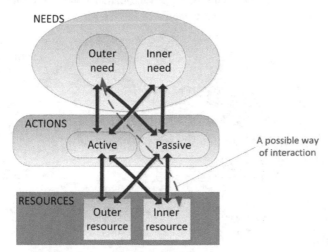

Figure 3. Interconnections between needs and resources

The actions – as transitions from the physical world to the psychical world – have a special role in the model through aiming to the maximization of the simultaneous occurrences of needs and resources over a given period of time or the maximization of collecting resources for sufficing further needs expectable in the future (buffering). Actions – in that explanation – are experiments on sufficing special type of needs by special type of resources. The number of the mentioned synchronism can be regarded as the indicator of "success" (S). The *S-indicator* is the key of surviving of a given social structure under the changing boundary conditions in the environment.

The *S-indicator* has two types:

- Success of sufficing inner needs *(INS)*
- Success of sufficing outer needs *(ONS)*

Both of the success indicators have to be of high value in order to gain balanced and stable state of the psychosocial entity.

During the process of model synthesis a very important analogy cannot be left out of consideration. The eight value system defined in Table 1 and the 3-tuple structure of the basic psychosocial-environmental relation assuming the binary values of the tuple-elements indicates the $2^3=8$ decomposition of the Fűr-Ijjas levels. That kind of interpretation is well-known in the philosophy of TaijiQuan. Fig. 4. shows the development of behaviors. In the first column the whole system of balanced psychosocial entities is represented therefore this system can be regarded as neutral from the outside. In the second column the system is split into two main parts causing the stress and giving the base of model. Two main qualities of the psychosocial entities can be differentiated that is analogous with the two main forces of TaijiQuan: Yin is the acceptor and Yang is the donor. This is called the base binary value.

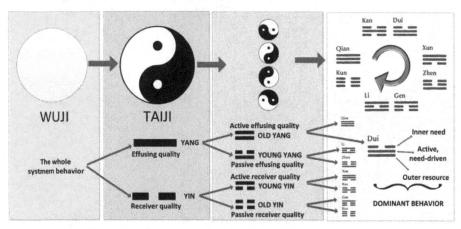

Figure 4. Interpretation of the TaijiQuan philosophy within the Fúr-Ijjas model

The base binary value (second column of Fig. 4.) represents the direction how the psychosocial entity relates to the environment (effusing or receiving quality). The second binary value has a comprehensive meaning of purity of the base behavior in the dual interpretation (third column of Fig. 4.). Young psychosocial entity does not own a clear self-concept includes disturbances from the other behavior, contrarily the old entity is cleared perfectly and it can show well defined functioning.

The most complex interpretation (fourth column of Fig. 4.) is based on the Trigrams that are combinations of three independent elements given by the base binary values. The bottom binary value represents the resources (outer resource=Yang, inner resource=Yin). The middle binary value represents the action (active, need-driven behavior=Yang, passive, resource-driven behavior=Yin). The top binary value means the type of needs (outer need=Yang, inner need=Yin)

The next table shows these levels linked to the corresponding binary values we have adopted from the Book of changes. These binary values designate different bio-psychophysical qualities, forces and movements. The lowest line in the Trigram represents the aimed resources by the psychosocial entity (inner/outer resources) the line in the middle determinates the way of acquiring the resource (actively or passively) and the top line stands for the prevailing needs of the psychosocial entity (inner/outer needs).

Fűr-Ijjas levels	Book of changes - binary value	Book of changes - image in nature	Gender quality	Outer/inner needs the active/passive way outer/inner resources
Surviving individuals	☰	The Creative, heaven	male individual	outer needs active outer resources
Superstitious clans/tribes	☵	The Abysmal, water	female collective	inner needs active inner resources
Egocentric warriors	☱	The Joyous, lake	male individual	inner needs active outer resources
Conformist groups	☴	The Gentle, flood	female collective	outer needs active inner resources
Creative hedonists	☳	The Arousing, earthquake	male individual	inner needs passive outer resources
Communities of human beings	☶	The Keeping Still, mountain	female collective	outer needs passive inner resources
System-thinking humans	☲	The Clinging, fire	male individual	outer needs passive outer resources
Holistic communities of human beings	☷	The Receptive, earth	female collective	inner needs passive inner resources

Table 2. Levels of psychosocial development 2

4. Knowledge attributed petri net based discrete simulation model

In this chapter a Knowledge Attributed Petri Net based discrete simulation model is described that is suited to map the previously highlighted eight value systems into reasonable experimental frames. The model aims the examination of struggling of entities for resources based on the level of psychosocial development.

4.1. The methodology and tool used

Proper describing of the physical reality in general has always been standing amongst the relevant questions of science. Several methodologies were developed based on classical mathematics, or statistics and also new disciplines – such as soft-computing techniques – appeared (Russel, S.J. – Norvig, P., 2002). Each methodology owns advantages in some fields of modeling, but none of them is adequate to describe complex processes in general. Multi-facetted problems require methodologies that are able to integrate high-level mathematical concepts in a natural way. An obvious solution of mapping reality to a well-structured form is given by the concept of Petri Nets (Petri, C.A. 1962).

Petri Nets follow an elementary abstraction of physical reality by describing containers as places, mobile entities representing temporal states as tokens, and rules – transitions – that determine the generation and elimination of tokens in space and time. These basic elements correspond to real or virtual elements: e.g. tokens to information, money, materials, living beings, and places to physical locations, or virtual containers – such as bank accounts, data storages or indicators – transitions can represent the static knowledge or rules of physical laws, economic or legal regulations.

During the last decades several extensions have been suggested to the original concept of Petri Nets in order to raise its describing power. Inhibitor arcs (Inhibitor Petri Nets, IPN), colored tokens (Colored Petri Nets, CPN), stochastic delayed streaming of mobile entities (Stochastic Petri Nets, SPN), object oriented architecture (Object Oriented Petri Nets, OOPN), numerical (Numerical Petri Nets, NPN) and linguistic attributes (Fuzzy Petri Nets, FPN) (Peterson, J.L., 1981, Jensen, K. Rosenberg, G., 1991, Balbo, G., 2002, Carl G. Looney 1994) broaden the range of capabilities.

In some fields of problem solving usage of static and mobile knowledge bases is needed: e.g. modeling of flexible manufacturing systems (Jávor, A., 1993), routing and treatment of patients in hospitals (Jávor, A. Benkő, M., Leitereg, A., Moré, G., 1994), or intelligent traffic simulation (Jávor, A., Szűcs, G., 1998). These problems to be investigated involved new conceptual developments of Petri Nets and led to the introduction of Knowledge Attributed Petri Nets (KAPN) (Jávor, A. 1993–2). At the same time artificial intelligence (AI) and distributed control in simulation appeared, intelligent demons (agents) (Jávor, A. 1992, 2006) supported the connection of mobile knowledge bases and static inference engines in an effective way.

In our research we decided to use the KAPN methodology that owns all the properties of High Level Petri Nets mentioned before and that is also able to host AI within the model. As simulation tool CASSANDRA (Cognizant Adaptive Simulation System for Application in Numerous Different Relevant Areas) simulation system was chosen. CASSANDRA is a KAPN based system with the accomplishment of the methodology of model identification by reconstruction (Fűr, A., Jávor, A., 2007). CASSANDRA was developed by the McLeod Institute of Simulation Sciences Hungarian Center where the authors contribute to the methodological research.

4.2. Model identification by reconstruction

During the process of model synthesis there are several cases where simulation experts can face incomplete knowledge on structure, parameters or the algorithms describing the operation of the system to be investigated. Sometimes even the behavior of the system itself is not strictly defined. In these cases the usage of model identification by reconstruction can be suggested. This methodology is based on a special simulation entity (agent) that aims to the modification of the model based on observing the trajectory of its behavior. Fig. 5. illustrates the closed-loop controlled like modification of the simulation model by an intelligent agent.

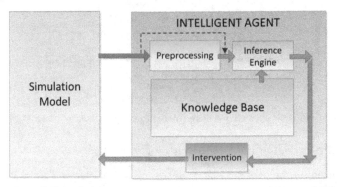

Figure 5. Agent controlled simulation

The identification process starts with the definition of the initial model that gives a soft delineation of the system based on the available theories and assumptions. The initial model – often built as a network of objects – should give a rough, but functionally approximating description.

The next step is triggering the model input by historical data and monitoring the output by the agent that compares it with the historical behavior of the system. In case when the agent finds significant deviations it may change the model until acceptable correspondence between reality and the model is achieved. The changes which the intelligent agent can execute are the following (Fűr, A., Jávor, A. 2007):

- Change the topology of the model network
- Change the parameters of the model elements
- Change the functions describing the effects of one model element on an other

There are several strategies how model reconstruction can be carried out, and also the complexity of the iterating algorithms can vary in a wide range from the multi-criteria estimation of parameters to the structure synthesis. After having reconstructed the initial model in such way that the output behavior corresponds to the reality the model can be used for further virtual experiments (e.g. prediction, or analysis) affording acceptable reliability regarding the simulation results (see Fig.6.).

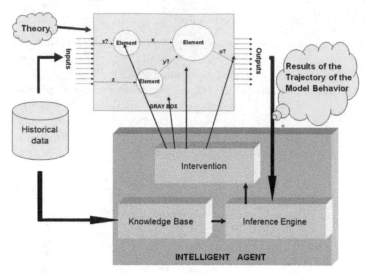

Figure 6. Identification by reconstruction using an intelligent agent (Fűr, A., Jávor, A. 2007)

4.3. The KAPN model of the Fűr-Ijjas levels

The basic concept of the model is based on two natural phenomena of the Petri Nets. The first one is the way of resource allocation. If there are two places given (the one representing the container of needs, the other the container of resources) and these are assigned to a transition object that has a place (container of results, state or success) at the output, the whole structure can be interpreted as a simple Petri Net model of resource allocation. In that case both input places (needs, and resources) has to contain tokens (representing the presence of need and resource at the same time) in order to trigger the event (this is called firing). After the process of resource allocation (firing) is finished the next state of the Petri Net model is that the resource (token) is destroyed in order to eliminate the need (token) and a result state (satisfaction, success) is generated (in also form of a token).

Figure 7. Basic Petri Net model of the process of resource allocation

The model delineated in Fig.7. shows deterministic functioning. In order to turn the model behavior more realistic an extension to the transition firing condition can be suggested. At

each simulation step to each transition a random number can be generated (of a given
distribution, within the range of [0,1]) and it has to be compared with the threshold of the
transition (constant ∈ [0,1] ∪ {-1}) and the firing is executed depending on the result of this
comparison. The threshold parameter (TRP) controls the permeability of the transition
regarding the tokens during a given time interval. TRP=-1 "closes" the transition (see Fig.8).

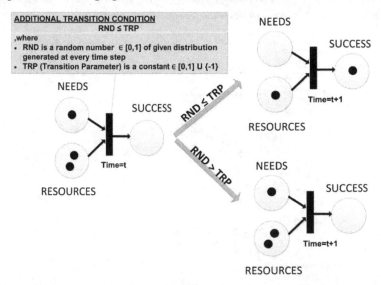

Figure 8. Knowledge Attributed Petri Net model of the process of resource allocation

The other main advantage of the Petri Net (or KAPN) based description lays in the power of
handling the conflict situations (see Fig.9.). In the case when e.g. needs of two different
entities require the same type of resource and in the container of the common resources
there is only one resource (token) present – although both of the competitors are marked to
be satisfied – in reality only one of them can access the resource (the one who fires before).

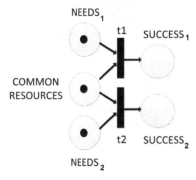

Figure 9. The common resource conflict situation

The firing order can be random or can follow priority considerations. (In our model the firing order is chosen to be random.)

In the following part a short description is given about the basic processes and their KAPN models that are subnets of the Fűr-Ijjas model. The basic consumption chain of a psychosocial entity (according to Fig.3.) can be need-driven (active) or resource-driven (passive). The main difference between the two approaches is the interpretation of the state of "success".

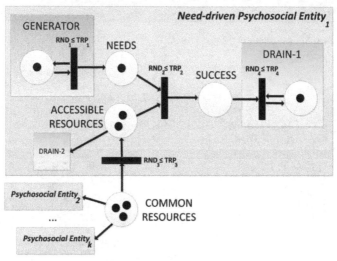

Figure 10. The model of a Need-driven Psychosocial Entity (NPE)

In the case of the Need-driven Psychosocial Entity (NPE) the success is defined as the simultaneous occurrences of needs and resources at a given time instant. The detailed network (see Fig. 10.) shows that the NPE collects tokens from the place of common resources transforming them into accessible resources (that can be utilized at any time instant by the NPE). This can correspond to the transportation or the exploitation of the resource. Each NPE possesses its own efficiency (TRP3) on accessing the common resources consequently the available base of sufficing needs is different for each of them. The accessible resources have natural diminution (represented by DRAIN-2) that is independent of the consumption intents of NPE. Reasons for the diminution can follow from the nature of the resource (if it is hard to store over a long period of time – e.g. food, electricity, water under given conditions, etc.) or can be caused by disasters, epidemics, or the unforeseen annihilation of the already acquired resource. The generator of needs awakes necessities during simulation run with given intensity (TRP1). The simultaneous occurrence of needs and resources triggers the transition causing the generation of a token in the place of SUCCESS. The efficiency of resource allocation can be adjusted by the parameter TRP2. If efficiency is low (e.g. thanks to backward governmental systems or infrastructure) then generated needs cannot be sufficed in time and thereunder natural diminution can happen. DRAIN-1 is responsible for not to get stuck the simulation because of the finite capacity of

SUCCESS place (infinite number of tokens in a place is not acceptable therefore each place possess given capacity – that corresponds to natural behavior).

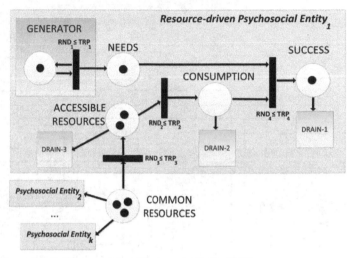

Figure 11. The model of a Resource-driven Psychosocial Entity (RPE)

Resource-driven Psychosocial Entities (RPE) appear to be different in their behavior. In the resource allocation process of these entities the annihilation of the accessible resources is independent of the presence of need. The consumption happens anyway hoping that in the future upcoming needs will be sufficed. That kind of buffering has a great advantage. This preventing behavior can possibly lower the number of cases when needs cannot be sufficed, but as a main disadvantage a new drain (see. Fig. 11. DRAIN-2) can appear that destroys that pre-buffered consumptions. Success is interpreted as simultaneous occurrence of needs and consumed resources. The main difference between NPEs and RPEs can be highlighted by the following metaphor: *"Need-driven Psychosocial Entities eat WHEN THEY ARE hungry, Resource-driven Psychosocial Entities eat NOT TO BE hungry"*.

As previously delineated the Fűr-Ijjas levels consider eight value systems with eight dominant behavior and relation to environmental resources. The given basic networks representing NPE and RPE (Fig. 10 and Fig. 11.) have to be integrated into a general KAPN model to ensure the possibility of modeling transitional (fuzzy) behavior between the clearly defined, dominant functioning of NPE and RPE. Also the two different types of needs and resources (inner/outer) have to be visualized in the model therefore the possible ways of resource allocation increases due to the different cross-couplings.

The following figure (see Fig.12.) outlines a possible KAPN implementation of the eight levels integrated into one psychosocial entity (PE). The aim of this general purpose entity is – beyond reproducing the original eight levels – the possibility of combination of the crisp behaviors. That corresponds to natural behavior of PEs because there are several cases when subdominant properties become also noticeable.

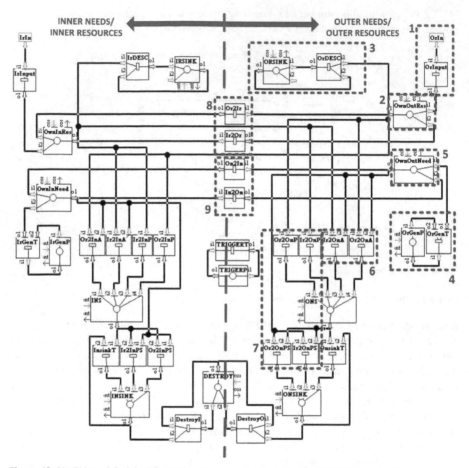

Figure 12. KAPN model of the Fűr-Ijjas levels (screenshot from the CASSANDRA system)

In the figure it can be seen that the model owns symmetric configuration with processes of inner needs and resources on the left side and outer needs and resources on the right side. The subnets marked by numbers on the picture correspond to the followings:

1. Input from the outer world for outer resources (connected to the place of common outer resources outside the PE)
2. Place of accessible (own) outer resources
3. Natural diminution of accessible resources (drain)
4. Generator of outer needs
5. Place of outer needs
6. Need-driven functioning (Or2OnA=Actively destroying outer resource on demand of outer need, Ir2OnA=Actively destroying inner resource on demand of outer need)

7. Resource-driven functioning (Or2OnP=Destroying outer resource anyway, consumption gets into buffer, then buffer is emptied when outer need is present, Ir2OnP=Destroying inner resource anyway, consumption gets into buffer, then buffer is emptied when outer need is present)

8. Cross-coupling between outer and inner resources. We assume that there are some cases possible when the resources of different types can be converted into each other. Outer resources can be converted into inner resources (e.g. the amount of water can have influence on the social, economic system) and inner resources can be converted into outer resources (with faith or a different approach inaccessible water resources can be accessed or even accessible water resources can be saved).

9. Cross-coupling between outer and inner needs. Several observations can give the ground to the assumption that inner and outer needs can substitute each other (within reasonable frames). In countries where the outer needs (e.g. need for nutriment) are hard to suffice PEs often turn to inner needs (e.g. need for family, need for express the existence by music, dance, art, etc.) in order to deflect attention from the real situation. In contrast developed countries often forget about their natural inner needs concentrating on sufficing their outer needs.

The figure below represents a possible test environment with different PEs accessing common outer and inner resource bases.

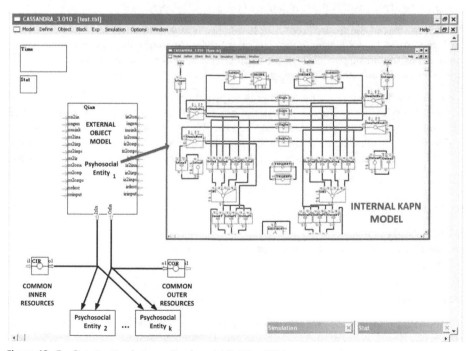

Figure 13. Conflict situation between Psychosocial Entities (PEs)

The KAPN model of the PE can be encapsulated into another object that hides the inner functioning from the higher level (see. Fig. 13.). Based on these hierarchical models multiple PEs and their conflicts can be examined under different boundary conditions.

Our aim is to simulate the behavior of different psychosocial entities living along e.g. Jordan River Basin, facing the growing water scarcity. In order to do that the parameters of the models have to be fine-tuned by the methodology of model identification by reconstruction.

After achieving the correct internal parameters of those countries that are involved in a resource allocation conflict we intent to forecast water usage at different climate scenarios. That should be our future work.

5. Conclusion

In this paper we identified eight levels of psychosocial development based on integrated theories. By the aid of TaijiQuan philosophy we found a way how to map these complex and fuzzy defined levels into concrete actions taken by an entity. The elaborated model gives well defined relation between the entity and the environmental resource considering the needs. This model is also able to combine the eight different levels with different intensity into a contemporary complex entity. Based on agent controlled simulation by the CASSANDRA system the changing behavior of entities during simulation run can become also possible. This ensures the simulation of the adaptation process of a psychosocial entity.

We believe that in the future cross-coupling between inner and outer needs and resources explored by the model become more important and narrowing outer resources or growing outer needs should be partially substituted by inner resources and inner needs. That is the key to successful adaption of future generations.

We also have a vision of our model in policy applications, especially in the identification of dominant psychosocial behavior patterns of different countries aiming to the recognition of adequate response in their adaptation process. We do hope that the model helps to understand – beyond the frequently communicated materialized problems – the inner aspects of the upcoming environmental-social-economical challenge caused by climate change.

Author details

Attila Fur and Flora Ijjas
Budapest University of Technology and Economics, Department of Environmental Economics, Hungary

Acknowledgement

We would like to express our acknowledgment to Prof. András Jávor who supported the research and development of this model within the McLeod Institute of Simulation Sciences Hungarian Center.

The work reported in the paper has been developed in the framework of the project „Talent care and cultivation in the scientific workshops of BME" project. This project is supported by the grant TÁMOP-4.2.2.B-10/1--2010-0009.

6. References

Alcamo, J., P.Döll, T. Henrichs, F.Kaspar, B. Lehner, T. Rösch and S. Siebert, (2003) Development and testing of the WaterGAP 2 global model of water use and availability. Hydrol. Sci. J., 48, 317-338.

APA (2011) Psychology and Global Climate Change: Addressing a Multi-faceted Phenomenon and Set of Challenges, Report of the American Psychological Association Task Force on the Interface Between Psychology and Global Climate Change, American Psychologist Special Issue, Vol. 66, Issue 4, 50-51.

Balbo, G. (2002) Introduction to Stochastic Petri Nets, Lectures on formal methods and performance analysis, Springer-Verlag New York, Inc. New York, NY, USA, ISBN: 3-540-42479-2

Beck, D.; Cowan, C. (1996) Spiral Dynamics: Mastering Values, Leadership, and Change, Malden, Blackwell Publishers

Blofeld, J. (1965). The Book of Changes: A New Translation of the Ancient Chinese I Ching. New York: E. P. Dutton.

Carl G. Looney (1994) Fuzzy Petri Nets and Applications, Fuzzy Reasoning in Information, Decision and Control Systems, Intelligent Systems, Control and Automation: Science and Engineering, Volume 11, Part 5, 511-527, DOI: 10.1007/978-0-585-34652-6_19

Cowan, C.C., Todorovic, N., (2007) Spiral DynamicsOverview.Retrieved28.7.2008 from www.spiraldynamics.org/aboutsd theory.htm. [Accessed 04.15.2011]

Dietz, T., Rosa, E. A. (1994). Rethinking the environmental impacts of population, affluence and technology. Human Ecology Review, 1, 277–300.

Fűr, A., Jávor, A. (2007) Intelligent Agent Controlled Simulation with the CASSANDRA System, EUROSIM 2007, Ljubljana, Slovenia, September 9-13.

Graves, C. W. (1970) Levels of Existence: An Open System Theory of Values. Journal of Humanistic Psychology, 10(2): 131-155.

Graves, C. W. (1974) "Human Nature Prepares for a Momentous Leap" The Futurist, April, 72-87.

Grumbine, R.E., (1994) What is ecosystem management? Conserv. Biol. 8, 27–38.

Ijjas F., Valko L. (2011) Integral concepts and virtual water trade for a peaceful Jordan River Basin, In: Proceedings of the international conference "Handshake across the Jordan – Water and Understanding in the Middle East. In: Forum Umwelttechnik und Wasserbau, Nr. 9. IUP – Innsbruck University Press, Innsbruck, 288.

Intergovernmental Panel on Climate Change Working Group II (2008) Climate change and water: technical paper of the Intergovernmental Panel on Climate Change, Geneva Intergovernmental Panel on Climate Change, 45.

Jávor, A. (1992) Demon Controlled Simulation, Mathematics and Computers in Simulation 34 3-4, 283-296.

Jávor, A (1993) AI Controlled High Level Petri Nets in Simulating FMS, Fourth Annual Conference: AI, Simulation, and Planning in High Autonomy Systems, Tucson, Arizona, USA, September 20-22,. 302-308.

Jávor, A. (1993) Knowledge Attributed Petri Nets, Systems Analysis, Modelling, Simulation, 13 1/2, 5-12.

Jávor, A. Benkő, M., Leitereg, A., Moré, G. (1994) AI Controlled Simulation of Complex Systems, Computing & Control Engineering Journal 5 2, 79-82.

Jávor, A., Szűcs, G. (1998) Simulation and Optimization of Urban Traffic using AI, Mathematics and Computers in Simulation 46 13-21.

Jávor, A. (2006) Demons as Forerunners of Software Agents in Simulation – Invited paper, 2006 SCS International Conference on Modeling and Simulation - Methodology, Tools, Software Applications (M&S-MTSA'06), Calgary, Canada, July 31- August 2. 151-155.

Jensen, K. Rosenberg, G. (1991) High-level Petri Nets, Springer Verlag

Myers, N., Kent, J. (2003) New Consumers: The influence of affluence on the environment. Proceedings of the National Academy of Sciences, 100(8), 4963-4968.

Peterson, J.L. (1981) Petri Net Theory and Modeling of Systems, Prentice Hall

Petri, C.A. (1962) Kommunikation mit Automaten, Bonn, Institut für Instrumentelle Mathematik, Schriften des IIM Nr. 2,

Russel, S.J. and Norvig, P. (2002) Artificial Intelligence: A Modern Approach. Prentice Hall

Stern, P. C., Dietz, T., Ruttan, V., Socolow, R. H., and Sweeney, J. (Eds.). (1997). Environmentally Significant Consumption: Research Directions. Washington, DC: National Academies Press.

Straatsma, M. W.; Schipper, A.; Van der Perk, M.; Van den Brink, C.; Leuven, R.; Middelkoop, H.; (2009) Impact of value-driven scenarios on the geomorphology and ecology of lower Rhine floodplains under a changing climate, Landscape and Urban Planning 92., 160–174.

WWAP (World Water Assessment Programme). (2012). The United Nations World Water Development report 4: Managing water under uncertainty and risk. Paris: UNESCO.

Wilber, K. (1997) An Integral Theory of Consciousness, Journal of Consciousness Studies, 4 (1): 71-92.

Zhao, X., & Belk, R. W. (2008). Politicizing consumer culture: Advertising's appropriation of political ideology in China's social transition. Journal of Consumer Research, 35, 231-244.

Chapter 5

Climate Change on the Urban Scale – Effects and Counter-Measures in Central Europe

Wilhelm Kuttler

Additional information is available at the end of the chapter

1. Introduction

According to the Intergovernmental Panel on Climate Change (IPCC, 2007), global climate change, which will lead to a temperature increase of the lower atmosphere, can be traced to an increased concentration of carbon dioxide (CO_2) and other infrared-active gases. Urban agglomerations are particularly affected because, in contrast to the surrounding rural areas, they are characterised by a high population density, a considerable extent of tightly sealed and rough surfaces and air pollution.

Of the consequences projected for central Europe, among which are changes in the precipitation patterns and the weather regime, this article focuses exclusively on the thermal and air-hygienic effects. These impacts pose a high risk for the inhabitants of every city, not only those of cities prone to flooding. Therefore, select measures are described that can counter the excess heating and the increase in temperature-dependent air pollution at the local level. The countermeasures presented in this chapter refer mainly to adaptation measures, although there exists a smooth transition between adaptation and mitigation.

2. Urban excess heating

Compared with non-built-up space outside conurbations, cities have higher air and surface temperatures, which occur particularly during local (autochthonous) meteorological conditions characterised by low winds and limited cloud cover. Among other factors (Table 1), high daytime radiation, a negative radiation balance (Q^*) in the evening and at night as well as a limited atmospheric exchange guarantee the development of a positive horizontal temperature difference between urban (t_u) and rural, non-built-up surroundings (t_r; $\Delta T_{u-r} > 0$ K). This phenomenon is called the urban heat island (UHI).

Influencing factor (IF)	Sign of correlation coefficient between UHI and IF
Cloud cover	−
Wind speed	−
Anthropogenic heat emission	+
Bowen ratio, β[1]	+
Population	+
Sky view factor, SVF[2]	−
Ratio building height/street width (H/W)	+
Surface sealing	+
Green- and water-surface area/total area	−
Latitude	+

[1] $\beta = Q_H/Q_E$; (Q_H/Q_E = turbulent sensible/latent heat flux density)

[2] SVF: degree to which the sky is obscured by the surroundings at a given location

Table 1. A selection of meteorological and structural factors influencing the UHI (source: Kuttler, 2009)

UHI can be categorised according to the types of vertical structures in an urban area as follows:

- below ground (subterranean UHI),
- on the surface (surface UHI),
- in the urban canopy layer (UCL UHI) and
- in the urban boundary layer (UBL UHI).

Subterranean and surface UHIs are predominantly caused by radiation and most frequently found in combination with sealed surfaces. The UCL and the UBL UHIs are caused by the heating of the air and are consequently much less congruent with sealed areas. Furthermore, the three-dimensional structure of the UCL and the UBL UHIs allows the wind to change their shape.

2.1. Factors influencing urban excess heating

In addition to the long-wave radiation fluxes ($L\uparrow$, $L\downarrow$), the dominance of either sensible or latent heat fluxes (Q_H, Q_E) is important to the heating of a city atmosphere. The Bowen ratio ($\beta = Q_H/Q_E$; Table 1) determines the respective proportion of Q_H and Q_E in specific cases. The surface type, i.e., whether an artificially sealed or a natural evaporation surface, is the decisive factor for the energy partitioning (Fig. 1). For example, if the plan area density (λ_P) is low, the major part of the energy resulting from the radiation balance (Q^*) in bright and sunny weather is transferred via the latent heat flux (Q_E) to the UBL without heating the layer significantly (Fig. 1 a). In such a case ($\lambda_P = 0.1$), the Bowen ratio reaches $\beta < 0.25$. However, where the surface is completely sealed ($\lambda_P = 1.0$), the sensible heat flux (Q_H) dominates (here, approximately 80 % of Q^*) with a Bowen ratio of $\beta > 2$.

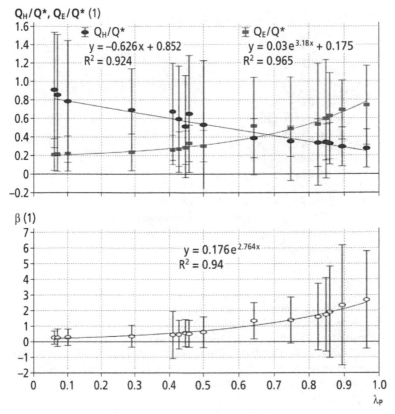

[1] Mean 30 min values and σ_1, $Q^* > = 0$ W/m²; sectors of wind direction and λ_P dependent calculation of urban/suburban data; 8/ 2010 – 4/ 2011; soil heat flux (Q_b) omitted

Figure 1. Ratios of Q_E/Q^* and Q_H/Q^* (a) and Bowen ratios ß (b) as a function of plan area density (λ_P) in Oberhausen, Germany (213,000 inhabitants; source: Goldbach & Kuttler, 2012)[1]

Therefore, weather conditions and the degree of sealing substantially influence the UHI intensity, which displays temporal and spatial dependencies.

As shown in Fig. 2, the air-temperature differences between a completely sealed urban location (t_u) and an undeveloped location in the rural environment (t_r) display typical diurnal and annual cycles in the UCL. Hence, the following conclusions can be made for central European metropolises:

- UHImax usually occurs in the evening or night and usually lasts until the early morning.
- During the day, no or little UHI effect develops.
- The dependency of the UHI on the weather is mirrored in the cellular distribution of air temperatures. However, in cloudy, rainy and windy (allochthonous) weather, the overheating is substantially reduced (October to February).

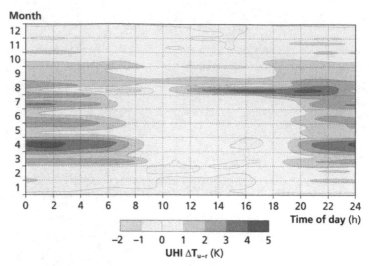

Figure 2. Hourly mean values of the UCL UHI (ΔT_{u-r}) in Bochum, Germany (375,000 inhabitants; 10/2006 – 10/2007)

Compared with the surrounding countryside, a spatially homogeneous overheating within the city limits seldom occurs and can display a highly heterogeneous pattern as a function of plan area density (λ_P). The expression "heat archipelago of a city" is a suitable term for the phenomenon (Hupfer & Kuttler, 2006). The example shown in Fig. 3 clearly emphasises the higher air temperatures of overheated, tightly sealed inner-city and industrial areas (red) compared with natural areas (blue) such as forests and undeveloped open countryside with lower air temperatures.

As a result, the most significant horizontal differences are found between the densely built-up areas of the city ($\lambda_P = 0.9$) and the unsealed and open countryside ($\lambda_P = 0.14$) with the above-mentioned peaks in the nocturnal hours (Fig. 4). The differences are far less significant between the inner city and the suburbs ($\lambda_P = 0.39$) because the degree of surface sealing in the two areas is more similar than in the first case. The thermal behaviour of an intra-urban park (1.2 ha) is particularly interesting. The park's air is cooler than that of the city. However, the temperature does not reach the low value of the countryside. The comparably higher air temperatures detected in the park are caused by the warmer atmosphere of the surrounding sealed areas, preventing the green area's air temperature from decreasing further (Kiese, 1995). In the end, the value of overheating in the park is approximately half the value measured in the densely built-up inner city.

To calculate the intensity of a UHI for urban agglomerations with an easily ascertainable parameter, the number of inhabitants is frequently used as a substitute value. In contrast to other size-related specifications, this value is usually easily obtainable. As the results of an exemplary evaluation for German cities demonstrate (Fig. 5), the urban excess heating can be positively correlated with the population figures.

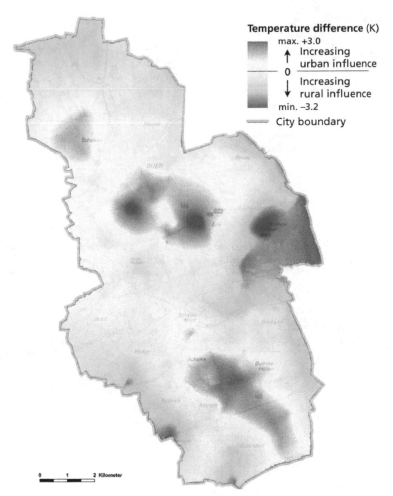

Temperature difference (K)

max. +3.0

↑ Increasing
urban influence

0

↓ Increasing
rural influence

min. –3.2

—— City boundary

¹⁾ "Calm and clear weather conditions" according to Dütemeyer, 2000, resp. Polster, 1969

Figure 3. UHI in the urban canopy layer during a night with clear and calm weather¹⁾ in Gelsenkirchen, Germany (298,000 inhabitants; 4/2011)

However, only 44 % of the values are represented through the exponential fit, i.e., the statistical connection is not very tight, because numerous other factors can exert an influence on the UHI in addition to the population figures. These other factors include, e.g., anthropogenic heat (Böhm, 1998; Quah & Roth, 2012), the traffic density, the city size, the allocation and mixture of sealed and unsealed surfaces, the proportion and distribution of green spaces, the vegetation (leaf-area index (LAI); leaf-area density (LAD)), the building materials, the building volumes, the proportion of high-rise buildings and topographical features.

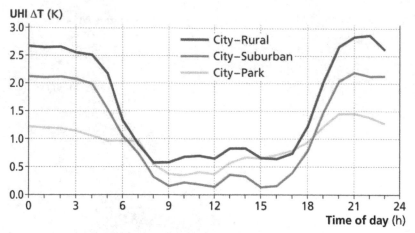

Figure 4. Hourly mean values of excess heating between the city and an urban park, the city and a suburban site and the city and a rural site in Oberhausen, Germany, for 76 calm and clear days (08/2010 – 07/2011)

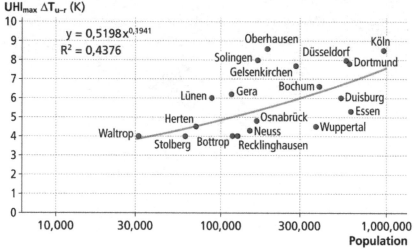

Figure 5. Maximal UHI intensity ($\Delta T_{(u-r)\,max}$) in the urban canopy layer as a function of population for selected German cities (data basis: stationary and mobile measurements; source: Kuttler, 2010a)

Generally, the intensity of a UHI is dominated by local and regional influences. However, large-scale influences can be identified, albeit in severely weakened form. If selected UHI intensities are assigned to three clusters (e.g., tropics, subtropics, mid-latitudes) according to the cluster's latitudes (Fig. 6), positive correlations can be observed. For instance, for the tropical cluster, a mean value of 4 K can be calculated for UHI_{max}, whereas 5 K can be calculated for the subtropical cluster and 6 K for the mid-latitudes (50 % quantile each; for more information regarding the basis for the cluster calculations, see Wienert & Kuttler, 2005, Fig. 6). This correlation, which was first assumed by Kratzer (1956) but never statistically proven, appears to be mainly caused by anthropogenic heat and the radiation balance (input and output parameters according to Terjung & Louie, 1972).

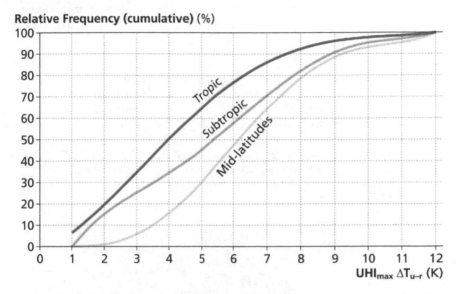

1) Basis for the cluster calculations: city population, height above sea level, topographic features (coast, plain, valley), climate classification according to Terjung & Louie, 1972

Figure 6. Cumulative frequency distribution of UHImax in the urban canopy layer for different climate zones (i.e., tropics, subtropics, mid-latitudes; source: Wienert & Kuttler, 2005)[1]

2.2. Urban excess heating and its effect on humans

Those city districts that develop heat islands under current climatic conditions can be expected to increase thermal stress for their inhabitants due to the projected increase in frequency, intensity and duration of heat spells under global climate change. This increased thermal stress will occur unless timely measures are taken in urban and environmental planning to counter these developments.

The comparison of climatic event days, meaning values of a meteorological variable fall below or exceed thresholds within one day, helps to quantify the degree to which the thermal climate will change. Thus, a *summer day* is defined by a maximum air temperature of $\geq 25\ °C$.

For instance, if the mean value of the daily maxima of air temperature in Essen, Germany, is 21.9 °C (Fig. 7) under the current climatic conditions in the summer months June, July and August, an average number of 26 summer days can be calculated. If, according to the IPCC

(2007), the mean values of the daily maxima of air temperature in Essen rise by 2.3 K until the year 2100, 18 additional summer days will occur, amounting to a total of 44 summer days. Under the current conditions, on 6 days the air temperature maxima reach values of ≥ 30 °C (*hot day*). Under climate change, this number will double to 12 days.

These numbers demonstrate that for estimating the consequences of thermal climatic change, observing the change of the annual mean temperature alone is insufficient. In the respective distribution statistics, the significant change will be in the quantity of peak values.

A distinction between humid (allochthonous) and dry (autochthonous) heat is required when analysing specific periods of high air temperatures or heat spells in central Europe. From the perspective of human biometeorology, humid heat creates a higher thermal stress for the human body than dry heat due to increasingly difficult transpiration. Humid heat is frequently caused by the advection of humid and warm air, for instance, air originating from the Mediterranean, whereas dry heat is developed locally by a combination of strong solar radiation and stagnating air masses under the influence of high pressure.

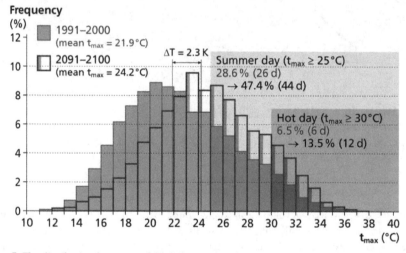

Figure 7. The distribution frequency of the daily maxima of air temperature in summer months (JJA) in Essen, Germany, in the past (1991 – 2000) and future climate (2091 – 2100) (source: Modell WETTREG/ECHAM5; IPCC-SRES-scenario: A1B, modified; here according to Kuttler, 2011a)

Repeated and distinct diurnal cycles of global radiation, wind speed, relative humidity and air temperature suggest an autochthonous heat episode (Fig. 8). The development of the air temperature from day to day displays an increase, even with relatively similar maxima of daily global radiation flux density. At the beginning, a maximum of 22 °C (22.06.2010) and a minimum of 10 °C in the night were observed. These values increased to 34 °C and 22 °C, respectively, at the end of the heat spell (04.07.2010). This increase is caused by the heat storage effect of the buildings. The thermal stress is not only caused by the high temperatures during the day but also by the daily reduction of the nocturnal cooling effect, which can have a detrimental effect on the recreational function of the sleep of the population (Höppe, 1984).

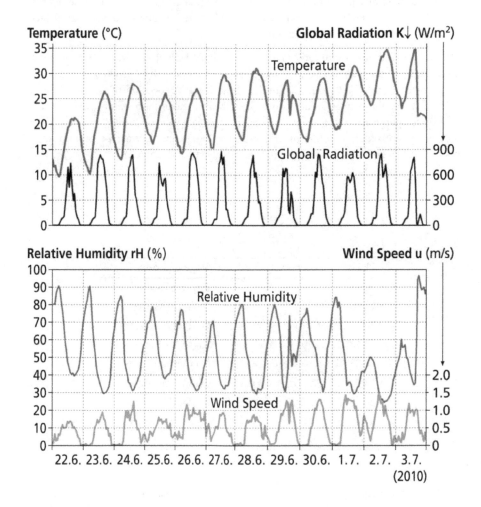

[1] Data of the University of Duisburg-Essen climate station, Campus Essen, 22.06.2010 – 04.07.2010; (source: P. Wagner, pers. comm.)

Figure 8. Hourly mean values of selected meteorological elements (t, K↓, rH, u) during a fair-weather episode in Essen, Germany [1]

Thermal stress, to which the population is exposed for considerable periods of time, can result in increased rates of morbidity and mortality, as the increase of mortality rates in central Europe during the two heat waves in August 2003 demonstrated (Jendritzky, 2007; Souch & Grimmond, 2004). To enable an objective classification of the thermal stress perceived by humans, human-biometeorology employs several thermal indices which include the physically measurable data of meteorological elements, such as air temperature, humidity, radiation temperature and wind speed as well as human physiological factors, such as physical activity, or clothing type.

Of the available thermal indices (Mayer, 2006), the physiologically equivalent temperature (PET), which is calculated based on the above-mentioned values, is chosen as a point of reference (for details, see Verein Deutscher Ingenieure, Association of German Engineers (VDI, 2008)). By the way all indices, including the PET values, are mainly governed by the radiation temperatures of the enclosing wall surfaces to which a person is exposed. The PET range is categorised into different classes according to the thermal perception and the physiological stage of stress of individuals and is given in °C (Table 2).

PET	Thermal perception	Stage of stress
< 4°C	Very cold	Extreme cold stress
4-8°C	Cold	Great cold stress
8-13°C	Cool	Moderate cold stress
13-18°C	Slightly cool	Slight cold stress
18-23°C	**Comfortable**	**No thermal stress**
23-29°C	Slightly warm	Slight heat stress
29-35°C	Warm	Moderate heat stress
35-41°C	Hot	Great heat stress
> 41°C	Very hot	Extreme heat stress

Table 2. The Physiologically Equivalent Temperature (PET), its thermal perception by and stage of stress for humans (source: Matzarakis & Mayer, 1996)

Fig. 9 shows examples of the varying thermal stress of an individual (PET) according to land use during a day of clear and calm summer weather for a rural, a suburban and an urban location, comparing day and night.

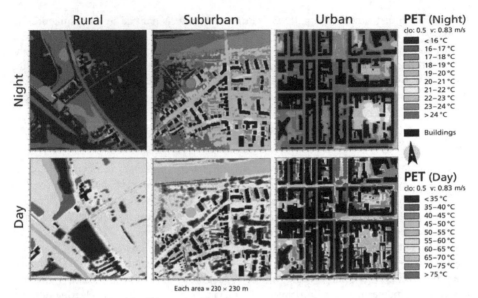

Each area = 230 × 230 m

[1] clo: heat transfer resistance of the clothing, v: wind velocity

Figure 9. ENVI-met simulation results: the PET current state for three plan areas (rural, suburban and urban), with a comparison of day and night conditions in Oberhausen (source: Müller, pers. comm.)[1]

The model simulation was executed with the model ENVI-met (Bruse & Fleer, 1998) and generally mirrors the lesser thermal stress in the open countryside in comparison with the two urban plan areas, although „extreme heat stress" also occurs in unshaded rural locations. However, the PET value is significantly higher in the city, particularly where no shade of buildings or trees reduces the radiation temperature. How significantly solitary canopy trees influence the PET values can be seen in the lower right image of Fig. 9. Approximately in the middle of the image, an east-west road displays a strong but localised reduction of the thermal stress due to solitary trees in its eastern section.

In sum, depending on the general weather conditions, cities are more or less overheated in contrast to the surrounding countryside. Therefore, even under the current climate conditions, urban agglomerations can be regarded as harbingers of the thermal aspects of global climate change. As noted, thermal stress in summer is caused by the UHI and has detrimental effects on the population.

3. Urban air quality

In the central European agglomerations, ozone (O_3), nitrogen dioxide (NO_2) and particulate matter (PM_{10}) are air pollutants of present and future significance that in higher concentrations have harmful effects on humans (e.g., Bell et al., 2010; Ebi & McGregor, 2009; Revich & Shaposhnikov, 2010). Nitrogen dioxide is created as a secondary pollutant through the oxidation of primary emitted NO. Additionally NO_2 is emitted in large quantities as primary air pollutant (Kourtidis et al., 2002). However, changes in the atmospheric concentration of NO_2 do not depend significantly on temperature. Therefore, the temperature increase caused by global climate change should have little impact on the increase of this air pollutant. For this reason, the behaviour of NO_2 concentrations will be disregarded in the following observations. Only the temperature dependency of the urban air constituents ozone and particulate matter will be analysed.

3.1. Ozone

Because high temperatures and intense solar radiation are important preconditions for the formation of the secondary trace substance ozone, ozone concentrations are expected to increase in the future due to the increasing number of heat waves (Kuttler & Straßburger, 1999; Lin et al., 2001). Fig. 10 shows an example of the temperature-dependent formation of ozone. The calculated regression curve allocates an air temperature of 20.5 °C to an ozone concentration of 50 µg/m³. Thus, a doubling to 100 µg/m³ is probable at 28.1 °C, and a tripling of the initial value can be seen at 32.5 °C. One reason for this exponential increase is that the precursor gas peroxyacetyl nitrate (PAN), which contributes to ozone formation, decomposes increasingly fast at rising air temperatures (> 25 °C), thereby releasing nitrogen dioxide, which is the basis for ozone formation in the troposphere.

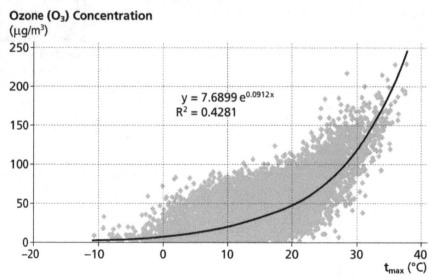

Figure 10. The correlation of the daily maximum ozone concentration and the air temperature at an industrial site in the Ruhr Area (Duisburg-Walsum, 1984 – 2007; source: Melkonyan, 2011)

The volatile organic compounds (VOC) are another group of ozone precursor substances. They can be categorised in two groups: anthropogenic VOCs (AVOCs; for instance, benzene, toluene and xylene) and biogenic VOCs (BVOCs; for instance, isoprene, α-pinene and limonene (Sharkey et al., 2008)). The latter VOCs are emitted by deciduous and coniferous trees and shrubs when temperatures are high. The plant physiology is not sufficiently understood, however, thermally triggered stress reactions of the vegetation have been discussed (Kesselmeier & Staudt, 1999; Sharkey et al., 2008). Even though BVOCs are expected in the urban atmosphere only at low concentrations due to the usually low vegetation density, the emission rates are highly temperature-dependent. Therefore, even in urban areas, summer weather with strong solar radiation will lead to additional ozone formation by biogenic trace substances. This increase occurs particularly if vegetation is dominantly composed of plants that emit BVOCs on a large scale, the so-called high-emitters (Benjamin & Winer, 1998). For this reason, at certain urban locations, BVOC concentrations (e.g., isoprene) can be produced (Fig. 11) that are several times higher than the AVOC concentrations (e.g., benzene). Additionally, the figure shows the high dependency of isoprene concentrations on the air temperature in contrast to benzene.

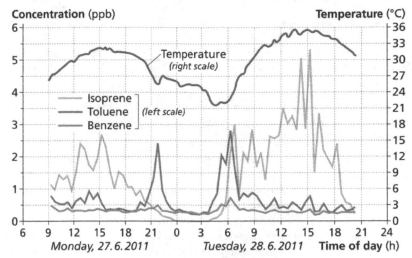

Figure 11. AVOC and BVOC concentrations near a university parking area in Essen, Germany, during hot weather (27.06.2011 – 28.06.2011; source: P. Wagner, pers. comm.)

To avoid increased ozone formation by biogenic VOCs, predominantly those plants should be used as greenery in the city which emit no or only low amounts of these trace substances and belong to the so-called low-emitters.

The number of days on which the ozone level exceeds the acceptable limit for individuals indicates to what degree ozone pollution will change for humans in the future climate. Fig. 12 shows a comparison of the annual number of days with ozone exceedance for the current and future thermal climate at an industrial site in the Ruhr Area, Germany. Accordingly, the mean number of days with ozone exceedance will increase from currently 8 days to 19 days in the future, more than doubling the overall number (Melkonyan, 2011).

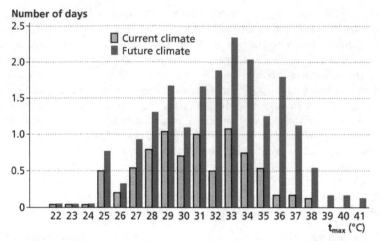

Number of days

[1] Target value for ozone since January 2010: highest 8-hour mean of one day ≥ 120 μg/m³; 25 days with ozone exceedance allowed per calendar year, averaged over three years

Figure 12. The annual number of days with ozone exceedance[1] for the current (1984 – 2007) and future (2100) thermal climate at an industrial site in the Ruhr Area, Germany (Duisburg-Walsum; source: Melkonyan, 2011)

The increase in ozone concentrations in Germany during calm and clear summer weather with intense solar radiation or during heat waves can be traced to four groups of determinant factors (Fig. 13).

1. Dry soil causes a lower atmospheric deposition velocity (v_d), aerodynamic resistances (r_{aero}) are relatively high and the plant's stomata are prematurely closed. Hence, less ozone reaches the plant's surface or is absorbed by the plant through the stomata.

2. Anthropogenic volatile organic compounds (AVOCs) and biogenic volatile organic compounds (BVOCs) contribute to the overall emission of VOCs in the summer at a ratio of 60:40 with isoprene being a major trace substance of the BVOCs. Although the BVOCs occur at a lower concentration in street canyons than the AVOCs, the reactivity of the BVOCs (e.g., isoprene) – measured in terms of reaction with OH radicals – is increased by a factor of approximately 100 compared with the AVOCs (e.g., benzene). The emission of BVOCs increases exponentially until approximately 40 °C (particularly the isoprene's). However, the emission of AVOCs shows little or no temperature dependency. Thus, ozone formation is expected to increase in city districts with a significant quantity of plants that are high-emitters of BVOCs.

3. Anthropogenic combustion processes lead to the emission of high amounts of NO_x, which results in a higher potential of photolysis in the near-surface air layer and an increased release of the oxygen atoms required for ozone formation. Furthermore, PAN is highly unstable at temperatures exceeding 25 °C, causing the release of ozone precursors.

4. Under polluted conditions in urban areas, the reaction of OH radicals with VOCs and CO leads to ozone formation.

In sum, the projected temperature increase in Germany caused by global climate change is expected to lead to a maximum increase in the ozone concentrations of up to 10 % compared with current peak concentrations.

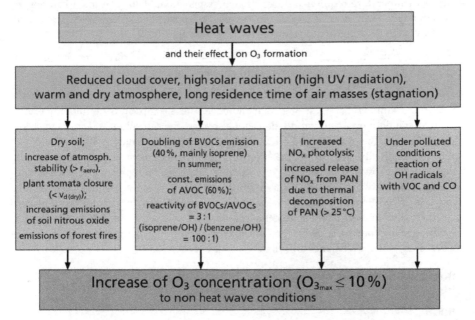

[1] Benjamin & Winer 1998; Chameides et al., 1988; Dawson et al., 2007; Demuzere & van Lipzig, 2010; Emeis et al., 1997; Forkel & Knoche, 2006; Held et al., 2004; Klemm et al., 2000; Mayer & Schmidt, 1998; Meleux et al., 2007; Meng et al., 1997; Narumi et al., 2009; Pöschl, 2005; Sillmann, 1999; Solberg et al., 2005; Solmon et al., 2004; Straßburger, 2004; Taha, 1996; Wu et al., 2008; Yashuda et al., 2009

Figure 13. Four groups of factors leading to ozone formation during calm and clear summer weather, compiled according to several authors[1]

3.2. Particulate matter

Particulate matter is defined as the particle content of the air, which can be categorised into three different modes of size and formation (GDCh, 2010):

- The nucleation mode (ultrafine particles, $\varnothing \leq 100$ nm) with little mass fraction and formation mostly by gaseous precursors, which agglomerate quickly;
- the accumulation mode (100 nm $< \varnothing \leq 1,000$ nm); formation mainly through the nucleation mode by adsorption of gases or by coagulation;
- the coarse mode ($\varnothing > 1$ μm); particles originating from deflation and erosion processes or mechanical abrasion and suspension.

The entire class of particles with a particle size of up to 10 μm (PM_{10}) is commonly called particulate matter.

Approximately half the particulate matter is directly emitted (primary particles), while the remainder is formed by gas-to-particle conversion (secondary particles). The question whether atmospheric concentration of particulate matter displays a statistical dependency on the air temperature requires a differentiated examination. Neither in the urban nor the rural boundary layer does a significant correlation appear to exist between the overall concentration of PM_{10} and the air temperature (Melkonyan, 2011).

However, a connection with temperature can be ascertained for certain modes, e.g., for ultrafine and very large particles (for instance, plant pollen).

Among the ultrafine particles, the BVOCs play a special role, being emitted by specific plants in the summer during strong solar radiation. These compounds do not only exert a sustained influence on ozone production, but also affect the aerosol mass. Due to their low vapour pressure, the organic compounds formed during the oxidation of BVOCs adhere to the particles already present in the atmosphere, increasing their overall mass. Several measurements prove (Plass-Dülmer, 2008) that the BVOCs add significantly to the aerosol mass in summer, whereas an influence on the formation of new particles could not be proven (Kiendler-Scharr et al., 2009) or their contribution should at least be viewed as doubtful (Plass-Dülmer, 2008). Under the increasing temperatures of global climate change, this problem is expected to grow. The summer increase of the particle mass likely has an impact on the radiation balance and cloud formation (Pöschl, 2005).

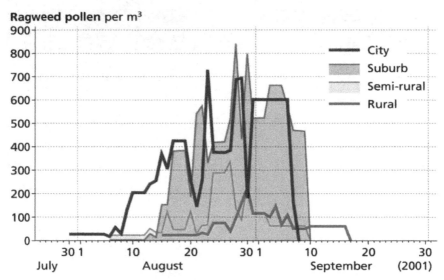

Figure 14. The summer ragweed pollen production for four sites along an urban transect in Maryland, USA, in 2001 (source: Ziska et al., 2003, modified)

In the case of plant pollen (< 250 μm), ragweed pollen (Ambrosia artemisiifolia) was used to demonstrate (Ziska et al., 2003) that, with higher temperatures and CO_2 concentrations, pollen dispersal begins significantly earlier in cities than in the surrounding countryside. In addition, a considerably higher pollen count can be measured in cities in comparison with rural areas (Fig. 14). The pollen contains an allergen that is detrimental to human health (Amb a1) and whose production is affected because ragweed is a C_3 plant. Because the growth of C_3 plants is limited by carbon, UHIs and higher concentrations of urban CO_2 boost the production of this allergy-triggering protein. Laboratory research has shown (Ziska & Caulfield, 2000) that pollen production and the number of floral spikes per plant are approximately doubled if atmospheric concentrations of CO_2 increase from 370 ppm (2000) to 600 ppm (2050).

Other types of plant pollen, such as that of the birch, are triggered in their allergen release (Bet v1) by air pollution (e.g., by NO or NO_2), increasing the intensity of allergic diseases (Pöschl, 2005). On the basis of the projected global climate change, particularly in cities, an increase in the pollen count, which is responsible for triggering allergies, is expected.

In regard to the influence of atmospheric particulate matter concentrations in Germany during calm and clear summer weather or heat waves, four main factor groups can be identified (Fig. 15):

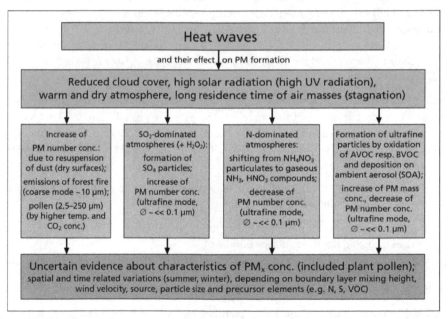

Figure 15. The influence of temperature on the formation of particulate matter (source: compiled according to Carvalho et al., 2010; Dawson et al., 2007; Jacob & Winner 2009; Pöschl, 2005; Wise & Comrie, 2005; Ziska et al., 2003)

1. Large (PM_{10}) and very large particles (< 250 µm) can be taken up from dry ground or surfaces in case of sufficiently high wind speed and shearing stress as primary or

resuspended dust, or be released from plants as pollen. Furthermore, forest fires can increase the particle concentration in the atmosphere.

2. Sulphate particles for instance can be formed in an SO_2-dominated atmosphere with the participation of H_2O_2, thereby increasing the particle number concentration in the ultrafine mode ($PM_{0.1}$).

3. An N-dominated atmosphere encourages, for instance, the conversion of NH_4NO_3 particles in the gaseous N compounds NH_3 and HNO_3, decreasing the particle number concentration in the ultrafine mode ($PM_{0.1}$).

4. AVOCs and BVOCs adhere to particles already present in the atmosphere due to their low vapour pressure. Therefore, no new formation occurs but, rather, an increase of mass.

In sum, both regional and seasonal aspects are found to exert diverse influences on the concentration of particulate matter.

4. Countermeasures on the local level

Countermeasures against global climate change being the sum of mitigation and adaptation measures should be taken particularly in cities. Although cities occupy only a relatively small proportion of the Earth's land mass, they are home to more than half the world's population. Consequently, cities are viewed as the strongest net sources of anthropogenic carbon dioxide (Büns & Kuttler, 2012; Grimmond et al., 2004; Velasco & Roth, 2010; Vogt et al., 2006).

Among other effects, the projected climate change (IPCC, 2007) will lead to an increase of temperatures during the entire course of the year in central Europe. The reduced requirement to remove snow and ice in the winter has to be emphasised as a positive effect, saving energy and costs (Brandt, 2007). Additionally, the heating costs of buildings and structures in the cold season should decrease, reducing CO_2 emissions.

These positive aspects of climate change are opposed by the known disadvantages in the summer, which are as follows:

*   higher thermal stress for individuals;
*   increased cooling requirements for buildings;
*   increased spread of pathogens (mosquitoes, ticks) and allergy-triggering plants (e.g., Ambrosia);
*   danger of flood and inundation caused by heavy rain.

Appropriate building-design and urban-planning measures are discussed in the following sections. These measures can counter the effects of overheating, which occur particularly in densely built-up urban districts. In this context it has to be considered that these temperature-reduction measures can prove effective only against autochthonous, dry heat. However, in allochthonous, humid and hot conditions, the measures are scarcely effective (Holst & Mayer, 2010).

4.1. Object-related means

Concerning central European climate, preventive object-related means to reduce CO_2 emissions are efforts to save energy in building air-conditioning. An example of such means would be to reduce the currently high specific heating-energy consumption of sometimes > 200 kWh/(m^2 a) in Germany. Better thermal insulation of buildings and resident consumption restrictions (Stadt Essen, 2009) could reduce this figure to ≤ 100 kWh/(m^2 a).

At 10 to 15 kWh/(m^2 a), significantly less energy is required by passive houses. Even better is the energy balance of plus-energy houses, which on a yearly basis generate more energy than they consume. Only when a household produces an energy surplus can an electric vehicle become feasible, because the vehicle's energy could be produced without the help of CO_2-emitting power stations (Hegger, 2009; in Germany, the CO_2-emission rate is approximately 0.6 kg CO_2/kWh).

Generally, the UHI effect means that the energy costs to keep inner-city buildings cool in the summer is higher than in the surrounding countryside. Consequently, office buildings in the City of London require 16 % more energy for cooling than comparable buildings in the surrounding countryside (Kolokotroni et al., 2006).

It is well known that brightly-coloured surfaces reflect more short-wave radiation than dark surfaces and therefore release significantly less long-wave radiation (L↑) into the vicinity and via heat conduction (λs) into the ground or the building.

However, absorption cannot only be reduced by increasing the amount of reflection in the short-wave range. Additionally, if surfaces are coated or coloured to reflect infrared radiation (Cool Colours) the amount of reflection in the long-wave range can be increased. Comparative measurements (Fig. 16) show that during highest solar radiation a reduced surface temperature of up to 6 K is possible for a surface coated with Cool Colours in comparison with a standard surface due to a stronger reflection in the near-infrared range ($\lambda > 750$ nm). Because a surface coating of this type creates lower temperatures than a standard surface during intense solar radiation, less energy is conducted into the material and a reduction in long-wave radiation flux density (L↑) is realised.

In the summer, highly reflective surfaces can lead to a noticeable reduction of temperatures and save cooling energy for buildings. However, in cold climates and during the heating period, building walls that are highly reflective are counterproductive in regard to the thermal balance because highly reflective walls do not help to achieve a solar radiation energy gain for the building when outside temperatures are low. Temperature-dependent, seasonally changing, thermochromic house colours (in summer: bright; in winter: dark) for a technical application do not yet exist. Thus, a compromise must be found that takes into account a house's yearly energy consumption for cooling and heating. Research conducted with model houses for different climate zones resulted in a highly differentiated picture. A location in the middle latitudes (Berlin, Fig. 17, left) shows that for annual energy consumption with respect to the short-wave albedo (α) and the long-wave emissivity (ε) of the surfaces, the lowest possible albedo values and equally low emission coefficients lead to

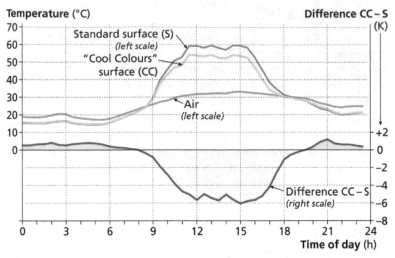

[1] Site: University of Duisburg-Essen climate station, Campus Essen

Figure 16. The diurnal air-temperature progression for a standard surface (S), a Cool Colours-surface (CC) and their difference (CC–S) on a day with clear and calm weather in Essen (26.06.2010)[1] (source: P. Wagner, pers. comm.)

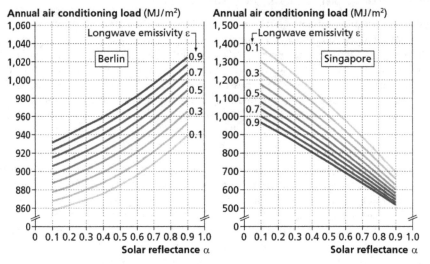

[1] one-storey house, flat roof; 20 m width, 20 m length, 4 m height; each wall has glass windows (16 m²) with a U-value of 3.2 W/(m² K); wall thickness: 0.28 m (bricks and cement mortar), R_{th} = 0.42 (m² K)/W; roof: 0.1 m reinforced concrete, 30 mm XPS board as insulation and 20 mm cement mortar, R_{th} = 1.12 (m² K)/W; ach (air change per hour): 0.3

Figure 17. Yearly energy consumption for heating and cooling a one-storey model house[1] as a function of the wall's short-wave reflection (α) and long-wave emissivity (ε) (source: Shi & Zhang, 2011, modified)

the lowest annual energy consumption. Here, a low emissivity value offers the highest saving potential. For comparison, the juxtaposed example taken from the tropics (Singapore, Fig. 17, right) illustrates that in this climatic zone a house's walls should have the highest possible values of both coefficients. Here, a high albedo has the strongest influence compared with a high emissivity, when targeting the saving of energy for heating and cooling.

If buildings are greened, numerous micro-climatic and air-hygienic advantages can be identified. Among these advantages are lowered building surface temperatures during high radiation with less diurnal fluctuations in comparison with a building without vegetation (Alexandri & Jones, 2008). Furthermore, evergreen vegetation can protect the building against heat loss in the winter.

In addition, the green surface leads, as a function of the LAD, to ad- and absorption of atmospheric trace substances, frequently improving the air quality in the building's vicinity (Köhler, 1993).

Fig. 18 shows the substantial differences between the surface temperatures of different types of roofing during clear and calm weather in summer. Black roofing felt reaches a temperature of more than 90 °C for a short time around solar noon. Considerably lower values are measured for bright and dry gravel as well as bright paint. However, the lowest values are produced by artificially moistened and, in particular, watered and planted roofs. Between a watered and planted roof and the black roofing felt, a temperature difference of up to 70 K can be measured at approximately noon. Consequently, planted and watered roofs are among the most efficient measures to reduce surface temperature if the strongest cooling effect is desired at maximum radiation. At night, the surface temperatures of all roofs reach a similarly low level.

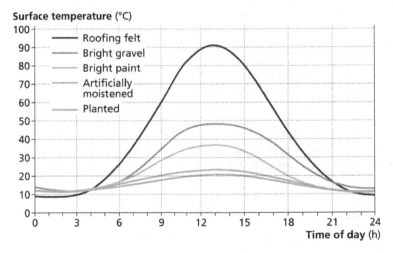

Figure 18. Surface temperatures of different roofs during clear and calm weather in summer (Berlin; source: Horbert, 2000, modified)

Planted roofs display similarly positive characteristics with respect to water budget when compared with roofs without vegetation (Table 3). The evapotranspiration of a green roof in regard to overall precipitation is substantially higher (72 %) than that of compared surfaces in a built-up area, even when in the case of a conventional roof, an additional partial evaporation of the run-off precipitation in swales and trenches (23 % and 10 %) is considered. Because the surface run-off of the green roof amounts to only 28 % in comparison with the conventional roof with 80 % and, furthermore, occurs after a delay, discharge peaks are reduced severely, leading to a balanced catchment yield factor. Because the frequency of strong rain is expected to increase in the future, a large-scale greening of roofs in developed areas would offer a possibility to mitigate the force of potential floods or prevent them.

| | Water budget (% of precipitation) | | |
	Evapotranspiration	Surface run-off (including sewage water)	Groundwater recharge
Natural predeveloped area			
Grassed area and cultivated land	55	9	36
Developed area			
Roof[1]	20	80	0
Green roof[1]	72	28	0
Traffic area	25	75	0
Water semi-permeable pavement[2]	30	55	15
Swale infiltration[3]	23	1	76
Trench infiltration[3]	10	1	89
Undeveloped area (mixed vegetation)	65	7	28

[1] 70 % roof area with 45°-slope, 30 % with 2°-slope

[2] Average over pedestrian way and terrace

[3] Roof run-off infiltration

[1)] Calculations were conducted with the analytical hydrological precipitation discharge model erwin 4.0 (2002). Green roof: sedum-grass-herb roof with an extensive substrate of 10 cm

Table 3. A water budget for representative land use with medium precipitation rates (799 mm/a)[1)] (source: Göbel et al., 2007)

Green roofs have undeniable climatic, air-hygienic and hydrological advantages for appropriately equipped buildings. For green roofs to exert a beneficial climatic influence on a larger scale, e.g., on the city-district level, as many houses as possible and in particular low houses should be greened.

4.2. Area-related means

Area-related means are interventions in solar-energy balances and the CO_2 budget of the UBL, here, however, relating to the scale of central city districts and urban residential quarters. A brief compendium of possible adaptation measures to climate change – of which some will be discussed later – is presented in Table 4. These and more measures correspond to recommendations made by federal state governments in Germany for example. For a successful implementation of measures sufficient legislation is a prerequisite.

Residential quarters scale
Built infrastructure
Preserve and create open space; minimise impervious surfaces
Create open water bodies
Optimise building orientation to the sun
Provide shading of relevant spaces
Use building material with reduced heat conduction and heat storage properties; cool colours
Green areas and vegetation
Create, maintain, redesign parks
Provide greening of streets; use of suitable plants
Guarantee sufficient irrigation of urban vegetation
Increase the number of trees and shrubs for providing shade
Use plants that increase the permeability of the soil due to root and water penetration
City scale
Restrict building land to prevent urban sprawl
Preserve and/or create surfaces which produce fresh air
Preserve and create ventilation channels as links between rural and urban environment

Table 4. Selected area-related adaptation measures to climate change on the residential quarters' and city scale.

An important means to reduce thermal stress in cities and thereby adapt to the climate change is the shading, evaporative cooling and ventilation of appropriate spaces. The reduction of radiation and air temperature of squares and streets depends on a variety of factors (Erell et al., 2011; Holmer et al., 2007). These factors include the geometric settings produced by the building structures, the influences of which can be determined by the ratio building height to street width (H/W), the sky view factor (SVF) (Blankenstein & Kuttler, 2004; Synnefa et al., 2006), the ventilation degree, the shading possibilities, the ground colour (Synnefa et al., 2006) and the ground type (sealed/natural). The amount of shade produced by large-canopy roadside trees depends on the shape, size and density (LAI, LAD) of their crowns, among other factors (Erell et al., 2011; McPherson & Simpson, 1995). In addition, the location of the tree plays a crucial role for the area to be shaded (Donovan & Butry, 2009). Furthermore, a canopy cover in the middle of the road should be avoided if trees are planted on both sides because the tunnel effect created could lead to street-level exhaust-gas accumulation. Approximately 80 % of the cooling effect of trees results from the shade, and, where soil moisture is optimal, approximately 20 % results from the latent heat flux (Q_E) of evapotranspiration (Shashua-Bar & Hoffman, 2000). In addition to improving thermal comfort, trees contribute to a pleasant light climate by reducing UV radiation and long-wave radiation fluxes under the canopy (Holst & Mayer, 2010).

Fig. 19 displays the cooling effect of roadside trees during hot weather. The thermal comfort perceived by a person located in the shade is increased significantly as PET values are reduced up to 22 K. This effect can be attributed more to lower radiation temperatures than to transpiration. Outside the shade, only a minor effect on thermal comfort can be observed. The direct solar radiation (K↓) is considerably reduced by trees during the bright daylight hours (in this case, up to 850 W/m²; Fig. 20, a) as is the long-wave radiation from the ground (L↑) (by up to 200 W/m²; Fig. 20, b), contributing essentially to improving thermal comfort.

A sunscreen can also be produced artificially, e.g., with covered footpaths or galleries (Ali-Toudert & Mayer, 2007), double-skin façades (Baldinelli, 2009), membrane structures covering semi-closed spaces (He & Hoyano, 2010) and canvas blinds or parasols. The reduction of radiation temperature can be substantial, improving the thermal comfort of the occupants and, by reducing the daily temperature range, countering the fatigue of material located in the shade. Additionally, light-induced ozone formation decreases.

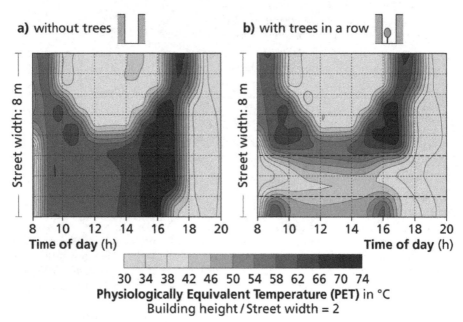

a) without trees   **b) with trees in a row**

Street width: 8 m

Street width: 8 m

8   10   12   14   16   18   20 8   10   12   14   16   18   20
Time of day (h)   **Time of day** (h)

30 34 38 42 46 50 54 58 62 66 70 74
Physiologically Equivalent Temperature (PET) in °C
Building height / Street width = 2

[1] (E-W oriented; H/W = 2; tree row on the north side of the street; tree height: 10 – 16 m, dense treetops; PET > 42 = very hot)

Figure 19. The effect of roadside trees on thermal comfort (PET) in a street canyon[1] in central-Saharan Ghardeia, Algeria (source: Ali-Toudert & Mayer, 2007, modified)

Green areas with grass only, often found as city greens, contribute little to the improvement of urban climatic conditions during the day because these areas provide little or no shade and exert their climatic influence largely through the latent heat flux (Q_E) caused by evapotranspiration, which depends on sufficient soil moisture.

As Fig. 21 illustrates, the surface temperature of grass-covered areas increases significantly with the passing of precipitation-free hours, which means that the climatic effects of grass-covered areas are closely linked to the soil's water supply. If the area dries out for a few days (here, a maximum 11 d), the desired effect of temperature reduction increasingly disappears. Intra-urban (grass-covered) green areas should therefore be optimally supplied with water to maintain and ensure their beneficial thermal effects on the local climate, particularly in hot weather.

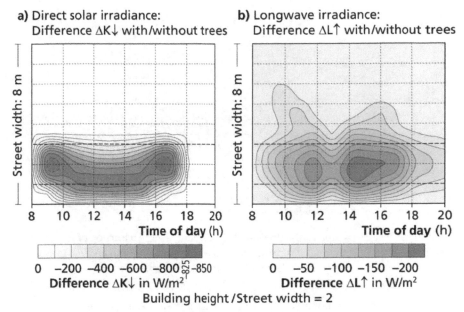

a) Direct solar irradiance: Difference $\Delta K\downarrow$ with/without trees

b) Longwave irradiance: Difference $\Delta L\uparrow$ with/without trees

0   −200 −400 −600 −800 −850
Difference $\Delta K\downarrow$ in W/m²

0   −50   −100 −150 −200
Difference $\Delta L\uparrow$ in W/m²

Building height / Street width = 2

[1] (E-W oriented; H/W = 2; tree row on the north side of the street; tree height: 10 – 16 m, dense treetops)

Figure 20. The difference between (a) direct solar radiation ($\Delta K\downarrow$) and (b) long-wave radiation of the street surface ($\Delta L\uparrow$) in a street canyon[1] with/without a tree row in central-Saharan Ghardeia, Algeria (source: Ali-Toudert & Mayer, 2007, modified)

Because their deposition velocities (v_d) reach only low values in comparison with other vegetation (Horbert, 2000; Litschke & Kuttler, 2008), grass-covered areas ad- and absorb relatively small amounts of air pollutants. In contrast, the combination of greater volume and the reduced wind velocity caused by trees in wooded green areas results in an effective filtration of the air beneath the canopy. For example, coniferous trees can filter a multiple of trace substances compared with an open area by dry and wet deposition (Wrzesinsky, 2004). The type and density of vegetation, particularly LAI, play a crucial role in the absorption of particles and gases (compilation in Jonas et al., 1985; Litschke & Kuttler, 2008).

For example, 200 m from an intensively used road, mitigation values between 40 % and 50 % could be determined as the annual mean values for the atmospheric trace substances

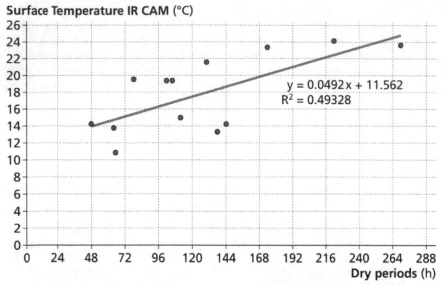

[1] 14., 16., 22., 23., 26., 28., 30.09.; 15., 17., 21., 24., 27., 31.10.2011

Figure 21. The surface temperature of a grass-covered area at the University of Duisburg-Essen climate station as a function of the number of precipitation-free hours[1]

NO and NO_2 in a green area (grass, trees and shrubs) and a 20 % reduction of the immission concentration of CO compared with concentrations measured at the roadside. However, the ozone concentrations in the green area were up to 20 % higher than at the street location (Ropertz, 2008).

The positive thermal effect of green spaces with trees and shrubs in comparison with sealed surfaces becomes clear when the turbulent heat flux densities (Q_H, Q_E) of an urban ($\Lambda_P = 0.82$) and a suburban location ($\Lambda_P = 0.31$) (Fig. 22) are juxtaposed. While the turbulent sensible heat flux (Q_H) dominates at the urban location in summer and in winter, the turbulent latent heat flux (Q_E) constantly reaches higher values than Q_H at the planted and irrigated suburban location (wooded grassland). At the urban location, the mean Bowen ratio ($ß = Q_H/Q_E$) amounts to $ß_u = 1.8$ in the summer and $ß_u = 21.4$ in the winter. The conclusion can be drawn that because of the limited availability or absence of water, far more energy is used for the heating of the air than for the evaporation. In contrast, the ß-values reach the expected significantly lower values of $ß_{sub} = 0.5$ at the suburban location in the summer and in the winter, as opposed to the urban location. Applying the proportion of Q_E to the radiation balance Q^*, it becomes clear that at the urban location on days with clear and calm weather in the summer only 23 % and in winter only 4 % of the energy can be dissipated by evaporation. At a suburban location, these proportions are higher: 54 % in summer and 41 % in winter. Additionally they fluctuate much less than in the city because of the optimal availability of water in the diurnal course. At a suburban location, the value of Q_E occasionally exceeds the value of the radiation balance Q^*, particularly in the afternoon hours (not shown here). Such episodes are called the oasis effect because a high evaporation potential cannot be covered by Q* and therefore draws energy from the surroundings to compensate the radiation balance, cooling the vicinity.

The comparison shown in Fig. 22 illustrates that green areas, which have a high evapotranspiration potential if supplied well with ground water, can contribute significantly to improved thermal comfort. From the urban-planning perspective, increased attention should be paid to this contribution because heat waves will occur more frequently in the future. However, a sufficient supply of irrigation water must be ensured (see above, Cleugh et al., 2005).

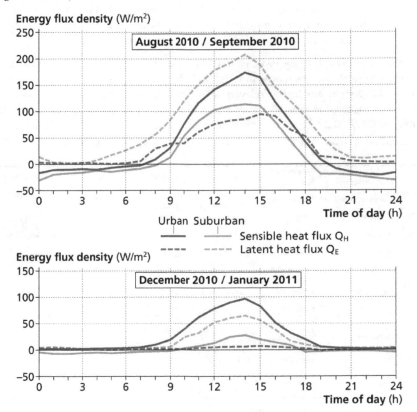

1) 08/2010 – 09/2010: 10 days with calm and clear weather; 12/2010 – 01/2011: 8 days with calm and clear weather

Figure 22. Average diurnal courses of turbulent heat flux densities (Q_H, Q_E) at an urban and suburban site[1] in Oberhausen, Germany (source: Goldbach & Kuttler, 2012)

A sufficient supply of water can be ensured by skillfully designing areas for the collection and storage of rainwater, thereby preventing a sewage-system overload in case of heavy rainfall.

Due to the mostly lower temperatures of urban green areas as compared with the warmer built-up areas in their vicinity, the green areas are called "urban cold islands" (Bongardt, 2006).

The lower temperatures of the green areas can generate a local circulation between the green areas and the built-up neighbourhood. Such local air movements, called park winds, reach only low speeds and occur more intermittently than continuously. The extent to which the cool air flows into the built-up surroundings depends on the design and enclosure of the green area and the type of the surrounding buildings and structures. For instance, if the green area is located in a ground depression or is surrounded by a high wall, the air-exchange rate will be reduced and the spread of cool air into the built-up area restricted. However, wall apertures in combination with streets at right angles can assume the functions of ventilation channels and lead cold air into the street canyons.

The questions of how large a green area must be to cause a temperature reduction and up to what distance from the green area into the built-up surroundings a temperature-reducing effect can be proven are highly relevant to climatological urban planning. Generally, a connection between the size of a green area and a higher thermal spread seems likely (Hamada & Ohta, 2010). To what degree this phenomenon occurs can be seen in Table 5.

Size in ha (rounded)	Location (UHI in K)	Park	Structure	PCI_{max} in K	Reach in m	Source
3	Kumamoto (Kyushu) (3 day, 2 night)	Kengun Shinto Shrine	Trees	2.5 (3 p.m.)	50	Saito et al., 1990
5	Vancouver (6 at dusk)	Trafalgar Park	Grass, tree fringe partly irrigated	5.0 (night) mobile measurements	200 - 300	Spronken-Smith & Oke, 1998
10	Dortmund	Westpark	Grass and trees	4.0 (night)	150	Bongardt, 2006
18	Berlin	Stadtpark Steglitz	-	1.0 (evening)	80 - 140	von Stülpnagel, 1987
30	Mainz	Stadtpark	-	2.0 (morning)	< 300	Naumann, 1981
44	Stuttgart	Schlossgarten	-	1.3 (yearly mean) 3.8 (daily mean)	200	Knapp, 1998
80	Copenhagen	Falledparken	Grass and trees	2.1 (10 p.m.)	100	Eliasson & Upmann, 2000
125	Berlin	Kleingärten Priesterweg	Garden	5.4 (evening)	250	von Stülpnagel, 1987
156	Gothenburg	Slottsskogen	-	3.3 (6 p.m.) measuring station	250	Eliasson & Upmann, 2000
212	Berlin	Tiergarten	Forest / grass	4.3 (evening)	200 - 1,300	von Stülpnagel, 1987
525	Mexico City	Chapultepec	Mixture (trees, grass); not irrigated	4.0 (dry season) 1.0 (wet season)	2,000 (one park depth)	Jauregui, 1990

Table 5. A compendium of size, surroundings, maximal cooling effect (PCI_{max}) and thermal reach of urban green areas (source: according to a compilation in Bongardt, 2006; here, according to Kuttler, 2010b, modified)

Because frequently higher summer temperatures and a lower amount of precipitation are expected due to the climate change, plants with high drought tolerance should be used in urban green areas. Additionally, which plants release significant quantities of BVOCs during high temperatures should be considered. Mainly those plants should be used that are low-emitters of isoprene (Benjamin & Winer, 1998; Taha, 1996), which are plants with isoprene emissions during thermal stress of not more than 2 µg per g dry matter per hour. Table 6 displays selected trees characterised by a low ozone-forming potential and optimal drought tolerance in case of limited water supply.

Scientific name	Common name	Low ozone-forming potential (OFP)	High drought tolerance (DT)
Acer campestre	Field Maple	++	++
Acer rubrum	Red Maple	++	++
Carya ovata	Shagbark Hickory	++	+
Carya tomentosa	Mockernut hickory	++	++
Fraxinus pennsylvanica	Green Ash, Red Ash	++	+
Ginkgo biloba	Ginkgo, Maidenhair Tree	++	++
Malus tschonoskii	Tschonoski Crabapple, Pillar Apple	++	+
Pinus ponderosa	Ponderosa Pine, Bull Pine, Blackjack Pine	+	++
Pinus sylvestris	Scots Pine	+	++
Prunus avium	Wild Cherry, Sweet Cherry, Bird Cherry, Gean	++	++
Pyrus communis	European Pear	++	+
Pyrus pyraster	European Wild Pear	++	+
Quercus rubra	Northern Red Oak, Champion Oak	+	+
Sophora japonica	Pagoda Tree	+	++
Ulmus parvifolia	Chinese Elm, Lacebark Elm	++	+
x Cupressocyparis leylandii	Leyland Cypress	++	+
Zelkova serrata	Japanese Zelkova, Keyaki	++	+

Low OFP: Isoprene emission 2µg/(g·h) DT; ++ = very good, + = good

Table 6. The ozone-forming potential (OFP) and drought resistance of selected trees and their suitability to higher temperatures (source: combined according to Benjamin & Winer, 1998; Roloff et al., 2008, modified; here, according to Kuttler, 2011b)

Cities require a great amount of parking space. For instance, parking space amounts to approximately 10 % of the city area in the USA and almost 7 % in Japan (Onishi et al., 2010). Because most of this space is covered with tarmac or paved and absorbs a great amount of heat due to the dark colour, the shading or greening of these surfaces would have a substantial positive effect on the energy balance. In comparison with an area of tarmac, a grassy parking area displays a maximal temperature difference of the surface of up to 15 K during calm and clear summer weather with maximum solar radiation values. During the daytime, the sensible heat flux can be reduced by 100 to 150 W/m² and in the night by approximately 50 W/m² based on investigations in Japanese cities (Takebayashi & Moriyama, 2009). This demonstrates that greened parking areas can limit or reduce the sprawl and intensity of the UHI effect even in the night, at least locally. However, this reduction is only possible for intact and irrigated stretches of grassy areas. If the turf is damaged, the cooling effects are severely limited (Kanemoto et al., 2009).

The following overview summarises the climatic and air-hygienic effects as well as the climate-related design recommendations for intra-urban green spaces (according to different authors):

- reduction of surface and air temperatures by means of shading and evapotranspiration ($q_{v,W\,20\,°C}$ ~2.4 MJ/kg water; negative correlation of surface temperature with the normalised difference vegetation index (NDVI))
- significantly higher avoidance of CO_2 production in the city with the help of city trees compared with similar forest trees (in its CO_2-reducing effect, one urban shade tree equals several forest trees)
- decrease of net CO_2
- creation of a pleasant light climate (shift of reflected and transmitted radiation to the long-wave section of the radiation spectrum, reduction of UV radiation)
- reduction of the wind speed and therefore the possibility of trace-substance deposition on plant surfaces, albeit a danger of atmospheric pollutant accumulation (tunnel formation in street canyons)
- reduction of discharge peaks because of interception and time-delayed percolation of precipitation into the ground
- decrease of ozone-formation potential by reduction of ambient temperatures
- large areas should be planted following the "savannah principle": grass and interspersed solitary canopy trees with shading effects
- green aisles should be designed with minimum roughness and connected linearly if possible
- intra-urban areas or areas close to the city should be planted with "energy plants" (e.g. poplars)
- green spaces in suburban areas should be preserved for the regeneration of cold and fresh air
- exploitation of the "shrinking cities problem" to transform sealed surfaces to green areas and water-retentive areas

Finally, it should be noted that shading, greening or surface brightening during periods of high summery radiation can have direct or indirect effects on buildings or city districts, leading not only to lower temperatures but also to a decrease in energy consumption by the reduced use of air-conditioning systems.

5. Energetic use of the urban subsurface

The excess heat of sealed surfaces in urban areas can be used to save energy expended on controlling the climate of buildings and thereby reduce urban CO_2 emissions. Because this overheating also occurs below the surface (subterranean UHI; see chapter 1), the intensity and frequency of the overheating must be established. As the exemplary data collected in Oberhausen, Germany, demonstrate (Fig. 23), the temperatures measured at up to 2 m depth (daily mean values) are predominantly higher (with the exception of April/May) than the air temperatures measured at the same location. If the average temperature difference (ΔT_{a-s}) between the air (t_a) and 2 m below ground (t_s) is approximately 2 K and the relevant urban area has a size of 2 km², this temperature difference would represent approximately $2.2 \cdot 10^6$ kWh/a. Assuming a ground thermal capacity density of $\zeta = 2 \cdot 10^6$ J/(m³ K) and a complete use of the temperature difference for a layer thickness of 1 m. If used via a heat pump with an efficiency of $\eta = 0.5$, this energy would amount to the annual energy consumption of approximately 275 German households (4 persons; 4,000 kWh/a) and, based on an emission of 0.6 kg CO_2/kWh (German power plant mix), lead to an avoided emission of approximately 660 t CO_2. Because the temperature difference between the air and the ground is not constantly 2 K but fluctuates, this energy should be used to heat water for domestic use with the possibility to be stored to bridge periods without heat-pump energy.

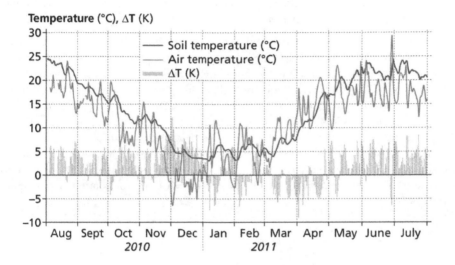

Figure 23. Daily mean values of air temperature (2 m above ground) and subterranean (2 m below ground) temperatures in Oberhausen, Germany, and their difference for the period 08/2010 – 07/2011 (source: H. Püllen, pers. comm.)

The energy yield of the subterranean UHI can be increased if the energy content of the groundwater could be used. As the exemplary temperature measurements in wells (Cologne, Germany) show (Zhu et al., 2010), water temperatures of 16 °C can be found below the inner city at a depth of 15 m, whereas under less densely built-up city districts, only 11 °C are reached (Zhu et al., 2010). Assuming a difference of approximately 5 K, a thermal capacity of at least $4.8 \cdot 10^{10}$ kJ can be calculated for a layer thickness of approximately 10 m and an area of 1 km². If the energy requirements for buildings are approximately $1.9 \cdot 10^{10}$ kJ/(km² a) in Cologne, the subterranean UHI can provide at least 2.5 times the energy required for heating and cooling buildings in Cologne. For this estimate, an energy supply for a building of approximately 50 kWh/(m² a) was assumed (Zhu et al., 2010). In addition to the energy savings, an emission reduction of approximately 3,000 t CO_2/(km² a) could be achieved in this case.

6. Outlook

Urban climate-change mitigation measures, such as surface unsealing, greening, preservation or designation of ventilation aisles, networking corridors or subterranean energy use, require space. However, space is often not available. In German cities, the current building-renewal rate is estimated at no more than 2 %/a. In the short term, the implementation of mitigation measures is rather improbable.

The former industrial cities are an exception. Here, structural change allows the reallocation of antiquated and vast industrial estates to new uses. In addition, the population decrease that can be observed in some cities enables the greening of open residential areas (Oswalt & Rieniets, 2006) and thereby the prevention of overheating.

Urban planning that takes climate change into account should favour compact but open building structures equipped with sufficient open spaces and green areas and offering possibilities for shading. Readily available public transport would obsolete the use of individual vehicle transport in the inner cities ("city of short distances"), thus reducing the emission of exhaust gases, particulate matter and CO_2. Building density and shading possibilities should be designed to provide sufficient protection against solar radiation in the summer while guaranteeing a maximum of radiation absorption in the winter. Suburban growth ("urban sprawl"), which has been monitored in numerous locations for some time now, should be superseded by suburban cold-air formation areas to provide the optimal ventilation of urban centres with fresh air from the surrounding countryside.

Author details

Wilhelm Kuttler
Applied Climatology and Landscape Ecology, University of Duisburg-Essen, Germany

7. References

Alexandri, E. & Jones, P. (2008). Temperature decreases in an urban canyon due to green walls and green roofs in diverse climates. *Building and Environment*, Vol.43, pp. 480-493

Ali-Toudert, F. & Mayer, H. (2007). Effects of asymmetry, galleries, overhanging façades and vegetation on thermal comfort in urban street canyons. *Solar Energy*, Vol.81, pp. 742-754

Baldinelli, G. (2009). Double skin facades for warm climate regions: Analysis of a solution with an integrated movable shading system. *Building and Environment*, Vol.44, pp. 1107-1118

Bell, M.L.; McDermott, A.; Zeger, S.L.; Samet, J.M. & Dominici, F. (2004). Ozone and Short-term Mortality in 95 US Urban Communities, 1987-2000. *Journal of the American Medical Association*, Vol.292, No.19, pp. 2372-2378

Benjamin, M.T. & Winer, A.M. (1998). Estimating the ozone-forming potential of urban trees and shrubs. *Atmospheric Environment*, Vol.32, pp. 53-68

Blankenstein, S. & Kuttler, W. (2004). Impact of street geometry on downward longwave radiation and air temperature in an urban environment. *Meteorologische Zeitschrift*, Vol.13, No.5, pp. 373-379

Böhm, R. (1998). Urban bias in temperature time series – A case study for the city of Vienna, Austria. *Climatic Change*, Vol.38, No.1, (January 1998), pp. 113-128

Bongardt, B. (2006): Stadtklimatische Bedeutung kleiner Parkanlagen – dargestellt am Beispiel des Dortmunder Westparks, In: *Essener Ökologische Schriften*, Vol.24, Kuttler, W. & Sures, B., Westarp-Wissenschaften, Hohenwarsleben

Brandt, K. (2007). Die ökonomische Bewertung des Stadtklimas am Beispiel der Stadt Essen, In: *Essener Ökologische Schriften*, Vol.25, Kuttler, W. & Sures, B., Westarp-Wissenschaften, Hohenwarsleben

Bruse, M. & Fleer, H. (1998). Simulating surface-plant-air interactions inside urban environments with a three dimensional numerical model. *Environmental Modelling and Software*, Vol.13, pp. 373-384

Büns, C. & Kuttler, W. (2012). Path-integrated measurements of carbon dioxide in the urban canopy layer. *Atmospheric Environment*, Vol.46, pp. 237-247

Carvalho, A.; Monteiro, A.; Solman, S.; Miranda, A.I. & Borrego, C. (2010). Climate-driven changes in air quality over Europe by the end of the 21st century, with special reference to Portugal. *Environmental Science & Policy*, Vol.13, pp. 445-458

Chameides, W.L.; Lindsay, R.W.; Richardson, J. & Kiang, C.S. (1988). The role of biogenic hydrocarbons in urban photochemical smog: Atlanta as a case study. *Science*, Vol.241, pp. 1473-1475

Cleugh, H.A.; Bui, E.N.; Simon, D.A.P.; Mitchell, V.G. & Xu, J. (2005). The impact of suburban design on water use and microclimate, *Proceedings of MODSIM 2005 International Congress on Modelling and Simulation*, Melbourne, Australia, December 12-15, 2005

Dawson, J.P.; Adam, P.J. & Pandis, S.N. (2007). Sensitivity of ozone to summertime climate in the eastern USA: A modeling case study. *Atmospheric Environment*, Vol.41, pp. 1494-1511

Demuzere, M. & van Lipzig, N.M.P (2010). A new method to estimate air-quality levels using a synoptic-regression approach. Part I: Present-day O_3 and PM_{10} analysis. *Atmospheric Environment*, Vol.44, No.10, pp. 1341-1355

Donovan, G. & Butry, D. (2009). The value of shade: Estimating the effect of trees on summertime electricity use. *Energy and Buildings*, Vol.41, pp. 662-668

Dütemeyer, D. (2000). Urban-Orographische Bodenwindsysteme in der städtischen Peripherie Kölns, In: *Essener Ökologische Schriften*, Vol.12, Burghardt, W.; Kuttler, W. & Schuhmacher, H., Westarp-Wissenschaften, Hohenwarsleben

Ebi, K. & McGregor, G. (2009). Climate change, tropospheric ozone and particulate matter, and health impacts. *Ciência & saúde coletiva*, Vol.14, No.6, pp. 2281-2293

Emeis, S.; Schoenemeyer, T.; Richter, K. & Ruckdeschel, W. (1997). Sensitivity of ozone production to VOC and NO_x emissions – A case study with the box-model BAYROZON. *Meteorologische Zeitschrift*, Vol.6, N.F., pp. 60-72

Erell, E.; Pearlmutter, D. & Williamson, T. (2011). *Urban Microclimate – Designing the Spaces between Buildings*, Earthscan Publications Ltd., ISBN 9781844074679, London, Washington

Forkel, R. & Knoche, R. (2006). Regional climate change and its impact on photooxidant concentrations in southern Germany: Simulations with a coupled regional climate-chemistry model. *Journal of Geophysical Research*, Vol.111, No.D12302

GDCh (= Gesellschaft Deutscher Chemiker). (ed.). (2010). *Feinstaub. Statuspapier*. Zusammen mit KRdL (Kommission Reinhaltung der Luft), ISBN 978-3-89746-120-8

Göbel, P.; Dierkes, C.; Kories, H.; Meßer, J.; Meißner, E. & Coldewey, W.G. (2007). Impacts of green roofs and rain water use on the water balance and ground water levels in urban areas. *Zeitschrift Grundwasser*, Vol.12, No.3, pp. 189-200

Goldbach, A. & Kuttler, K. (2012). Quantification of turbulent heat fluxes for adaptation strategies within urban planning, In: *International Journal of Climatology*, 28.02.2012, available from http://onlinelibrary.wiley.com/ doi/10.1002/joc.3437/pdf

Grimmond, C.S.B.; Salmond, J.A.; Oke, T.R.; Offerle, B. & Lemonsu, B.A. (2004). Flux and turbulence measurements at a densely built-up site in Marseille: Heat, mass (water and carbon dioxide), and momentum. *Journal of Geophysical Research*, Vol.109, No.D24101

Hamada, S. & Ohta, T. (2010). Seasonal variations in the cooling effect of urban green areas on surrounding urban areas. *Urban Forestry & Urban Greening*, Vol.9, No.1, pp. 15-24

He, J. & Hoyano, A. (2010). Measurement and evaluation of the summer microclimate in the semi-enclosed space under a membrane structure. *Building and Environment*, Vol.45, No.1, pp. 230-242

Hegger, M. (2009). Sonnige Aussichten, Verlag Müller und Busmann, Wuppertal

Held, A.; Nowak, A.; Birmili, W.; Wiedensohler, A.; Forkel, R. & Klemm, O. (2004). Observations of particle formation and growth in a mountainous forest region in central Europe. *Journal of Geophysical Research*, Vol.109, No.D23204

Holmer, B.; Thorrson, S. & Eliasson, I. (2007). Cooling rates, sky view factors and the development of intra-urban air temperature differences. *Geografiska Annaler*, Vol.89, No.4, pp. 237-248

Holst, J. & Mayer, H. (2010). „Verbundkoordination (KLIMES ALUF-1)" und „Planungsrelevante human-biometeorologische Bewertung von städtischen Strukturen bei thermischen Stressbedingungen hinsichtlich der Anpassung an Extremwetter (KLIMES ALUF-2)". Schlussbericht zum Teilvorhaben, In: *Berichte des Meteorologischen Instituts der Universität Freiburg*, Vol.22, Mayer, H., pp. 7-99, Eigenverlag des Meteorologischen Instituts der Albert-Ludwigs-Universität Freiburg, Freiburg

Höppe, P. (1984). Die Energiebilanz des Menschen, In: *Münchner Universitäts-Schriften*, Fachbereich Physik, Wiss. Mitteilungen, Vol.49, München

Horbert, M. (2000). Klimatologische Aspekte der Stadt- und Landschaftsplanung, In: *Schriftenreihe Fachbereich Umwelt und Gesellschaft, TU Berlin*, Vol.113, Berlin

Hupfer, P. & Kuttler, W. (Eds.). (2006). Witterung und Klima (12), B. G. Teubner Verlag, ISBN 3-8351-0096-3, Wiesbaden

Intergovernmental Panel on Climate Change (= IPCC). (2007). *Climate Change 2007: The Physical Science Basis. Contribution of Working Group I to the Fourth Assessment Report of the*

Intergovernmental Panel on Climate Change, Cambridge University Press, ISBN 978-0521-88009-1, Cambridge (UK), New York (USA)

Jacob, D.J. & Winner, D.A. (2009). Effect of climate change on air quality. *Atmospheric Environment*, Vol.43, pp. 51-63

Jendritzky, G. (2007). Folgen des Klimawandels für die Gesundheit, In: *Der Klimawandel. Einblicke, Rückblicke und Ausblicke*, Endlicher, W. & Gerstengarbe, F.-W., pp. 108-118, Potsdam-Institut für Klimafolgenforschung (PIK), Retrieved from http://www.pik-potsdam.de/services/infothek/buecher_broschueren/broschuere_cms_ 100.pdf

Jonas, R.; Horbert, M. & Pflug, W. (1985). Die Filterwirkung von Wäldern gegenüber staubbelasteter Luft. *Forstwissenschaftliches Centralblatt*, Vol.104, No.5, pp. 289-299

Kanemoto, S.; Matsushita, Y.; Moriyama, M. & Takebayashi, H. (2009). An observation study on the degradation of the heat island mitigation effect by damage to the grass in "Grass Parking", *Proceedings of the 5th Japanese-German Meeting on Urban Climatology*, Freiburg, October 6-8, 2008, In: *Berichte des Meteorologischen Instituts der Universität Freiburg*, Vol.18

Kiendler-Scharr, A.; Wildt, J.; Dal-Maso, M., Hohaus, T.; Kleist, E.; Mentel, T.F.; Uerlings, R.; Tillmann, R.; Schurr, U. & Wahner, A. (2009). New particle formation in forests inhibited by isoprene emissions. *Nature*, Vol.461, No.7262, pp. 381-384

Kiese, O. (1995). Die bioklimatische Funktion innerstädtischer, insbesondere baumbestandener Grünflächen, *Proceedings of the 24th Jahrestagung der Gesellschaft für Ökologie*, pp. 395-401, Frankfurt/Main, September 19-23, 1994

Klemm, O.; Stockwell, W.R.; Schlager, H. & Krautstrunk, M. (2000). NO_x or VOC limitation in East German ozone plumes. *Journal of Atmospheric Chemistry*, Vol.35, pp. 1-18

Köhler, M. (Ed.). (1993). *Fassaden- und Dachbegrünung*, Ulmer, Stuttgart

Kolokotroni, M.; Giannitsaris, I. & Watkins, R. (2006). The effect of the London urban heat island on building summer cooling demand and night ventilation strategies. *Solar Energy*, Vol.80, pp. 383-392

Kourtidis, K.; Ziomas, I.; Zerefos, C.; Kosmidis, E.; Symeonidis, P.; Christophilopoulos, E.; Karathanassis, S. & Mploutsos, A. (2002). Benzene, toluene, O_3, NO_2 and SO_2 measurements in an urban street canyon in Thessaloniki, Greece. *Atmospheric Environment*, Vol.36, pp. 5355-5364

Kratzer, A. (1956). *Das Stadtklima* (2), Vieweg, Braunschweig

Kuttler, W. & Straßburger, A. (1999). Air pollution measurements in urban parks. *Atmospheric Environment*, Vol.33, pp. 4101-4108

Kuttler, W. (2009). Zum Klima im urbanen Raum, In: *Klimastatusbericht 2008*, Deutscher Wetterdienst, pp. 6-12, Offenbach

Kuttler, W. (2010a). Urbanes Klima, Teil 1. *Gefahrstoffe - Reinhaltung der Luft, Umweltmeteorologie*, Vol.70, No.7/8, pp. 329-340

Kuttler, W. (2010b). Urbanes Klima, Teil 2. *Gefahrstoffe - Reinhaltung der Luft, Umweltmeteorologie*, Vol.70, No.9, pp. 378-382

Kuttler, W. (2011a). Climate change in urban areas, Part 1, Effects. *Environmental Sciences Europe (ESEU)*, Vol.23, No.11, pp. 1-12

Kuttler, W. (2011b). Climate change in urban areas, Part 2, Measures. *Environmental Sciences Europe (ESEU)*, Vol.23, No.21, pp. 1-15

Lin, C.Y.C.; Jacob, D.J. & Fiore, A.M. (2001). Trends in exceedances of the ozone air quality standard in the continental United States, 1980-1998. *Atmospheric Environment*, Vol.35, pp. 3217-3228

Litschke, T. & Kuttler, W. (2008). On the reduction of urban particle concentration by vegetation – a review. *Meteorologische Zeitschrift*, Vol.17, No.3, pp. 229-240

Matzarakis, A. & Mayer, H. (1996). Another Kind of Environmental Stress: Thermal Stress. *WHO Colloborating Centre for Air Quality Management and Air Pollution Control NEWSLETTER*, Vol.18, pp. 7-10

Mayer, H. & Schmidt, J. (1998). Problematik der Kennzeichnung von sogenannten „Ozon-Wetterlagen". *Meteorologische Zeitschrift*, Vol.7, N. F., pp. 41-48

Mayer, H. (2006). Indizes zur human-biometeorologischen Bewertung der thermischen und lufthygienischen Komponente des Klimas. *Gefahrstoffe – Reinhaltung der Luft*, Vol.66, No.4, pp. 165-174

McPherson, E.G. & Simpson, J.R. (1995). Shade trees as a demand-side resource. Home energy, Vol.2, pp. 11-17

Meleux, F.; Solmon, F. & Giorgi F (2007). Increase in summer European ozone amounts due to climate change. *Atmospheric Environment*, Vol.41, No.35, pp. 7577-7587

Melkonyan, A. (2011). Statistical Analysis of Long-Term Air Pollution Data in North Rhine-Westphalia, Germany, In: *Essener Ökologische Schriften*, Vol.30, Kuttler, W. & Sures, B., Westarp-Wissenschaften, Hohenwarsleben

Meng, Z.; Dabdub, D. & Seinfeld, J.H. (1997). Chemical Coupling between Atmospheric Ozone and Particulate Matter. *Science*, Vol.277, pp. 116-119

Narumi, D.; Kondo, A. & Shimoda, Y. (2009). The effect of the increase in urban temperature on the concentration of photochemical oxidants. *Atmospheric Environment*, Vol.43, pp. 2348-2359

Onishi, A.; Cao, X.; Ito, T.; Shi, F. & Imura, H. (2010). Evaluating the potential for urban heat-island mitigation by greening parking lots. *Urban Forestry & Urban Greening*, Vol.9, pp. 323-332

Oswalt, P. & Rienlets, T. (2006). *Atlas of Shrinking Cities*, Hatja Cantz Verlag, Ostfildern

Plass-Dülmer, C. (2008). Biogene Kohlenwasserstoffe und ihre Rolle in der Luftchemie – außer angenehmen Düften. *GAW Brief des DWD* (= Deutscher Wetterdienst), Vol.43

Polster, G. (1969). Erfahrungen mit Strahlungs-, Temperaturgradient- und Windmessungen als Bestimmungsgrößen der Diffusionskategorien. *Meteorologische Rundschau*, Vol.22, pp. 170-175

Pöschl, U. (2005). Atmosphärische Aerosole: Zusammensetzung, Transformation, Klima- und Gesundheitseffekte. *Angewandte Chemie*, Vol.117, No.46, pp. 7690-7712

Quah, K.L.A. & Roth, M. (2012). Diurnal and weekly variation of anthropogenic heat emissions in a tropical city, Singapore. *Atmospheric Environment*, Vol.46, pp. 92-103

Revich, B. & Shaposhnikov, D. (2010). The effects of particulate and ozone pollution on mortality in Moscow, Russia. *Air Quality, Atmosphere and Health*, Vol.3, No.2, pp. 117-123

Roloff, A.; Bonn, S. & Gillner, S. (2008). Baumartenwahl und Gehölzverwendung im urbanen Raum unter Aspekten des Kimawandels. *Forstwissenschaftliche Beiträge Tharandt/Contributions to Forest Sciences.* Vol.7, pp. 92-107

Ropertz, A. (2008). Transport atmosphärischer Spurenstoffe in eine innerstädtische Grünfläche - Eine Analyse mittels optischer Fernmessverfahren, In: *Essener Ökologische Schriften*, Vol.26, Kuttler, W. & Sures, B., Westarp-Wissenschaften, Hohenwarsleben

Sharkey, T.D.; Wiberley, A.E. & Donohue, A.R. (2008). Isoprene Emission from Plants: Why and How. *Annals of Botany*, Vol.101, pp. 5-18

Shashua-Bar, L. & Hoffman, M.E. (2000). Vegetation as a climatic component in the design of an urban street. *Energy and Buildings*, Vol.31, pp. 221-235

Shi, Z. & Zhang, X. (2011). Analyzing the effect of the longwave emissivity and solar reflectance of building envelopes on energy-saving in buildings in various climates. *Solar Energy*, Vol.85, pp. 28-37

Sillmann, S. (1999). The relation between ozone, NO_x and hydrocarbons in urban and polluted rural environments. *Atmospheric Environment*, Vol.33, pp. 1821-1845

Solberg, S.; Coddeville, P.; Forster, C.; Hov, Ø.; Orsolini, Y. & Uhse, K. (2005). European surface ozone in the extreme summer 2003. *Atmospheric Chemistry and Physics Discussions*, Vol.5, pp. 9003-9038

Solmon, F.; Sarrat, C.; Serça, D.; Tulet, P. & Rosset, P. (2004). Isoprene and monoterpenes biogenic emissions in France: modeling and impact during a regional pollution episode. *Atmospheric Environment*, Vol.38, No.23, pp. 3853-3865

Souch, C. & Grimmond, S. (2004). Applied Climatology: heat waves. *Progress in Physical Geography*, Vol.28, No.4, pp. 599-606

Stadt Essen. (2009). Der Essener Heizspiegel 2009, In: *Stadt Essen*, 01.03.2012, available from http://media.essen.de/media/wwwessende/aemter/59/energie/Essener_Heizspiegel_web 1.pdf

Straßburger, A. (2004). *Analyse atmosphärischer Spurengase zur Bestimmung des lufthygienischen Erholungswertes eines urbanen Parks.* Dissertation (unveröffentlicht) Universität Essen, Essen

Synnefa, A.; Santamouris, M. & Livada, I. (2006). A study of the thermal performance of reflective coatings for the urban environment. *Solar Energy*, Vol.80, pp. 968-981

Taha, H. (1996). Modeling impacts of increased urban vegetation on ozone air quality in the South Coast Air Basin. *Atmospheric Environment*, Vol.30, No.20, pp. 3423-3430

Taha, H. (1996). Modeling impacts of increased urban vegetation on ozone air quality in the South Coast Air Basin. *Atmospheric Environment*, Vol.30, No.20, pp. 3423-3430

Takebayashi, H. & Moriyama, M. (2009). Study on the urban heat island mitigation effect achieved by converting to grass-covered parking. *Solar Energy*, Vol.83, pp. 211-1223

Terjung, W.H. & Louie, S.S.-F. (1972). Energy Input-Output Climates of the World. *Archiv für Meteorologie, Geophysik und Bioklimatologie*, Vol.20, pp.129-166

Verein Deutscher Ingenieure (= VDI). (2008). *VDI-Richtlinie 3787, Blatt 2 – Umweltmeteorologie – Methoden zur human-biometeorologischen Bewertung von Klima- und Lufthygiene für die Stadt- und Regionalplanung. Teil I: Klima*, VDI, Düsseldorf

Velasco, E. & Roth, M. (2010). Cities as net sources of CO_2: Review of atmospheric CO_2 exchange in urban environments measured by eddy covariance technique. *Geography Compass*, Vol.4, No.9, pp. 1238-1259

Vogt, R.; Christen, A.; Rotach, M.W.; Roth, M. & Satyanarayana, A.N.V. (2006). Temporal dynamics of CO_2 fluxes and profiles over a Central European city. *Theoretical and Applied Climatology*, Vol.84, pp. 117-126

Wienert, U. & Kuttler, W. (2005). The dependence of the urban heat island intensity on latitude – A statistical approach. *Meteorologische Zeitschrift*, Vol.14, No.5, pp. 677-686

Wise, E.K. & Comrie, A.C. (2005). Meteorologically adjusted urban air quality trends in the southwestern United States. *Atmospheric Environment*, Vol.39, No.16, pp.2969-2980

Wrzesinsky, T. (2004). Direkte Messung und Bewertung des nebelgebundenen Eintrags von Wasser und Spurenstoffen in ein montanes Waldökosystem. Dissertation Universität Bayreuth, In: *Universität Münster*, 28.02.2012, available from http://www.uni-muenster.de/imperia/md/content/landschaftsoekologie/klima/pdf/2004_wrzesinsky_diss.pdf

Wu, D.; Wang, Y.; Lin, X. & Yang, J. (2008). On the mechanism of the cyclonic circulation in the Gulf of Tonkin in the summer. *Journal of Geophysical Research*, Vol.113, No.C09029

Yashuda, R.; Nakagawa, F. & Yoshida, A. (2009). Numerical study on summer-time photochemical pollution in Osaka area, *Proceedings of the 7th ICUC*, Yokohama, Japan, June 29–July 3, 2009

Zhu, K.; Blum, P.; Ferguson, G.; Balke, K.-D. & Bayer, P. (2010). The geothermal potential of urban heat islands. *Environmental Research Letters*, Vol.5, No.4

Ziska, L.H. & Caulfield, F. (2000). The potential influence of rising atmospheric carbon dioxide (CO_2) on public health: Pollen production of common ragweed as a test case. *World Resource Review*, Vol.12, pp. 449-457

Ziska, L.H.; Gebhard, D.E.; Frenz, D.A.; Faulkner, S.; Singer, B. & Straka, J.G. (2003). Cities as harbingers of climate change: Common ragweed, urbanization, and public health. *Journal of Allergy and Clinical Immunology*, Vol.111, No.2, pp. 290-295

Quantification of the Urban Heat Island Under a Changing Climate over Anatolian Peninsula

Tayfun Kindap, Alper Unal, Huseyin Ozdemir, Deniz Bozkurt,
Ufuk Utku Turuncoglu, Goksel Demir, Mete Tayanc and Mehmet Karaca

Additional information is available at the end of the chapter

1. Introduction

More than half of the world's population (i.e. approximately 3.5 billion) has been living in urban areas in 2010 and by 2030 this number will rise to almost 5 billion (UNFPA, 2011). In the meantime, global average surface temperature has been increasing significantly: last century saw an approximately 0.7 °C warming (IPCC, 2007; WMO, 2009). Urban growth and climate change are two major forcings on local climate (IPCC, 2007). Urbanization reshapes the surface of the earth causing changes in the energy budget at the ground surface while altering the surrounding atmospheric circulation characteristics leading changes in local climate (Huang et al., 2009; Oke et al., 1992). Urban heat island (UHI) refers to warmer air temperatures observed in urban areas as compared to those over surrounding non-urban regions. When naturally vegetated areas (e.g., grass and trees) are replaced with impervious surfaces having relatively low reflectivity and evapotranspiration rates, additional energy heats the atmosphere causing the phenomena. After sunset, non-urban areas cool more rapidly than urban regions resulting in a temperature differential. The UHI is presented as the difference between temperatures recorded within and outside the urban settlement. It has been suggested that UHI is influenced by population; topography; level of industrialization as well as the regional climate (Oke, 1987; Rosenzweig et al., 2005; Gill et al., 2008; Kolokotroni & Giridharan, 2008).

Anatolian peninsula, (i.e. Asian part of Turkey), is lying in the Eastern Mediterranean, and has seven distinct geographical regions: Eastern Anatolia; Central Anatolia; Black Sea Region; Mediterranean Region; Aegean Region; Marmara Region; and Southeastern Anatolia (Unal et al., 2003; Kindap, 2010). The Aegean and Mediterranean coastal regions have cool rainy winters, and hot moderately dry summers. Mountains along the coast prevent the Mediterranean influences from extending inland, giving interior of Anatolian Peninsula a continental climate and distinct seasons.

As urbanization rate has increased significantly, UHI has become a significant issue in the Anatolian Peninsula. There have been a few studies focusing on UHI effect over Anatolian Peninsula (Karaca et al., 1995; Tayanc & Toros, 1997; and Ezber et al., 2007). For example, Karaca et al. (1995) investigated the effects of urbanization on climates of two cities, Istanbul and Ankara in the Anatolian Peninsula with varying periods (1912-1992) and reported significant upward trend for the urban temperatures when compared to the rural temperatures in the southern part of Istanbul. Tayanc & Toros (1997) studied 4 urban areas (i.e., Adana, Bursa, Gaziantep, Izmir) suggesting that temperature is more sensitive to UHI than precipitation and their results showed that the Anatolian Region is under a cooling trend after the period ending by 1990. Ezber et al. (2007) used statistical and numerical modeling tools to investigate the climatic effects of urbanization in Istanbul from the period of 1951 to 2004, and they found statistically positive trends in the urban stations of the city. These studies focused on cities individually and have not conducted a comprehensive evaluation of all urban environments in Anatolian Peninsula. The objective of this study, hence, is to quantify the UHI effect of the all major cities (population>1,000,000) with quality meteorological data extending back to 1965. Climate forecasts using a regional climate model output are also analyzed to understand the effect of UHI under a changing climate.

2. Material and methods

2.1. Temperature data

Minimum daily temperature is the primary indicator of UHI (Rosenzweig et al., 2005). We have, therefore, compiled a database of minimum daily temperatures over urban cities in Anatolia. As a first step, we have developed and implemented the following criteria: Urban stations are selected in regions having population over 500,000, and rural stations with population of less than 100,000 (Hua et al., 2007); high quality temperature data (i.e., passes homogeneity test, have representative urban and rural stations); continuous data from 1965 to 2006. After comprehensive evaluation, especially for representativeness, we have identified 8 cities to analyze, a total of 25 meteorological stations (9 urban and 16 rural). These cities are highlighted in Table 1.

Selected cities have significant increase in urban population between 1935 and 2011, are shown in Fig. 1. Ratio of the center population is calculated by dividing the center population (urban districts of the city, not the rural) of the selected cities to the total population of the city. Dark colored bar in the figure is the ratio of the year 1935, and the total bar including the shaded and the dark is the ratio in 2011. For example, in 1935 approximately 50% of Istanbul's population is urbanized, which increased to approximately to 100% in 2011. In 1935, number of cities that have urban ratio of 50% or more was only one (i.e. Istanbul), whereas in 2011 all of the selected cities have urban population ratio of more than 50%.

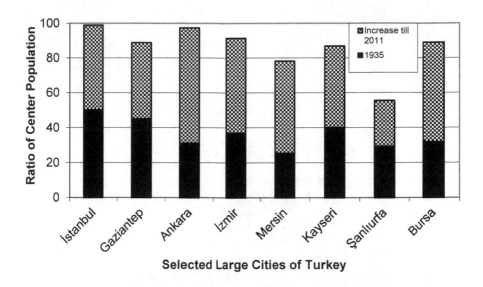

Figure 1. The rate of urbanization in terms of urban over total population (%) in 1935 and 2011

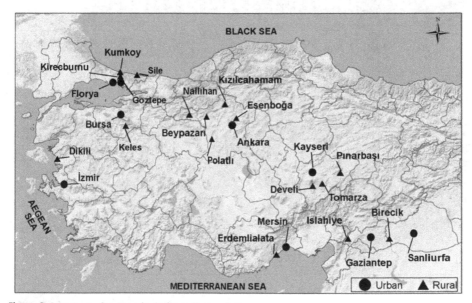

Figure 2. Locations of meteorological monitoring stations

Northern Marmara		Aegean and Western Mediterranean		Black Sea		Central Anatolia		Eastern Anatolia		Southeastern Anatolia		Eastern Mediterranean	
City	Pop.	City	Pop.	City	Pop.	City	Pop.	City	Pop.	City	Pop.	ties	Pop.
İstanbul	13,624,240	İzmir	3,965,232	Samsun	1,251,729	Ankara	4,890,893	Van	1,022,532	Gaziantep	1,753,596	Adana	2,108,805
Kocaeli	1,601,720	Bursa	2,652,126	Trabzon	757,353	Konya	2,038,555	Erzurum	780,847	Şanlıurfa	1,716,254	Antalya	2,043,482
Sakarya	888,556	Manisa	1,340,074	Ordu	714,390	Kayseri	1,255,349	Ağrı	555,479	Diyarbakır	1,570,943	Mersin	1,667,939
Tekirdağ	829,873	Balıkesir	1,154,314	Zonguldak	612,406	Eskişehir	781,247	Muş	414,706	Mardin	764,033	İskenderun	1,474,223
Edirne	399,316	Aydın	999,163	Giresun	419,498	Afyon	698,626	Bitlis	336,624	Malatya	757,930	Kahramanmaraş	1,054,210
Kırklareli	340,199	Denizli	942,278	Düzce	342,146	Sivas	627,056	Kars	305,755	Adıyaman	593,931	Osmaniye	485,357
Yalova	206,535	Muğla	838,324	Rize	323,012	Tokat	608,299	Erzincan	215,277	Elazığ	558,556		
		Kütahya	564,264	Karabük	219,728	Çorum	534,578	Iğdır	188,857	Batman	524,499		
		Çanakkale	486,445	Sinop	203,027	Yozgat	465,696	Ardahan	107,455	Şırnak	457,997		
		Uşak	339,731	Bartın	187,291	Isparta	411,245	Tunceli	85,062	Siirt	310,468		
				Artvin	166,394	Aksaray	378,823			Hakkari	272,165		
				Gümüşhane	132,374	Kastamonu	359,759			Bingöl	262,263		
				Bayburt	76,724	Niğde	337,553			Kilis	124,452		
						Amasya	323,079						
						Nevşehir	283,247						
						Kırıkkale	274,992						
						Bolu	276,506						
						Burdur	250,527						
						Karaman	234,005						
						Kırşehir	221,015						
						Bilecik	203,849						
						Çankırı	177,211						

* Cities highlighted are analyzed in the study

Table 1. Cities with their populations in each climatological region of the Anatolian Peninsula (TUIK, 2012)

2.2. Study area

Regarded by many as the cradle of civilization of the world, the Anatolian Peninsula is located at the confluence of Europe, Asia and Africa. This region has 75 million inhabitants (TUIK, 2012). The population has grown almost 4.7 times between 1935 and 2012, from 16 million to 75 million (TUIK, 2012). Ratio of rural to urban population has changed dramatically especially after 1980s and urban population ratio reached to about 80% in 2010 (Fig. 3). Table 1 lists the cities in each climate regions with their population. Approximately, 74% of the cities have the population between 100,000 and 1 million; 22% between 1 million and 5 million. Only 2 cities are below 100,000, and Istanbul is the only megacity (i.e., the city which has a population of five million or more) in the region having a population of over 13 million. Table 2 gives a classification about the site characteristics of the selected meteorological stations.

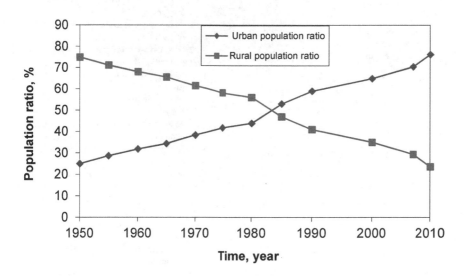

Figure 3. The population ratio of urban and rural parts of the region over total population from 1950 to 2010 (TUIK, 2012)

Station	Latitude	Longitude	Altitude (m)	Land-use	Location and siting
Ankara	39° 57′	32° 53′	891	Urban	In a residential area at the city center on the ground
Polatli (Ankara)	39° 35′	32° 09′	886	Suburban	70 km away from the city center on the ground
Kizilcahamam (Ankara)	40° 28′	32° 39′	1033	Rural	50 km away from the city center on the ground
Beypazari (Ankara)	40° 10′	31° 56′	682	Rural	75 km away from the city center on the ground
Esenboga (Ankara)	40° 07′	33° 00′	959	Urban	At the airport in the city center on the ground
Nallihan (Ankara)	40° 11′	31° 22′	650	Rural	120 km away from the city center on the ground
Bursa	40° 23′	29° 01′	100	Urban	5 km away from the city center at the airport on the ground
Keles (Bursa)	39° 54′	29° 13′	1063	Rural	60 km away from the city center on the ground
Gaziantep	37° 08′	37° 37′	900	Urban	In a residential area at the city center on the ground
Islahiye (Gaziantep)	37° 02′	36° 61′	706	Rural	50 km away from the city center on the ground
Istanbul (Goztepe)	40° 97′	29° 06′	33	Urban	In a residential area at the city center on the ground
Istanbul (Florya)	40° 97′	28° 79′	37	Urban	In a residential area at the city center on the ground
Kirecburnu (Istanbul)	41° 15′	29° 05′	58	Rural	25 km away from the city center on the ground
Kumkoy (Istanbul)	41° 25′	29° 04′	38	Rural	30 km away from the city center on the ground
Sile (Istanbul)	41° 17′	29° 60′	83	Rural	50 km away from the city center on the ground
Izmir	38° 39′	27° 08′	29	Urban	In a residential area at the city center on the ground
Dikili (Izmir)	39° 07′	26° 89′	3	Rural	75 km away from the city center on the ground
Kayseri	38° 77′	35° 49′	1092	Urban	In a residential area at the city center on the ground
Develi (Kayseri)	38° 28′	35° 54′	1180	Rural	40 km away from the city center on the ground
Pinarbasi (Kayseri)	38° 77′	36° 61′	1500	Rural	75 km away from the city center on the ground
Tomarza (Kayseri)	38° 27′	35° 48′	1397	Rural	35 km away from the city center on the ground
Mersin	36° 78′	34° 60′	3	Urban	In a residential area at the city center on the ground
Erdemli Alata (Mersin)	36° 38′	33° 94′	9	Suburban	35 km away from the city center on the ground
Sanliurfa	37° 16′	38° 79′	547	Urban	In a residential area at the city center on the ground
Birecik (Sanliurfa)	37° 02′	37° 96′	345	Rural	65 km away from the city center on the ground

Table 2. Site characteristics of selected meteorological stations

2.3. Urban Heat Island

The UHI effect refers to an increase in urban air temperatures as compared to surrounding suburban and rural temperatures (Oke, 1982; Quattrochi et al., 2000).

UHI effect is defined as:

$$\Delta T_{u-r} = T_u - T_r \tag{1}$$

where T_u is the urban station temperature, T_r is the rural station temperature, $\Delta T_u - r$ is the effect of UHI.

Heat islands develop in areas that contain a high percentage of non-reflective, water-resistant surfaces and a low percentage of vegetated and moisture-trapping surfaces (Rosenzweig et al., 2005). In particular, materials such as stone, concrete, and asphalt tend to trap heat at the surface (Landsberg, 1981; Oke, 1982; Quattrochi et al., 2000) and a lack of vegetation reduces heat lost due to evapotranspiration (Lougeay et al., 1996). The addition of anthropogenic heat and pollutants into the urban atmosphere further contributes to the intensity of the UHI effect (Taha, 1997). The pollution created by emissions from power generation increases absorption of radiation in the boundary layer (Oke, 1982) and contributes to the creation of inversion layers. Inversion layers prevent rising air from cooling at the normal rate in urban areas.

Globally average surface warming is projected to increase for the end of the 21st century (2090 to 2099), relative to 1980 to 1999 within the six different scenarios between 1.1 - 6.4 °C (IPCC, 2007). Climate change has the potential to significantly alter the intensity and increase the spatial extent of heat islands in urban environments. As temperature warms, the frequency with which UHI conditions occur could grow (Rosenzweig et al., 2005).

2.4. Mann-Kendall trend test

Trend analysis can be used to assess the climatic variations of the atmosphere and the Mann-Kendall trend test (Mann, 1945; Kendall, 1975) is one of the widely used non-parametric tests to detect significant trends in time series. The Mann-Kendall trend test is not affected by the actual distribution of the data and is less sensitive to outliers rather than parametric trend tests, which are more powerful, but more sensitive to outliers. Therefore, Mann-Kendall test is more suitable for detecting trends in temperature time series, which may have outliers (Hamed, 2008). Climate change can be detected by the Kendall coefficient t (Mann test) and when a time series shows a significant trend, the period from which the trend is demonstrated can be obtained effectively by this test. In a time series, for each element y_i, the number n_i of elements y_j preceding it $(i > j)$ is calculated such that $y_i > y_j$.

The test statistic t is then given by,

$$t = \sum n_i, \tag{2}$$

and is distributed very nearly as a Gaussian normal distribution with an expected value of $E(t) = n(n-1)/4$ and a variance of $\text{var}t = n(n-1)(2n+5)/72$. A trend can be seen for high values of $|u(t)|$ where,

$$u(t) = \frac{\left[t = E(t)\right]}{\sqrt{\text{var}t}}, \tag{3}$$

This principle can be usefully extended to the backward series and $u_i = -u(t_i)$ can be obtained. The intersection of the $u(t)$ and $u'(t)$ curves denotes approximately the beginning of the trend. This is called the sequential version of the Mann-Kendall test (Goossens & Berger, 1986). If a Mann-Kendall statistic of a time series is higher than 1.96, there is a 95% significant increase in that particular time series. If the result is just the reverse; lower than -1.96, there is a 95% significantly decreasing trend in the series. Also, results between 0.5 and 1.96 indicate increasing, -0.5 and -1.96 indicate decreasing trends, and 0.5 and -0.5 indicate no trend. The Mann-Kendall statistics are then plotted on a map in order to show the spatial distribution of both the significant and non-significant temperature trends in Turkey.

3. Results and discussion

Observation data over the Anatolian Peninsula were analyzed to understand trends in average temperature. As can be seen in Fig. 4a, as of 2009, average temperatures in all 8 cities are higher than temperature in 1960s. For example, average temperature increase in Kayseri, Gaziantep, and Mersin is over 2 °C. Fig. 4b presents anomaly in mean temperatures in Anatolian Peninsula as estimated using gridded dataset obtained by the Climate Research Unit (CRU - TS3.0). The results suggest an increase of 0.5 °C starting in 1990s.

Trends in daily minimum temperatures at individual meteorological stations were investigated with the Mann-Kendall trend test (Mann, 1945; Kendall, 1975). As an example, annual time series of minimum temperatures for the urban stations of Istanbul (Goztepe & Florya) and its sequential version of the Mann-Kendall test graph is presented in Fig. 5. In Mann-Kendall plots, as of 1965, Goztepe station has a minimum temperature of 10 °C, which increased to 11.5 °C in 2006. Similar trend is seen for Florya station. Significance of the trend has been identified by Mann-Kendall statistics where the area above the line passing 1.96 represents the 95% significance level. Both Goztepe and Florya stations show a significant increase starting by mid to late 1990s.

The Mann-Kendall statistic of the annual minimum temperatures for the urban and rural stations are presented in Table 3. In all urban stations, Mann-Kendall statistics suggest statistically significant increase. In rural stations, however, out of 16 stations, 13 stations do not show any significant trend; only 1 of them have statistically significant increase and 2 of them show decreasing trend. After 1990s, the population of Beypazari and Tomarza are on the decline, and this may be the reason for the negative trends in these two rural stations. These results suggest that an additional factor: possibly UHI, is in motion to cause significant increase in minimum temperatures over urban areas. In order to further investigate this hypothesis, we have studied trends in temperature differences between urban and rural environment for the selected cities.

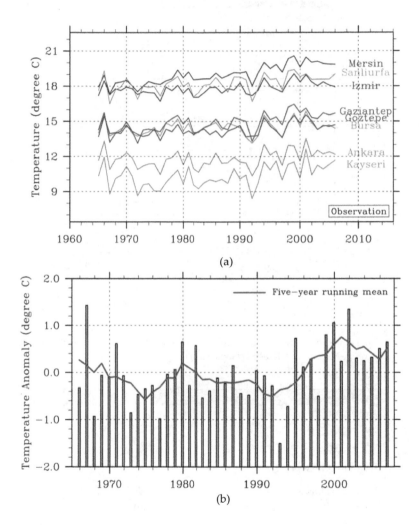

Figure 4. (a) Observation mean temperature variation in large Anatolian Cities and (b) anomaly in yearly mean temperatures in the Anatolian Peninsula from 1965 to 2006

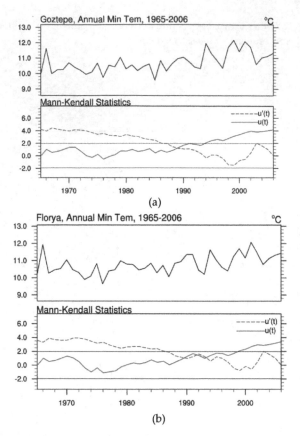

Figure 5. The annual time series of minimum temperature for the urban stations of Istanbul and its sequential version of the Mann-Kendall test; a) Goztepe, b) Florya

Urban Station	T$_{min}$ Statistics (1965-2006)
Ankara*	2,93
Bursa*	2,20
Gaziantep*	4,87
Istanbul (Goztepe)*	4,19
Istanbul (Florya)*	3,46
Izmir*	2,61
Kayseri*	5,08
Mersin*	7,06
Sanliurfa*	4,35

(a)

Rural Station	T$_{min}$ Statistics (1965-2006)
Polatli (Ankara)	0,38
Kizilcahamam (Ankara)	0,99
Beypazari (Ankara)	-2,87
Esenboga (Ankara)	0,55
Nallıhan (Ankara)	-0,73
Islahiye (Gaziantep)*	2,18
Kirecburnu (Istanbul)	1,81
Kumkoy (Istanbul)	1,22
Sile (Istanbul)	-0,14
Dikili (Izmir)	1,46
Develi (Kayseri)	-1,51
Pinarbasi (Kayseri)	0,49
Tomarza (Kayseri)	-2,29
Keles (Bursa)	-0,64
Erdemli Alata (Mersin)	1,48
Birecik (Sanliurfa)	0,40

(b)

*Shows statistically significant increasing trend

Table 3. Annual minimum temperature statistics of stations; a) urban, b) rural

The Mann-Kendall results of the other urban-rural pairs are given in Table 4. Example of the results are given in Fig. 6. Time series and the statistics for the city of Ankara, the capital of Turkey show strongly increasing warming trend (Fig. 6a). Ankara is the most crowded city of Turkey after Istanbul and located in the center of Anatolia. Because of the migration from rural sites to the center of the city (Fig. 2), Ankara has become a highly populated city, although there is not a significant industrial activity in the area. Eventually, urban-rural pairs of Ankara have high Mann-Kendall statistics over 1.96 (Table 4), which demonstrates the UHI effect in the capital of Turkey.

Bursa, which is located in the south of the Marmara Sea, has a growing population (Fig. 2) with a highly industrialized area, producing the urbanization phenomenon in the city with the human migration to the city like Istanbul. The minimum temperature difference series for Bursa-Keles station pair shows significantly increasing warming trend (Fig. 6b) with a Mann-Kendall statistic of 4.30 (Table 4). The city of Gaziantep is located in the southeastern part of Anatolia and has a growing population like the other large cities in the region and has a developing industry. The Mann Kendall statistics of the urban-rural pair shows a significantly increasing trend (Fig. 6c) with 5.82.

Istanbul and Bursa are the cities located in the north-west of Anatolia. This region is the most industrialized part of the country. Istanbul is the largest city of Turkey with over 13 million population (TUIK, 2012). Due to the cultural and financial features of the city, migration is generating the urbanization and making the Istanbul a mega city. It is located on the Bosporus and extends both on the European and Asian sides; therefore to investigate

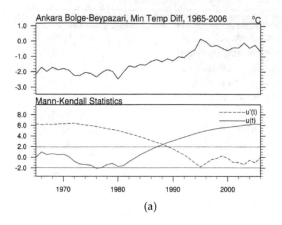

(a)

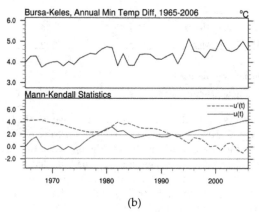

(b)

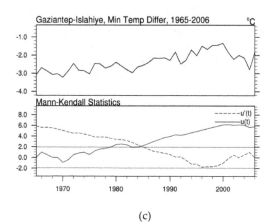

(c)

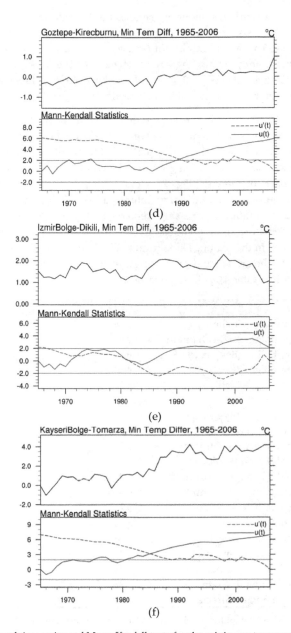

Figure 6. The annual time series and Mann-Kendall tests for the minimum temperature differences between; a) Ankara Bolge – Beypazari, b) Bursa – Keles, c) Gaziantep – Islahiye, d) Istanbul (Goztepe) – Kirecburnu, e) Izmir Bolge – Dikili, f) Kayseri Bolge – Tomarza

the UHI for Istanbul, two urban stations are selected on the both sides for representing the all city. The urban stations used in the European and the Asian sides are Florya and Goztepe, respectively. The annual minimum temperature statistics of the urban stations of the city show significantly increasing trend (Table 3). The urban-rural pairs show significantly increasing UHI effect (>2.5) in the city (Fig. 6d and Table 4), which is an accepting result for a highly urbanized city of Istanbul.

Izmir is the third largest city in Turkey with respect to population and comprising significant amount of economic activity in its region. The city is located in the west coast of Anatolia (Aegean Region) and has a typical Mediterranean climate. The city has encountered important amount of urban growth and with the 2007 above the 70% of the city live in the urban parts of the city (Fig. 2). Urban station, Izmir Bolge has a significant increasing trend in annual minimum temperature statistics (Table 3a). The Izmir Bolge-Dikili difference has a significant increasing trend above 95% (Fig. 6e). Kayseri is the other city studied in the same region of Anatolia along with Ankara. Kayseri is another big and also industrialized city in the region. Three rural stations are selected for the Kayseri with a one urban station. All the urban-rural pairs show a positive trend above the 95% with Mann Kendall values over 5.0 (Table 4), pointing out the UHI effect strongly in Kayseri (Fig. 6f).

Station pairs (urban-rural)	T_{min} Statistics (1965-2006)
Ankara Bolge – Polatli	5.88
Ankara Bolge - Kizilcahamam	2.92
Ankara Bolge – Beypazari	6.25
Ankara Bolge - Esenboga	4.30
Ankara Bolge - Nallihan	5.13
Bursa - Keles	4,30
Gaziantep - Islahiye	5.82
Istanbul (Florya) - Kirecburnu	2.66
Istanbul (Florya) - Kumkoy	3.74
Istanbul (Florya) - Sile	4.63
Istanbul (Goztepe) - Kirecburnu	6.01
Istanbul (Goztepe) - Kumkoy	4.63
Istanbul (Goztepe) - Sile	5.75
Izmir Bolge - Dikili	2.13
Kayseri Bolge - Develi	5.06
Kayseri Bolge - Pinarbasi	5.86
Kayseri Bolge - Tomarza	6.93
Mersin - Erdemli Alata	6.82
Sanliurfa Bolge - Birecik	5.54

Table 4. Mann Kendall statistics of the urban and rural minimum temperature differences

In 1965, practically there was no difference in daily minimum temperatures between urban and rural stations. However, as of 1985, urban station is approximately 2 °C warmer than the rural station. This trend accelerated and as of 2006, the urban station is over 4 °C warmer than rural. Mann-Kendall plot verifies this finding; since by 1985 the trend is estimated to be significant. Results for other urban-rural pairs are given in Table 4. All of the 19 pairs show a statistically significant increase in urban temperatures as compared to rural values.

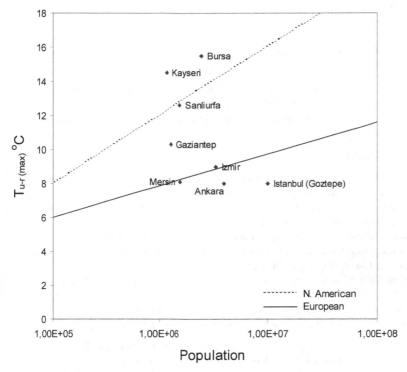

Figure 7. The intensity of maximum UHI of the selected cities with respect to the logarithm of 2007 populations

Fig. 7 presents the intensity of maximum UHI in relation with the logarithm of the population. Magnitude of maximum UHI effect is calculated by subtracting the minimum rural temperature from the minimum urban temperature and the maximum difference between them is taken as the maximum UHI intensity (Tayanc & Toros, 1997). The linear curves of Oke (1973) for the maximum UHI intensity of European and North American cities are illustrated in the figure. Maximum UHI intensities of the stations are almost positively correlated with Oke's fit for European cities; Istanbul (Goztepe - Sile), Izmir (Izmir – Dikili), Ankara (Ankara - Beypazari), Mersin (Mersin - Erdemli Alata), and Gaziantep (Gaziantep - Islahiye). But the cities; Kayseri, Sanliurfa, and Bursa are correlated with the North America line. This difference may be about the lower population densities or different sizes of settlements, which may generate higher ΔT_{u-r} values.

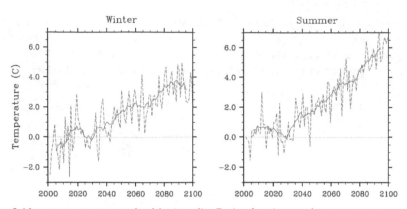

Figure 8. Mean temperature anomaly of the Anatolian Peninsula; winter, and summer season

4. Conclusion

Urbanization makes significant changes in the surface of the earth, and this change makes variations in the trends of the temperatures. In this study, urbanization effects on the temperature trends are investigated at the selected stations in the Anatolian Peninsula, which is located at the confluence of Europe, Asia, and Africa. Interactions between the climates of these continents, which have different characteristics, affect the Mediterranean region including the Anatolian Peninsula causing temporal and spatial variations (Giorgi & Lionello, 2008). Complex terrain of the study area, presence of water bodies as well as urbanization effects on local climate further complicates the system. In this study, we aim to quantify the UHI effect by contrasting the temperatures between urban-rural areas. Study area includes 81 cities of which 79 of them have population over 100,000. We have chosen 9 urban and 16 rural meteorological stations (in the close proximity of the city) in 8 cities.

The findings of this study suggest that there is no statistically significant increase in rural daily minimum temperatures between 1965-2006. However, all the urban sites show significant increase, which is a strong indication for the existence of UHI effect over this region. These findings are different from the previous studies (Karaca et al., 1995; Tayanc & Toros, 1997), which suggest either no significant or is a cooling trend existing for this region. This is mainly due to the fact that our study includes the period between 1960 and 2006, where a clear upward trend is seen especially after 1990s. Similar to our findings, Kataoka et al. (2009) demonstrated the UHI in several Asian cities.

IPCC has identified Eastern Mediterranean covering Anatolian Peninsula as one of the most vulnerable zones in terms of Climate Change (Stern Review, 2006). Our group conducted research studies (Scientific and Technical Research Council of Turkey-TUBITAK Project No: 105G015) to investigate climate change over Anatolia via regional climate models. Fig. 8 presents anomaly of temperature for Anatolian Peninsula, which were estimated by regional model in a 27 km resolution for the period 2000-2100 as compared to reference period 1961-1990, following the International Panel on Climate Change Special Report on Emission Scenarios (IPCC SRES) A2 forcing. In winter, temperatures do not show increasing trend until

2030. However, there is almost linear increase reaching up to 3°C at the end of the century. In summer the trend is much more significant and reaches up to 5 °C (Fig. 8b). It should be pointed out that most of the climate models do not take UHI effect into account, since climate models use fixed Land Use/Land Classification (LU/LC) for the simulation period. Therefore, there is a possibility of greater increase in temperatures over urban areas, when UHI effect is considered. Such analysis is definitely required to better understand future climate.

Author details

Tayfun Kindap, Alper Unal, Deniz Bozkurt and Mehmet Karaca
Istanbul Technical University, Eurasia Institute of Earth Sciences, Turkey

Huseyin Ozdemir and Goksel Demir
Bahcesehir University, Environmental Engineering Department, Turkey

Ufuk Utku Turuncoglu
Istanbul Technical University, Informatics Institute, Turkey

Huseyin Ozdemir and Mete Tayanc
Marmara University, Environmental Engineering Department, Turkey

Acknowledgement

This study is partly supported by a grant (108Y064) from TUBITAK (The Scientific and Technological Research Council of Turkey).

5. References

Ezber, Y.; Sen O.L.; Kindap, T. & Karaca, M. (2007). Climatic effects of urbanization in Istanbul: a statistical and modeling analysis. *International Journal of Climatology*, 27, 667-679

Gill, S.E.; Handley, J.F.;, Ennos, A.R.; Pauleit, S.; Theuray, N. & Lindley, S. (2008). Characterising the urban environment of UK cities and towns: A template for landscape planning. *Landscape and Urban Planning*, 87, 210-222

Giorgi, F. & Lionello, P. (2008). Climate change projections for the Mediterranean Region. *Global and Planetary Change* 63, 90-104. Doi:10.1016/j.gloplacha.2007.09.005

Goossens, C. & Berger, A. (1986). Annual and Seasonal Climatic Variations over the Northern Hemisphere and Europe during the Last Century. *Annal. Geophys*. 4, 385–399

Hamed, K.H. (2008). Trend detection in hydrologic data: The Mann-Kendall trend test under the scaling hypothesis. *Journal of Hydrology*, 349, 350-363

Hua, L.J.; MA, Z.G. & Guo, W.D. (2007). The impact of urbanization on air temperature across China. *Theoretical and Applied Climatology*. Doi:10.1007/s00704-007-0339-8

Huang, S.; Taniguchi, M.; Yamano, M. & Wang, C.; (2009). Detecting urbanization effects on surface and subsurface thermal environment – a case study of Osaka. *Science of the Total Environment*, 407: 3142-3152

IPCC, Climate Change (2007). Synthesis Report. Contribution of Working Groups I, II and III to the Fourth Assessment Report of the Intergovernmental Panel on Climate Change

[Core Writing Team, Pachauri, R.K and Reisinger, A. (eds.)]. *IPCC*, Geneva, Switzerland; 2007.

Karaca, M.; Tayanc, M. & Toros, H. (1995). Effects of urbanization on climate of Istanbul and Ankara. *Atmospheric Environment*, 29: 3411-3421

Kataoka, K.; Matsumoto, F.; Ichinose, T. & Taniguchi, M. (2009). Urban warming trends in several large Asian cities over the last 100 years. *Science of the total environment*, 407: 3112-3119

Kendall, M.G. (1975). Rank Correlation Methods. *Griffin*, London

Kindap, T. (2010). A severe sea-effect snow episode over the city of Istanbul. *Natural Hazards*, 54(3), 707-723

Kolokotroni, M. & Giridharan, R. (2008). Urban heat island intensity in London: An investigation of the impact of physical characteristics on changes in outdoor air temperature during summer. *Solar Energy*, 82: 986-998

Landsberg, H.E. (1981). *The Urban Climate*. National Academy Press, New York 275pp

Lougeay, R.; Brazel, A. & Hubble, M. (1996). Monitoring intraurban temperature patterns and associated land cover in Phoenix, Arizona using Landsat thermal data. *Geocarto International* 11 (4), 79-90

Mann, H.B. (1945). Non parametric tests against trend. *Econometrica* 13, 245-259

Nasrallah, H. A. & Balling, R.C. (1993). Spatial and Temporal Analysis of Middle Eastern Temperature Changes. *Climate Change* 25, 153-161

Oke, T.R. (1973). City size and the urban heat island. *Atmospheric Environment* 7, 769-779

Oke, T.R. (1982). The energetic basis of urban heat island. *Journal of the Royal Meteorological Society* 108 (455), 1-24

Oke, T.R. (1987). *Boundary Layer Climates*. 2nd ed., Routledge

Oke, T.R.; Zeuner, G. & Jauregui, E. (1992). The surface energy balance in Mexico City. *Atmospheric Environment*, 26(B): 433-444

Quattrochi, D.; Luwall, J.; Rickman, D.; Estes, M.; Laymon, C. & Howell, B. (2000). A decision support information system for urban landscape management using thermal infrared data. *Photogrammetric Engineering and Remote Sensing* 66 (10), 1195-1207

Rosenzweig, C. Solecki, W.D.; Parshall, L.; Chopping, M.; Pope, G. & Goldberg, R. (2005). Characterizing the urban heat island in current and future climates in New Jersey. *Environmental Hazards*, 6: 51-62

Stern Review. (2006). *The Economics of Climate Change*. UK Treasury

Taha, H. (1997). Urban climates and heat islands: albedo, evapotranspiration, and anthropogenic heat. *Energy and Buildings* 25, 99-103

Tayanç, M. & Toros, H. (1997). Urbanization effects on regional climate change in the case of four large cities of Turkey. *Climatic Change*, 35: 501-524

TUIK. (2012). *Turkish Statistics Institute*. http://www.tuik.gov.tr. Accessed March 19, 2012

Unal, Y.; Kindap, T. & Karaca, M. (2003). Redefining the climate zones of Turkey using cluster analysis. *International Journal of Climatology*, 23: 1045-1055

UNFPA (United Nations Population Fund). State of world population (2011). People and possibilities in a World of 7 Billion, 2011. ISBN: 978-0-89714-990-7

WMO (2009). World Meteorological Organization statement on the status of the global climate in 2008. WMO-No. 1039; 2009

Chapter 7

Targeting the Future: Smarter, Cleaner Infrastructure Development Choices

Andrew Chen and Jennifer Warren

Additional information is available at the end of the chapter

1. Introduction

Asia's giants China and India are pivotal changelings for the world's ecosystem and economic system. These two countries, with high economic growth trajectories, offer the planet a significant opportunity to reduce carbon emissions provided they invest in climate-smart infrastructure. Choosing greener and more efficient energy infrastructure offers the chance for generations to reduce carbon emissions, a key contributor to climate change. Additionally, attention to more efficient use of water resources and infrastructure will be needed in tandem. Understanding the energy-water-climate nexus is critical to achieving sustainable development. The challenges of climate change and minimizing its negatives impacts will require bold choices by governments, private sector actors, consumers, and significant capital investments.

Climate change is a *force majeure* that afflicts developed and developing countries alike but some countries get hit harder than others. An Economics of Climate Adaptation Working group indicates that some regions could lose 1- 12% of GDP because of existing climate patterns [1]. In September 2006, the Chinese government calculated its economic growth within an environmental context. Economic losses to the economy from environmental impacts amounted to 3.05% of GDP; water pollution cost the economy $37 billion, seconded by air pollution of $28 billion.[1] In a May 2010 Economist survey report, water pollution and scarcity were cited as costing the Chinese economy 2.3% of GDP. In a 2009 economic survey, India's government reported that it was already spending over 2.6% to adapt to climate change in regard to agriculture, water resources, health and sanitation, forests, coastal-zone infrastructure and extreme weather events [2].

---

[1] Climactico, "China & Green GDP," Feb 3, 2010; chinadialogue, "China Issues First Green GDP Report," Sep 7, 2006.

These economic and cost equations interact and change with ever-evolving variables —how water is used in conjunction with energy; growth rates changes unforeseen; and even the climate change-related effects and real costs of pollution and health expenditures that conventional country economic-growth forecasts lack calculations for.

As more countries move along the development spectrum, the growth of energy demand is a key culprit for increasing CO_2 emissions. Global energy demand is expected to be 53% higher up to year 2035 according to the Energy Information Administration's 2011 International Energy Outlook [3]. China and India alone would account for over half of the demand. In these projections, carbon emissions rise 43% by 2035, from a 2008 baseline. The International Energy Agency projects China consuming 70% more energy than the U.S. in 2035.[2] By 2015, China accounts for 28% of total global emissions as Figure 1 shows. This trend of China's continues, even slightly upward to 2035. India replaces Russia as the third largest emitter by 2015, and even surpasses the entire Middle East in carbon emissions by 2025.

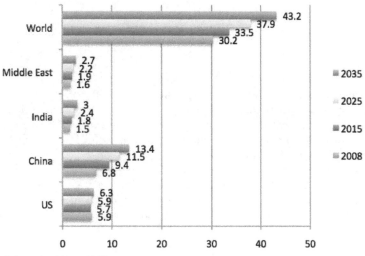

U.S. Energy Information Agency, 2012

Figure 1. CO_2 Emissions, gigatons

To finance the large-scale "greener" infrastructure needed for China and India's expected growth and emissions trajectories, new infrastructure funding approaches will be needed to bridge the gap between climate-friendly intentions and actions that benefit both economy and environment. China's pledges toward a lower-carbon economy could bring about 25% of the reductions needed to steer the globe toward the desired 2º C temperature increase limit, also known as the 450 Scenario.[3] India claims a low-carbon agenda as well. The challenges ahead are complex and expensive.

---

[2] International Energy Agency (IEA), "World Energy Outlook 2011 Factsheet."

[3] The 450 Scenario refers to greenhouse-gas concentrations being stabilized at 450 ppm (parts per million) to limit the most severe weather, sea-level rise, and temperature increases. From IEA's World Energy Outlook, 2009.

First some of the key challenges and drivers related to climate change and infrastructure will be explored. Growing energy demand parallels rising emissions and is therefore a major focal point for intervention. Water resource and infrastructure issues are an area of vulnerability for growing economies, especially in an increasingly climate-stressed world. Following that, Section II offers an overview of the infrastructure development environment and why past approaches are lacking. A new "market-finance" approach can help initiate and sustain the low-carbon and efficient infrastructure projects that will be needed. A discussion about the environment for implementation follows in Section III. Conclusions are drawn in the last section.

2. Challenges of climate change

Choices made now serve as an insurance policy against climate change effects both seen and unforeseen. A fundamental conclusion about climate change was declared by scientists world-wide in a May 2010 *Science* article: "The planet is warming due to increased concentrations of heat-trapping gases [such as carbon emissions] in our atmosphere." [4] This warming is occurring due to human activities, especially the burning of fossil fuels. With increased warming, the hydrologic cycle will be altered; oceans are acidifying; sea levels are rising. A ton of CO_2 emitted today is worse than a ton emitted a decade ago with the earth's lessened capacity to absorb the emissions, according to reporting by Union of Concerned Scientists.[4] Thus energy infrastructure choices today can impact global warming, and simultaneously, water challenges need to be addressed.

The costs of climate change are high in a resource and infrastructure context. Delaying action and investment on climate change becomes more expensive in the future. A climate change opportunity exists to develop and fund the new "green" energy and water infrastructure needed to meet developing countries continuing economic growth and development. Overall estimates vary greatly about the costs of climate change globally: the United Nations Framework Convention on Climate Change (UNFCC) says its $49 -$171 billion annually, while an Imperial College London report says it's two to three times those amounts.[5] They cite that extreme weather damage alone costs more than $50 billion annually. In India's Maharashtra state, drought could cost their agricultural productivity up to $570 million annually by 2030. Emerging markets with reliance on hydropower that are susceptible to drought put 50% of their energy budget at risk to the detriment of their economy [1].

2.1. Energy variables

The challenge of meeting energy demand in the future requires large-scale investments. The International Energy Agency estimates that $38 trillion will be needed to meet future energy

---

[4] Union of Concerned Scientists, Climate Science Update, Feb 2009. Also at
http://www.ucsusa.org/global_warming/science_and_impacts/science/global-warming-faq.html
[5] "Costing Catastrophe," The Economist.com. Dec. 8, 2009.

supply to 2035, two-thirds of which goes to developing countries.[6] To deploy clean energy and reduce emissions, a McKinsey 2011 resources report projects another $315 billion on average will be needed to meet the 450 Scenario, or $7.88 trillion to 2035 [1].

A large contributor to emissions growth world-wide, China's demand for electricity escalates to 2030. By 2015, the demand for electricity consumption of China exceeds that of the U.S., and forecasts even two years earlier had China not doing so until 2030 (see Table1). With China's continued heavy reliance on coal-burning power plants[7] into the future and ever-increasing demand for electricity, energy source choices and power demand combine for unfavorable emissions calculations. India's coal consumption also increases dramatically, doubling use by 2035 and displacing the U.S. as the second-largest coal consumer by 2035 [5].

	2007	2015	2030	2007-2030
U.S.	3826	3986	4679	0.9%
China	2717	4723	7513	4.5%
India	544	892	1966	5.7%
Middle East	575	790	1382	3.9%
Asia	4108	6777	11696	4.7%
OECD	9245	9792	11596	1.0%

Source: World Energy Outlook, 2009

Table 1. Electricity Demand (TWh)

Increased demand for electricity serves as a proxy for development. Developing rapidly, the new investment needed for China to 2030 to meet electricity demand is estimated at $3.12 trillion. Of total world demand for power generation, China consumes 30% of the world's total energy budget over the period 2008-2030, also revised upward since earlier calculations. India's demand exceeds that of the Middle East region. The sole country of China takes in more investment than the entire developed country region of Europe and the U.S. as Table 2 illustrates.

	Capacity Additions (Gigawatts)	2008-2015	Capacity Additions (Gigawatts)	2016-2030
U.S.	148	$554	420	$1515
Europe	220	762	492	1672
China	530	1209	795	1912
India	117	332	338	1016

Source: World Energy Outlook, 2009
*Includes power generation, transmission, and distribution

Table 2. Power Generation Infrastructure Investment 2005-2030 (in billion $)

---

[6] IEA 2011 World Energy Outlook Factsheet.
[7] China accounts for nearly 75% of the global increase in coal generation (EIA 2011, Global Econ Outlook article, aol.com).

How the energy mix changes for these growing economies can create a different and more positive outcome for generations to come. The choices of China and India, such as including more renewables (wind, solar, and hydropower), zero-carbon nuclear and movement away from the heaviest polluting fossil fuels are key to outcomes. If China decides to fully embrace a glide path toward a 450 scenario, in 2030 coal generation would be 30% over 69% in a business as usual case according to McKinsey study scenarios (Figure 2). Hydropower use remains virtually the same for China at 13% of its energy mix. But solar and nuclear power grow considerably from a currently low base. India following a 450 scenario would need to make significant solar power contributions, reduce coal-based power, add wind (14%) and other renewable forms. The IEA expects investment of $15.2 trillion by both supply-side and consumers to meet the 450 Scenario by 2035.[8] Both countries need to use more natural gas with its lower carbon content for power generation and potentially transportation. While pivotal countries address energy demand, energy mix, and investment issues, another critical resource warrants analysis.

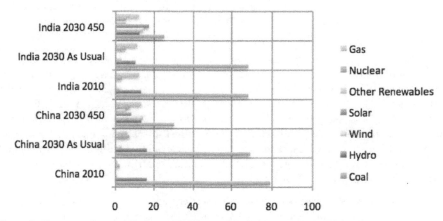

Figure 2. Changes to Energy Mix Scenarios China and India, 2010-2030 (% TW hours)

2.2. Water resources

The United Nations reported that by 2030 nearly half of the globe will be water stressed [6]. Water resource security has become a top global policy issue, and can be more problematic than energy issues. Private companies, governments and civil society institutions realize that inadequate water resources can impede the best-laid plans for economic and business development. The high risk that water scarcity poses for economic growth and stability makes this policy area a priority.

A number of drivers impact the future of water for China and India, though these two countries are not alone. Economic growth, urbanization, industrialization and increasing affluence will shift demand for water and sanitation purposes even higher. For example,

---

[8] IEA, 2011, FactSheet.

China's middle class is expected to grow from 4% of the population in 2005 to 56% by 2030.[9] Population growth makes declining water supplies spread among more people, agriculture, industry and businesses challenging, even posing competitive threats to one another.

Climate change is expected to worsen water problems by increasing the frequency and severity of floods and droughts. According to climate change projections, the changes in snow and glacier melt of the Himalayas could affect hundreds of millions of India's and China's population [7]. Several of India's river systems rely upon Himalayan glacier melt. In the immediate term, snowmelt increases their flows; in the long run, the impact is expected to be a decrease of 30-50% [8]. Projections indicate that the freshwater availability in India will drop from 1820 cubic meters to less than 1000 m by 2025.[10]

The policy areas of energy, health, food security and environment intersect water demand and supply issues. In Kashmir and the borders between Assam and Bangladesh, national security risks arise because of conflict over resources and the environmental refugees that flood into neighboring countries. Owing to the effects of drought, hydropower generation in India declined 10% in 2009 over the year before.[11] The World Preservation Foundation stated that aquifers under Beijing and Delhi are drying up.

China also has the potential for security conflicts both outside and within its own borders, owing to water resource conflicts. Water refugees within China are driven from their homes. Ten provinces in China, accounting for 45% of GDP, are considered water poor by the World Bank.[12] The Mekong River runs through the Yunnan province, Myanmar, Laos, Thailand, Cambodia and Vietnam. Fifty years of drought and the Chinese government's building of dams have caused insecurities, even unrest in cases, for the 60 million who depend on the river. Worsening water resource challenges lie ahead for China, India, and other countries, with both developed and developing countries facing different sides of climate change-related water challenges.

The case of China warrants a brief focus as its water challenges reveal a slow-moving crisis. For its fast-growing economy, water challenges are intertwined in meeting its energy demand. Rapid urbanization and industrial growth are main drivers for its water-energy demand challenges. Agriculture makes up 50% of water demand, and industrial demand comprises the other 32%, which is largely driven by thermal power generation [8]. Water shortages cost China about 1.3% of its annual economic output, with a further 1% lost to water pollution, says the World Bank.[13] Significant industrial and domestic wastewater pollution makes the quality gap larger than the quantity gap, according to the Water Resources Group. Urban sewage, refuse, and industrial waste have polluted over 90% of groundwater.[14]

---

[9] Water Resources Group report, p.56

[10] IPCC Report 2007, Ch 7.

[11] "India Braces For Drought..." Reuters, Aug 18, 2009.

[12] Economy, Elizabeth, "China's Global Quest for Resources and the Implications for the Unites States, " Testimony before the U.S. Congress, Jan26, 2012.

[13] *Economist* survey report, May 2010. Also note 25.

[14] Ibid. "China's Global Quest for Resources and the Implications for the Unites States, " Testimony before the U.S. Congress, Jan26, 2012.

Industrial users dominate overall water demand growth in China. With industrial and urban water users being the fastest growing segment, water-efficient infrastructure programs and a focus on conservation could result in net annual savings of $22 billion even after capital costs are considered [8]. Productivity gains could also accrue by implementing greater industrial efficiency measures in water usage. China knows how to squeeze out energy efficiencies in power production to reduce carbon emissions as it did across two earlier decades. Water is a key industrial input for China and India's power base, since coal is a heavy user of water. With its recent policy shift to more renewable power generation, water efficiencies in solar power should be a priority as well. India has a similar dynamic with increasing coal usage in the future.

Sector	2005	2015	2030	Cumulative annual growth rate
Municipal/ Domestic	68	88	133	2.7%
Industry	129	194	265	2.9%
Agriculture	358	385	420	0.6%
Totals	555	667	818	1.6%

Source: Water Resources Group, 2009

Table 3. China Water Demand by Sector, Withdrawals, billion m3

In 2009, India was especially hard hit with blistering droughts afflicting 177 districts, owing to poor monsoons in the year.[15] Agricultural production, a significant part of the Indian economy that impacts 2/3 of Indian livelihoods, was curtailed; India's economic growth rate had also taken a tumble when all was accounted for.[16] India with water needs of 1.5 trillion m^3, is projected to face a water supply deficit of 50% or 754 billion m^3 (cubic meters) by 2030, according to a study on global water resources.[17] Current supply is 740 billion m^3. Increases in demand for water in agriculture, alongside a limited supply infrastructure will contribute to this gap. India's water demand from agriculture is expected to double from 2005 levels, and comprise 80% of total water demand. Demand from water users in India, besides agriculture, is also expected to growth rapidly. Municipal and domestic water demand doubles by 2030, and industrial users demand four times the amount as 2005 levels.

As *Figure* 3 indicates, a supply deficit of 201 billion m^3 is estimated up to 2030 in China given water demand. Certain regions will suffer more severe shortages than others. The Yangtze Basin is expected to face the largest size water gap of 70 billion m^3 or a 25% gap.

---

[15] Prime Minister and Agriculture Minister declaration from August 17, 2009 news story on FEER.
[16] "India Braces For Drought..." Reuters, Aug 18, 2009.
[17] Water Resources Group, "Charting Our Water Future," 2009.

China has water storage capacity per capita of 2,220m³ — less than half that of the U.S. and 100 times that of India's. However, China's water *availability* per capita is approximately one-quarter of the global average.[18]

In developing countries, municipal and domestic water demand will grow significantly. Figure 3 illustrates how China and India's sectors grow, and where the efficiency gains could be targeted. China's industrial water demand accounts for 40% of the *additional* growth of global water demand for industrial users, largely as a result of power generation. China's increased water demand from 2005 to 2030 is 61% versus India's increase of 58%. China and India could be considered benchmark cases, as they represent developing countries having common urbanization and water-energy challenges, which could impact development.

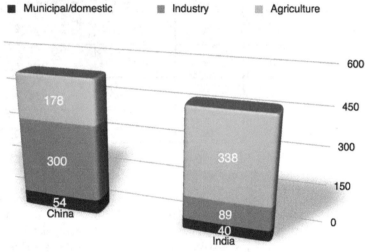

Source: Water Resources Group, 2009

Figure 3. Comparison of China and India's water demand, 2005-2030 (billion m³)

2.3. Managing water

Water managers may not be fully aware of climate change impacts related to the infrastructure they manage, or they have other priorities. Climate change needs to be factored in relative to an area's particular challenges like flooding, drought, or low-lying area problems such as salt-water intrusion. The UNFCC estimates that additional investments for climate change adaptation will be $28 billion-$67 billion (even up to $100 billion) per year several decades from now.[19] Water infrastructure, usage patterns and institutions have developed in the context of past and present conditions, according to the

---

[18] Congress Report of Second China Water Congress 2008, Beijing, *International Journal of Water*.
[19] U.N. World Water Development Report 3 (Factsheet on Water), 2009.

Intergovernmental Panel on Climate Change [9]. Water managers in numerous countries – the Netherlands, US, UK, Germany, Bangladesh, Australia and others—have begun to address climate change in their planning. While China faces water shortage issues, India's challenge partly surrounds managing its supply better. In fact, India can learn from the mistakes of China in upstream and downstream management practices.

China and India can learn from the lessons of advanced economies and their outcomes. A recent study by the Pacific Institute found that one way to manage scarce supply of water is simply by curbing waste. The state of California illustrates developed-country water challenges and how to begin managing them among stakeholders. The authors suggest that some combination of irrigation technologies and management practices can save 17% of all the water used by California farmers and more than twice the total of the state's millions of city dwellers. They add that spending on capital-intensive projects like de-salination plants and solar power plants make less sense when you can gain efficiencies in this manner.[20] With most of India's water budget spent by agriculture, efficiencies in this sector should be one, among other priorities.

By making appropriate investments in infrastructure and changes in land use and management structures, the impacts of floods and droughts could be abated. For example, India's Konkan Railway suffers approximately $1 million annually in damages because of landslides during the rainy season [9]. Twelve percent of India's land is prone to flooding and 80% could be abated by providing reasonable protections.[21] China's Yunnan Province officials studied the temperature over the last 30 years in an attempt to understand the droughts of the last half a century. Temperature increases between 1950 to 2003 caused losses to agriculture costing the economy 10 billion yuan (approximately $1.5 billion). Its Water Resources Department recently outlined a number of drought-related infrastructure undertakings (though it had focused on floods earlier and missed the drought abatement opportunities that are now critical). Pushing for more private investment will exacerbate demand. A senior engineer at a research institute in Kunming, China said by 2020, urbanization and industrialization will take its toll on water supply, with an annual shortfall that amounts to the city's current water demand.[22]

While India has a substantial water resource base, current infrastructure to buffer its variability is low. India has only 200 m³ of water storage capacity per capita compared to China's 2,220 and the U.S.'s 6000m³. The Water Resource Group's (WRG) assessment says India's "accessible, reliable supply of water amounts to 744 billion m³, or 29% of its total water resource."[23] Water quality is a major issue for India and a lack of wastewater treatment plants in the middle and lower parts of most Indian rivers causes surface water quality degradation.

---

[20] "Waste Not: A Demand-side Solution for California's Water Troubles," Wall Street Journal, July 24, 2009.
[21] Food & Agriculture Organization of the United Nations (FAO), Aquastat country profile (India). http://www.fao.org/nr/water/aquastat/countries/india/index.stm
[22] Hujun, Li. "No Easy Fix After Decades of Droughts In Yunnan," Caing.com (English), Mar 11, 2010. http://english.caixin.com/2010-03-11/100125559.html
[23] Water Resources Group, 2009. p 56.

More than half of China's 660 cities suffer from water shortages, says the Institute of Public and Environmental Affairs. To accommodate continued urbanization, both agriculture and industrial users will have to reduce consumption. The government has plans for more wastewater treatment facilities to deal with water pollution. In anticipation of water shortages, one of China's most technically challenging infrastructure projects, the South-North transfer, will divert water from the Tibetan Plateau to western regions through a 300-mile network of tunnels. At one point, siphoning water from the Brahmaputra River, vitally important to India, was considered but the Chinese backed down after controversy ensued. The project's costs are an estimated $62 billion, surpassing that of the Three Gorges Dam project of over $20 billion.[24] Additionally, the project's intent to re-supply the dry north may not be keeping up with nature's changing courses. China will have to re-think future water diversion projects.

China will have to simultaneously consider reducing the carbon and water footprints of their new infrastructure additions in the years ahead. Water and energy cost savings can accrue by focusing on water efficiencies in thermal power processes and energy choices. To close China's 201billion m^3 water supply gap, investment capital of $7.8 billion is needed annually to fill the deficit to 2030, or $156 billion in total. However, when for operational expenditures are considered, net annual savings of $21.7 billion could accrue according the WRG. These savings get distributed among thermal power, wastewater reuse, pulp and paper, textile, and steel industries. The water efficiency measures required to create these savings will divert resources to water efficiency that may slow growth in the short-term but create sustainable business practices and technical innovation over the long haul. Limited water supply and environmental pressures in numerous basins indicate that wastewater treatment and wastewater reuse are critical challenges ahead for China.

Greater private sector participation is being encouraged in China, according to an official at a recent water congress in Beijing. However the Chinese government is believed to lack the capacity to implement the many intended reforms of the water sector.[11] Raising water tariffs are expected to bring more private sector participation. Shenzhen planned to increase water prices 30% for households and 60% for businesses; cities in China that have already raised tariffs or plan to include Beijing, Shanghai, Tianjin, Shenyang, Guangzhou, Nanjing, and Chongqing.[25] Of World Bank infrastructure private-public partnerships (PPPs), China accounted for more than 70% of projects established in developing countries; 60% are for sewage treatment plants, reflecting a government priority of dealing with urban wastewater.[26] Though India will come to the same conclusions as China, India has a rougher road ahead with its policy reversals being greater in water infrastructure development than in China.

Water savings can accrue to India, and other developing countries, by managing water leakage better. In India's municipalities, a 26% savings is possible. The World Bank found

---

[24] http://www.water-technology.net/projects/south_north/
[25] "China Cities Raise Water Price in Bid to Conserve," *Wall Street Journal*, July 31, 2009. And
http://news.xinhuanet.com/english/2009-12/27/content_12711285.htm
[26] World Bank PPI database, June 2009

inefficiencies in Indian cities of more than 40%, partly because of water leakages and a lack of invoicing customers.[27] In Brazil's Sao Paulo an effort toward reducing waste has resulted in losses falling from 32% of revenue to 24%, with a goal of 13%; Brazilian losses average 40%.[28]

Water is a complex, multi-faceted infrastructure and resource challenge for countries developed and developing alike. Resource challenges however have also been managed from a national policy standpoint — Germany's current zero-nuclear movement to 2022 and China's energy efficiency and conservation efforts in the period 1980-2000. Australia implemented water sector reform, resulting in agricultural productivity gains of 36% between 2000-05 and a financial water market worth $1.7 billion in 2007-08, the Water Resources Group notes. Across the globe, water-oriented institutions are informing the dialogue on the need for better management of water resources. A key challenge remains: How will the green and efficient infrastructure for sustainable growth and development be funded?

3. Infrastructure needs and funding

Developing world economic growth is a driver of needed infrastructure. As the growth paths of China, India, Africa and South America will contribute to increased carbon emissions, countries will need a new set of tools to finance the much needed climate-sensitive infrastructure. Over the next twenty years, to deliver renewables at scale, $500 billion could be needed annually, according to an expert in a World Economic Forum symposium.[29] Research shows that the ways in which infrastructure has been financed in the past has wasted resources. Given resource constraints in both developed and developing countries, new approaches are needed that are more conducive to modern economic and political realities.

Demand for urban infrastructure will continue to increase dramatically. Research from the McKinsey Group projects 136 new cities entering the top 600, all in developing countries with 100 new cities in China alone. With trends in urbanization, the opportunity for economies of scale exists in infrastructure. Cities generally have denser infrastructure and policies directed toward low-carbon and water resource efficiency offer a unique opportunity to counter the energy-water-climate nexus. Also what happens in one city bleeds into the comprehensive carbon emissions count, ie., the pollution of Shanghai reaches the atmosphere of Japan, even the West Coast of the U.S.

Economies of scale in infrastructure exist in the growing urban centers of China, India and other developing countries. "The doubling of a population of any city requires an 85% increase in infrastructure," according to research on scaling laws. There are similar savings

---

[27] McKinsey "Resource Revolution" p.95. Citing Pronita Chakrabarti Agrawal, "Designing an effective leakage reduction and management program," World Bank, April 2008.
[28] Ibid, p.138.
[29] Kerr, Thomas. "Renewable energy: developing solutions to deliver low-carbon investment," World Economic Forum presentation, 2011 July 6.

in carbon footprints. Most large, developed country cities are greener in terms of per capita emissions though they are centers of concentrations of emissions as well [12]. Rapidly developing cities in China and India have inconclusive results to date, though researchers are pursuing the data trail.

However, creating and operating the same infrastructure in higher densities is more efficient, economic and can lead to better innovations. Because of denser settlement, the per capita space required shrinks resulting in a more intense use of infrastructure. The notably accelerated pace of life in cities with their universal features leads to higher productivity, which is also characteristic of a more efficient and economic use of infrastructure. Finally, owing to the nature and commonalities of cities, the intertwining of more diverse economic and social activities heightens economic specialization and expression of social networks. Innovation is a byproduct of the wealth and creativity effects of urbanization and agglomeration that does not typically happen in smaller places. Infrastructure is a natural beneficiary of urban scaling laws and cities as complex adaptive systems. To make better use of finite resources and growing, wealthier populations, innovation, economy and efficiency are imperatives to deal with the energy-water-climate nexus.

With the region of Asia needing around $8 trillion over the next ten years in infrastructure investment, new approaches to leverage capital are needed. In a recent high-profile Asian water summit, the director of the Asian Development Bank called for extensive use of new technologies, large-scale reuse of wastewater, and plugging in the gaps with small-scale efforts. India's power sector requires $600 billion by 2017 to meet demand, according to a 2011 report by McKinsey. Private-public partnerships are expected to fill in the infrastructure gaps the public sector cannot fund or manage.

In the traditional water sector for 'downstream' water supply such as sanitation and industrial use, spending across this sector was $485 billion in 2005. It is expected to grow to $770 billion by 2016, mainly in the water and sanitation areas. Utilities account for 70% of the total spending. In comparison, utilities in other sectors spend $770 billion in natural gas sectors and in electricity some $1.5 trillion.[30] Upstream water supply, often government-funded, is expected to become more expensive in time as cheaper supply is superceded by demand. In developing countries, spending on water infrastructure is a small part of total global spending as mentioned above. Given their respective states of development, expected growth, and resource dynamics, China and India will be spending vast sums in the future.

To fill the large gap in funding infrastructure development, private capital will need to be mobilized. Many governments around the globe acknowledge the limits of their ability to fund projects. With greater integration of capital markets, globalized markets offer opportunities that benefit both developed country investment searching for returns and diversification, and developing countries seeking funds. Financing infrastructure projects by way of global capital markets' invisible hand can infuse economic development alongside the policy imperatives of lower-carbon energy and water security.

---

[30] Datamonitor, "Electricity: Global Industry Guide," 2009.

3.1. Some pitfalls of existing approaches

In the past two decades, the primary approaches of Build-Operate-Transfer (BOT) and Private-Public Partnerships (PPP) have been widely applied for hundreds of large-scale infrastructure projects worldwide. Both BOT and PPP approaches can be considered "contract finance" approaches. These approaches were encouraged to counter the shortcomings in the older build-own-operate approach and create a pathway for private participation. Throughout numerous cities in China—Pudong BOT project (Shanghai), Chengdu, Tianjin, Lanzhou, and many more— the government is privatizing its water and wastewater infrastructure to add capacity and modernize operations. But these existing approaches of BOT and PPP reveal shortcomings as well, and have led to substantial amounts of wasted resources across countries and sectors.

Infrastructure project financing is structured in a way, which creates the problem of "plums" in contrast to Akerlof's illustrious 'lemons' problem [13]. The plums problem arises when the buyer (bidder or firm providing capital) knows more about the quality and economic value of the project than the seller (government agencies). Under the existing approaches of BOT and PPP, project companies have incentives to play political games which gives rise to corruption and waste. Project sponsors and investors may then be deterred from future projects in the host country or even the region [14].

Under existing approaches, infrastructure project financing is structured in a way which creates flaws: inefficiencies and added costs, greater political (policy) risk, and a lack of diverse ownership needed for transparent incentives. India's failed $2.9 billion Enron-Dabhol project highlights some of the issues surrounding infrastructure projects. In this case, there was a lack of competitive bidding, unfair contracts, and limited knowledge by the seller (the government) in terms of project scale, technologies and complexity [15]. Unfair competition for contracts will not yield the longer-term goals of sustainable growth and development and better governance records. Consequently, investors may then be deterred from future projects as happened in India post-Enron for many years.

Both BOT and PPP projects in China have a checkered history. In 1988, China's first BOT project with private participation was the Shaijiao power plant in Shenzhen. Early PPPs revealed signs of the plums problem by foreign investors, and at a later time, state-owned enterprises displayed operational and management inefficiencies. In Ke's study of sixteen Chinese PPP projects [16], the classic shortcomings of the "contract finance" approach were observed: a) inefficient bidding processes, b) imperfect project contracts, and c) lack of diversification and liquidity in project finance. The sixteen projects studied either failed to bring reasonable returns to investors; were suspended or purchased by the government during the concession period; or were forced to re-negotiate with the government. These projects were predominantly water and energy infrastructure PPPs for which policy risk ran high across the spectrum of projects.

Political or policy risk is a significant concern for investors considering developing country projects. In a study of the political risks associated with BOT projects in China, several obstacles were found as China embarked on its campaign to encourage private participation. The top five critical risks faced by foreign investors in order of threat were: changes in legal risk (law and regulatory changes by government), corruption risk, delay in approval risk, expropriation risk, and Chinese entities reliability risk [17]. This story repeats itself in India as well. Research by Wilkinson revealed that corruption in infrastructure projects in India often has political roots.[31]

In China, most wastewater treatment plants with private investment are BOT, and foreign investors have grown wary. The government controls the water prices, thus creating return on investment or market risk for the private firm. Other foreign firms have complained that local governments give preference to domestic firms, though their technology is in fact superior. The government is acting as provider, regulator and customer, with the conflicts of interest implied—political and policy risk for the foreign firm/investor.[32] Inconsistent laws and regulations and their irregular application to foreign firms are classic BOT shortcomings that plague other developing countries besides China and India.

3.2. Market approach for progress

Governments around the globe are seeking the experience and capabilities of the private sector to both fund and provide necessary services to their citizens. Given the emergent lower-carbon and water-related infrastructure needs in China and India, a progressive approach is needed that levels the playing field, deters political and policy risk, and develops more efficient, transparent market mechanisms. Rather than initiating a government-controlled infrastructure project, which may then be privatized, the market-based approach can allocate capital and resources more efficiently from the project's onset. A government can choose to be the majority shareholder under this approach, but the risks and incentives will be transparent when shareholder-stakeholders are responsible for the project's sustainability. This is a truer form of private-public partnership.

A "market finance" approach, which creates immediate private ownership of public investment projects among diverse groups of investors, may lead to more efficient and successful infrastructure development. In a study of seven provinces in northern China facing water scarcity, groundwater markets through the privatisation of tubewells re-organized water usage and management for farmers. Water was managed more efficiently and cropping patterns became more productive and profitable, without a lack of access for poorer farmers. India's large agriculture withdrawals and waste are the result of a lack of market pricing mechanisms.

Project securitizations or initial public offerings of project securities can be designed with financial innovations for any new large-scale infrastructure project, or projects linked by theme, sector or region. These could effectively be "green zones." A group of green projects— a large biomass plant, a solar plant, and a water-efficient utility—could be

---

[31] Chen and Warren (Kubik), 2007.
[32] *China Economic Review*, "Drinking Buddies," Dec 2009.

funding targets for capital markets. This approach would complement the low-carbon centers that China and India's governments intend, but do so in a more financially-sustainable manner.

Securitizations would create diversification, liquidity, and mitigate many of the problems that accompany existing approaches in financing infrastructure. Managerial incentives could be more aligned with productivity, thus reducing the widespread problems of cost overruns and inefficiency in traditional BOT and PPP. It could also unravel the perverse incentives pervading infrastructure spending in China, India and other developing countries.

Financial innovations in the securities offering can serve as both a deterrent and an incentive. For example, including event-risk provisions in project bonds can deter politicians' attempts to make undesirable policy changes. This can ultimately foster a more investment-friendly environment that developing countries need to attract. Sound decisions and proper management will bring its own reward through enhanced project value and the value it brings to the community and economy at large. The invisible hand may prove more capable in setting infrastructure project agendas spanning varied political tenures and agendas.

Governments—central, provincial, and local—could be allocated project securities to achieve true public-private ownership. In market-based PPPs, governments can play a role as needed according to their capacity. Market-based PPPs can address investors' reluctance due to political risk and profitability concerns, bring projects online more quickly, and attract longer-term institutional players. Conventional PPP projects and deals line the news media pages, but another approach will be needed to marshall the financial resources for the challenges ahead.

Water resource issues and price increases often lead to public backlash. Under this new approach, citizens can participate as shareholders and stakeholders, and therefore participate in governance and oversight issues. The bankruptcy of U.S.-government backed solar firm Solyndra might have had a different outcome for solar power development in the U.S. were an alternate approach taken. In Asia, the Asian Development Bank can play an important role as project guarantor in water infrastructure PPPs as well. Additionally, utilizing capital markets offers the potential to scale up projects that might otherwise receive funding on a smaller scale. This is important with economies of scale needed to deal with the significant carbon emissions reductions required of China and India and with respect to the scope of wastewater and efficient water systems.

Much of the world's infrastructure revolves around cities —highways, ports, power plants, and water systems — with fossil fuels at their epicenter of operations. Owing to age and a preference of cleaner energy, a massive capital stock turnover is coming across the next few decades. China's wind power and hydropower generation comprise 25% of their total capacity.[33] China plans wind power to grow a staggering 14.2% annually to 2035. While China has shown strides as major green player, in order to grow and power their economy, more attention to infrastructure choices and sustainable financing will be required.

---

[33] *China Economic Review*, "89.7 Million kW of Power in 2009," 11 Jan 2010.

India plans to double its hydropower capacity by 2030, but environmental concerns have already led to rejections of several projects. India will need to manage its water resources more stringently with hydropower and agriculture demands, coupled with expected changes owing to climate. India's National Solar Mission plans to expand solar power from 20 GW to 200 GW by 2050,[34] but again government may not be capable of financing their clean energy ambitions.

The time for action is now as delaying investments mean less chance to reach a 450 Scenario. According to the IEA, if coordinated international action is not taken by 2017, all permissible emissions would come from the then existing infrastructure and therefore all new infrastructure (includes power plants, factories, and buildings) from 2017 to 2035 would need to be zero carbon unless older emitting infrastructure is retired early. Much of the power sector infrastructure stock in existence today accounts for half of the emissions locked-in to 2035. The IEA calculates for every $1 of investment avoided before 2020 in the power sector, an additional $4.3 would need to be spent after 2020 to compensate for the higher emissions [5]. The stakes are high— and developing countries, particularly high current and future emitters such as China and India— are a targeted response to prevention.

4. Practical and policy implications

Numerous opportunities exist to address the energy-water-climate nexus through cleaner energy infrastructure and by utilizing efficiencies in water in developing countries. However energy and water resources are often subsidized by governments, distorting pricing signals. In fact, according to the McKinsey resource study, between 70-85% of opportunities to boost resources productivity are in developing countries. And importantly, in order to engage the private sectors, reducing subsidies in energy and water make more projects attractive for private sector engagement through higher rates of return. If carbon were priced and subsidies reduced, water projects with rates of return of 10% or greater increase from 76% to 90%.[35] In general, resource competition from developing country growth will require new approaches and interventions to provide power, water and modern lifestyles.

A Chinese official recently stated that they would be pursuing and competing for natural resources globally alongside other major economies. Coal, for example, is one resource that China will heavily influence in the future, as it adds 550 gigawatts (GW) of coal capacity between 2010 - 2030, from 50 GW in 2005-2010.[36] China and India can further leverage their domestic agricultural resources through biomass power plants. Power from biomass is more reliable than wind and solar and has the ability to provide base-load capacity when old coal plants are de-commissioned. But the challenges in project financing are proving to be an obstacle for biomass expansion. Market prices for the agricultural waste are not set like coal inputs or natural gas. This type of 'greenfield' project would be ideal for the market-based approach. It could help determine a market price, and therefore further biomass's expansion, which supports China and India's green agendas.

---

[34] U.S. EIA, 2011, p. 97-8.
[35] McKinsey Global Institute, 2011, p.17.
[36] Ibid. p 108.

In emerging markets, many of the new greenfield opportunities are "sustainable" and green types of infrastructure. A fund managed by PIMCO, one of the largest bond investors in the world, will purchase fixed income securities in infrastructure sectors of importance to emerging market governments such as energy, transport, telecoms, water and treatment. They are targeting both retail and institutional investors; this is effectively an outsourcing of infrastructure investing owing to demand for the asset class and the knowledge it takes to manage these types of assets. These opportunities are not being ignored in progressive developed countries. A U.K. start–up firm is scaling up waste-to-energy power plants. A key reason for underdevelopment in this novel area has been financing, even in an advanced economy.

In India, the past five years has seen considerable growth in infrastructure investment, which is a stated government goal. Over the next five years, infrastructure investment is expected to reach a new high relative to GDP. The private sector is expected to make $1 trillion in investment. According to a recent Economist article, the infrastructure firms developing roads, power stations and airports are heavily indebted however. The top 70 Indian stock exchange-listed (BSE) firms were roughly $12 billion cash flow negative. The time for more equity, and shareholders (over oligarchs), is ripe in India to sustain its economy's growth, which 'slowed' to 7%.[37] This market-based approach could jettison greener infrastructure development in India.

Institutional investors like infrastructure investing because these investments often mirror the long-term nature of their portfolio needs— and they are "real" assets, which are more attractive post-financial crisis. Specific to climate-change infrastructure, power generation and their grids, energy storage, and water infrastructure rank high on the list as climate change-targeted infrastructure opportunities. Upgrades to power plant infrastructure — from subcritical technology to ultra-supercritical coal— is one area where resource and carbon emissions savings can accrue. China's coal plants are 80% subcritical, with about one-third of conversions taking place toward greater efficiency.[38] More efficient gas-fired plants using combined-cycle gas turbines can also offer savings, especially with the projected switching to more natural gas from coal in the future. Greatly increased supply has come online from unconventional resources, such as shale gas, tight gas and coal-bed methane sources.

Within this space of infrastructure investing, climate change adaptation and mitigation efforts also apply to the areas of corporate governance and social responsibility. Many large institutional investors—banks, insurance companies, or pension funds—are integrating sustainable business practices into their lending and/or investment criteria. For example, India's YES Bank has a dedicated Sustainable Investment Group, specializing in alternative energy and other environmentally-focused sectors.[39] Players in global supply chains are becoming better environmental stewards. A number of coalitions, such as Ceres, the World Economic Forum, NGO groups and supra-national organizations, have been formed that

---

[37] "Infrastruggles," The Economist, Dec 31, 2011.
[38] McKinsey Global Institute, 2011.
[39] Ceres and RiskMetrics Group, "Addressing Climate Risks: Financial Institutions in Emerging Markets," Sept 2009.

produce meaningful research and new forms of assessment tools. In essence, business, government and varied organizations are raising the bar of infrastructure development, in which "greener," smarter, and more efficient forms of infrastructure are emerging.

In addition to larger infrastructure projects, small-scale water infrastructure opportunities have emerged such as the World Bank and ADB-backed Aakash Ganga project, which originated in India. This private-public partnership outperforms typical public works projects. The economic impact and quality of life has substantially improved the selected rural sites' because it is a community-driven initiative. Water conservation is at the heart of the process. A large-scale study of China's rural areas reveals a preference of villagers for infrastructure projects that raise standard of living and/or improve the environment, even beyond employment opportunities. These have included projects in irrigation, water supply, and environmental protection of forests. With China's decentralized approach in rural development, progressive local–level leaders can attract private investment for their green growth goals as well.

Developed countries will need approaches to maintain and upgrade their infrastructure. The most developed PPP markets of the UK and Australia are attempting to motivate new investment through the private sector to share the burden of funding. Governments are finding the investment capital needed to keep their infrastructure modernized do not exist in public coffers. In the UK, the government plans to attract pension funds (institutional investors), sovereign wealth funds, and investors in Latin America and China to invest in modernizing its infrastructure. They also have a Green Investment Bank, targeting low-carbon investments, with L18 billion available by 2014-15.[40] The UK government sees this strategy of modernization critical for adding growth prospects into a slowing-moving economy.

5. Conclusions

The climate change story has energy and water as its primary protagonists and antagonists. By enhancing the ability of market-based PPPs to operate within China and India's borders, new cleaner infrastructure can be developed based on a sustainable model. Financial institutions and other stakeholder groups have been leading the way in lower carbon footprints in energy projects, but the water footprint is becoming a more pressing issue as well. Attention to sustainable development and finance will matter even more to India and China as their economy's grow and develop. A flexible, market approach which runs in tandem with the varieties of ways in which infrastructure investment exists, offers a new source of capital at a time when government resources are under pressure and other priorities exist.

Citizens, numerous governments, financiers and investors are favoring "green" infrastructure as it mitigates many types of risks seen and unforeseen. If China or India are losing 3% of GDP to environmental impacts, it is a *de facto* halving of 6% growth rate. This is likened to

---

[40] Institutional Investor, Sept 17, 2011

inflation or a debt, which will erode true growth and real wealth and health over time. With climate change these debts will be paid by someone. Isn't it time to literally engineer this better? Entrepreneurs however want and need consistent policies from government to facilitate exchange. Barriers, policies, and practices that blunt clean energy and water security ambitions should be analyzed within the context of sustainability. Rather than the extreme step of diverting water resources as a first choice, aggressive national conservation measures can be policy instruments that complement what the private sector is incentivized to do.

The governments of China and India can implement progressive policies to become cleaner, greener and more efficient [18]. No objectives can be realistically met unless policy incentives are aligned; capital markets are more developed; a level playing field exists for investors; and steady governance exists. Germany has new energy policies underway to eliminate nuclear power, scale up renewables, and advance their low-carbon and energy efficiency export markets. Incentives are in place, and they will adjust subsidies accordingly as they did with their solar power push. But governments also lose credibility as in the case of Spain's solar and wind policy reversals and subsidy rollbacks. Governments need to make realistic promises to win the trust of investors. India had a long period of *mea culpa* with investors post-Enron.

Water and energy challenges are hard to disentangle. Water security problems could have a domino effect on energy development, and subsequently economic growth prospects. Global awareness about these issues is needed, and best practices in need of being shared. Investing in environmentally-sensitive forms of infrastructure is no longer an outlier. It is a mainstream trend—given the limits of the earth's resources and its ability to integrate particular types of man-made pollution into its cycles. Smarter and more sustainable ways to finance the development of Asia's giants complement the ideal of a cleaner, more efficiently managed use of the planet's resources.

Author details

Andrew Chen
Cox School of Business, Southern Methodist University, Dallas, Texas, USA

Jennifer Warren*
Dallas Committee on Foreign Relations, Dallas, Texas, USA,
Cox School of Business, Southern Methodist University, Dallas, Texas, USA,
Concept Elemental, Dallas, Texas, USA

6. References

[1] McKinsey Global Institute (2011). "Resource Revolution: Meeting the world's energy, materials, food, and water needs," November.

[2] IANS (2009), "Climate change costs India over 2.6% of GDP: Economic Survey," Thaindian News, July 2. Available: http://www.thaindian.com/newsportal/enviornment/climate-

* Corresponding Author

change-costs-india-over-26-percent-of-gdp-economic-survey_100212459.html.   Accessed 2012 Feb 5.

[3] Shifra, Mincer (2011). "Global Energy Outlook Puts China and India at the Fore," AOL Energy, Sep 20. Available: http://www.thaindian.com/newsportal/enviornment/climate-change-costs-india-over-26-percent-of-gdp-economic-survey_100212459.html. Accessed 2012 Feb 5.

[4] National Academy of Sciences Letter (2010). "Climate Change and the Integrity of Science." *Science, May 2010.*

[5] International Energy Agency (2011), "World Energy Outlook 2011 Factsheet."

[6] "Water Shortage," World Preservation Foundation. Available: http://www.worldpreservationfoundation.org/topic.php?cat=water#.T1F3dGDnvEV. Accessed 2012 Mar 2.

[7] Kundzewicz, Z.W., L.J. Mata, N.W. Arnell, P. Doll, P. Kabat, B. Jiménez, K.A. Miller, T. Oki, Z. Sçen and I.A. Shiklomanov, 2007: Freshwater resources and their management. Climate Change 2007: Impacts, Adaptation and Vulnerability. Contribution of Working Group II to the Fourth Assessment Report of the Intergovernmental Panel on Climate Change, M.L. Parry, O.F. Canziani, J.P. Palutikof, P.J. van der Linden and C.E. Hanson, Eds., Cambridge University Press, Cambridge, UK, 173-210.

[8] 2030 Water Resources Group (2009). "Charting Our Water Future."

[9] United Nations (2007). United Nations Framework Convention on Climate Change. "Investment and Financial Flows to Address Climate Change," October.

[10] Bates, B.C., Z.W. Kundzewicz, S. Wu and J.P. Palutikof, Eds. (2008) "Climate Change and Water," Technical Paper VI: Intergovernmental Panel of Climate Change, June.

[11] Varis, Olli (2008). "Congress Report: 2nd China Water Congress, 2008, Beijing, China, 14-16 May 2008, International Journal of Water, Mar 2009: p. 175-77.

[12] Bettencourt, Luis and GeoffreyWest. *Nature*, Oct 27 2010, p 912, vol 467.

[13] Akerlof, G. (1970). "The Model for 'Lemons': Quality Uncertainty and the Market Mechanism," Quarterly Journal of Economics, Vol. 89, 1970, 488-500.

[14] Chen, Andrew (2009). "Rethinking Infrastructure Project Finance," working paper, November.

[15] Chen, Andrew and Jennifer Warren (Kubik) (2007), "Complementing Economic Advances in India: A New Approach to Financing Infrastructure," *Journal of Structured Finance*, July/Aug.

[16] Ke, et al (2009) "Public-Private Partnerships in China's Infrastructure Development: Lessons Learned," *Proceedings of International Conference on Changing Roles: New Roles and New Challenges*, Netherlands, October.

[17] Wang, S. Q, et al (2000). "Evaluation and Management of Political Risks in China's BOT Projects," *Journal of Construction Engineering and Management*, May/June.

[18] Warren, Jennifer (2008). "China's Green Future," *Far Eastern Economic Review*, December, Vol 171, No.10.

Food Security in a Changing World

Land Grab, Food Security and Climate Change: A Vicious Circle in the Global South

Kihwan Seo and Natalia Rodriguez

Additional information is available at the end of the chapter

1. Introduction

The commercialization of public land in the Global South, which refers to the medium and low human development based on the United Nations development program report 2005 [1], has increased dramatically in recent years due to the wide spread leasing and sale of land to foreign companies and governments. The main goal of these investors is to secure food and energy production for their populations as multiple factors threaten their food security at home. On the one hand, the current population will increase worldwide from nearly 7 billion to over 9 billion by 2050 [2], a growth that would require the increase of food production from nearly 6 billion tons (gross) to 9 billion tons by 2050 [3]. Furthermore, competition for land, water, and energy will only intensify along with the need to reduce the many negative impacts of agriculture to the environment [4, 5]. Global food security has been further strained, notably during 2007-08 [6], by the growing volatility of the food market and the political controversy surrounding the use of grain to produce biofuels [7, 8]. Any one of these factors will likely pose significant challenges, but the sum of all of them could constitute a major threat to land ownership.

Overall, these factors have driven a change in perspectives of land ownership. Recent trends indicate that the need to provide food and energy security at home has led international corporations, sovereign wealth funds, foreign governments, private equity firms and domestic actors to buy or lease large tracts of land outside their national borders [9]. These land deals or "land grabbing", as labeled by many Non-Governmental Organizations and the media, are certainly to be considered a prominent factor in the list of significant drivers of land change in certain parts of the globe, especially in southern hemisphere continents. The need to secure food supply comes as a result of the many international pressures that took place in the mid-2000s; and initiated the rise of food prices by 2006. Among these pressures some that stand out are [10]:

1. Extreme weather events, such as droughts and floods that affected cereal exports in 2005-06 and decreased cereal production worldwide by 10 percent.
2. World cereal reserves fell as major cereal producers such as the USA, EU and China reduced holdings of food stocks.
3. Fertilizer prices and transportation costs increased due to the oil price hike from 2003 to 2008.
4. The increased demand for the production of biofuels conflicted with food crops as land was diverted for the production of monocultures such as sugar, oilseeds, palm oil and maize.

These, along with other international and local pressures caused concern in the international market leading to an increase in market volatility as speculations of food supply where unfavorable [10]. In response to these price hikes many food-importing countries found a long term strategy to outsource their food production and guarantee their food supply at low costs in the leasing and purchasing of foreign land [11]. For example, China holds approximately 20 percent of the world's population but possesses no more than seven percent of the world's arable land [12]. For many years this Asian country has been a net exporter of agricultural goods. In recent years due to its rapid economic growth, higher population income, changes in diets, and limited arable land, among other factors, China has become a net importer of agricultural goods since the beginning of the 2000s. In order to ensure its food security and promote its current economic growth the Chinese government and private corporations are investing in land suitable for agriculture outside its national borders. In the same way some wealthy import-dependent countries, such as Japan and South Korea being directly affected by the 2007-08 food crises, have initiated policies along this line. On the other hand, arid, oil-rich countries from the Gulf States under "harsh climatic conditions, poor soils and scarce land and water" among other limitations [11], such as Saudi Arabia, do so in an attempt to reduce its domestic water use [13]. In recent years the scale of this type of business has increased dramatically with millions of hectares being bought or leased outside their borders. [11]

Due to the increasing global demand for food-stuffs production and alternative energy development, the southern hemisphere is portrayed as an idoneous reservoir of arable land capable of satisfying the international needs, a particular example constitute African countries due to their relatively low population density [9] and cheap land. As mentioned in De Shutter (2009) southern countries in Africa and Latin America are the main targets for investors seeking farmland as it is scarce in Asia [14]. This race to buy land has been described by many as a new neo-colonial approach by wealthy countries to take over the key natural resources of poor countries [15]. Some analysts perceive these land deals as a threat to the livelihood of local communities while others stress the positive effects derived from the income generated in these deals [9]. Such benefits could be perceived as the injection of the much-needed capital to sustain agriculture [16] and therefore the creation of on-farm and off-farm jobs, the development of rural infrastructure such as irrigation canals, and the construction of schools and health clinics that will improve local livelihoods. Along this line of thought, many of the host countries of land deals have encouraged this type of investment and are keen to develop it as a potentially lucrative activity [6].

2. Global land grabs

2.1. Land grabs undermine food security

Global land grabs have recently become a major point of international discussion [17] due to the global struggle to ensure food security [18]. "Land grabbing, generally referred to the mass purchase of agricultural lands by transnational companies or foreign countries" [19], refers to the lease (often for 30–99 years) or purchase of vast areas of land outside their national borders [20] mainly for agricultural production. One of the main drivers of this practice is the current international demand for cheap food after the food price hikes of 2007-08 [13]. During these years the dramatic increase in basic food prices reduced the access to food of millions of people as they reached the highest levels in 30 years [21]. According to global estimates this price hike brought around 915 million people to undernourished levels worldwide, and additional hundreds of millions were added to the count due to the effects of the global financial and economic crisis [10]. Although the highest levels of food insecurity where reached in developing nations many food-importing countries felt the effects of food prices in their own population.

According to Brown (2011), wealthy but food-insecure countries worried about tightening markets [20] are seeking to ensure their food production by leasing and buying land overseas (e.g. the Gulf States). By controlling farm land beyond their national borders these countries are gaining control of the international supply-chain of food-stuffs [22]. This practice is perceived as an innovative, long-term strategy to ensure the food security of its population at cheap prices [20]. The majority of the investors are Asian countries such as China and India, which according to the Food and Agriculture Organization's (FAO) 2009 report, are currently food self-sufficient. Likewise import dependant countries, particularly affected by the food crisis, such as Saudi Arabia, Japan and South Korea, are also in the search for fertile farmland in African countries like Uganda, Madagascar, Mali, Somalia, Sudan and Mozambique, as well as in other developing countries such as the Philippines, Indonesia, Laos, Thailand, Vietnam, Cambodia, Pakistan, Burma, Brazil, Argentina, Kazakhstan, Ukraine, etc. [16, 20, 23]. (Table1. Shows the most recent estimates of land deals worldwide linked to the countries that are the major sources of land grabbers).

The governments of 'host' countries, such as Madagascar, Sudan and Cambodia, generally welcome foreign investment [24], even though much of their own population lacks sufficient food [17]. Large-scale land acquisition for food security by richer countries is increasingly contested, since it is not considered ethical to export food from countries in which there is widespread hunger. For example, Daewoo Logistics, the South Korean commercial group, failed its attempts to acquire 1.3 million hectares (over half the arable land of the country) of land in Madagascar for the production of maize for human consumption food and palm oil used in biofuels [16]. By doing this the company would have ensured future fuel stocks and guaranteed the countries' food security "by providing half of its maize imports from Madagascar alone" [25]. Ultimately the deal ran into trouble and was a direct factor in the overthrow of the country's government in 2009 [16].

Countries involved in land grabbing	Land purchased or leased (including deals still in process in ha)	Number of deals
UK	4,941,765	40
US	4,162,394	42
UAE	3,182,950	19
India	2,101,400	28
China	1,953,527	36
South Korea	1,412,394	16
Saudi Arabia	1,132,945	20
Germany	525,345	22

Note: Estimates have been calculated from GRAIN (2012) "GRAIN releases data set with over 400 global land grabs" Available at http://www.grain.org/article/entries/4479-grain-releases-data-set-with-over-400-global-land-grabs

Table 1. 2012 global estimates of major land deals carried out by governments and private companies.

In developing countries, land deals result most of the time in the displacement, dispossession and disenfranchisement of local communities. Most of the land utilized by small farmers in local communities is used under customary tenure arrangements; as a consequence, they often lack formal property titles over the land and can easily risk losing access to it [9]. In addition, most of the deals between foreign investors and local governments are arranged outside the public scope and therefore, smallholders may not even know they are losing their land. Women, who make up 70 percent of farmers in the developing world, are often the most vulnerable to this practice as they may not be able to protect their own land tenure claims in court due to local laws.

The general perception under the land deals scope is that most of the land available for buyers is abundant and underutilized; although in many cases it is already being used [6]. Existing land use it usually overlooked due to the lack of formal land rights of smallholders or their access to proper legal assistance [16]. For example, in Gambela, Ethiopia, the Ethiopian government has signed deals with investors from India, Saudi Arabia, China and other countries since 2008 for large-scale agricultural projects in the region (see Table 1). The deals give foreign investors control of half of Gambela's arable land [26]. All land allocations recorded are classified as involving 'wastelands' with no pre-existing users. As the Anywaa Survival Organization was able to verify, these are ancestral lands from which indigenous communities such as the Anuak have been dislocated. Without any information or consent for the sale and purchase of such territories, the surrounding communities have lost from these forests their refuge in times of violence, an excellent source of medicinal plants, and a valuable reserve of food during famines [26].

Land grabber	Base	Sector	Hectares	Production	Projected Investment	Status of the deal
Hunan Dafengyuan	China	Agribusiness	25,000	Sugar Cane	-------------	Completed
ARS Agrofoods	India	Agribusiness	3,000	Cotton, groundnut, sesame, soybean	US $5 million	In process
BHO Agro	India	Agribusiness	27,000	Cereal, oil seeds, pulses	US $8/ha/yr (lease)	Completed
Karuturi	India	Agribusiness	311,000	Maize, palm oil, rice, sugar	US $1.2/ha/yr (after first 7 years)	Completed
Ruchi Group	India	Agribusiness	50,000	Soybeans	US $4 million (lease cost for 25,000 ha)	Completed
Al Amoudi	Saudi Arabia	Finance	140,000	Livestock, maize, oilseeds, rice sugar cane, teff	US $2,500 million	Completed

Note: Estimates have been calculated from GRAIN (2012) "GRAIN releases data set with over 400 global land grabs"

Table 2. Examples of land deals initiated in Gambela since 2008.

These large-scale land deals increase local food insecurity as the export of locally produced agricultural products force farmers to purchase agricultural goods elsewhere as opposed to benefiting from the harvest of their own lands [27]. The country of Ethiopia claimed as the epicenter of current land deals [29], shows the direct relationship between food insecurity and land grabs. Since 1984, Ethiopia has been well known for its extreme food shortages [22]. In 2010 ten percent of its population relied on food aid [27], and in 2011, due to the dearth of rain in the Somali and Oromiya regions, the nation appealed for emergency food aid at the United Nations. Betting on economic growth projections, the Ethiopian government promised that the country would be food self-sufficient within five years. Although the economic speculation is promising it comes at the expense of the displacing and dispossession of the population as the government is closing deals with private investors over the citizen's lands [22]. The detrimental effects of these land deals were evident during the 2008 famine in which food instability levels increased among the population while food was being exported [30]. The USAID, which has been one of Ethiopia's largest aid donors, is strong critic to this practice, and argues that the right way to ensure the country's food security is by guaranteeing the complete ownership of land by its citizens and to stimulate the local consumption [31]. Nevertheless, the irony of these land deals continued as a $116-million food aid package is planned to reach the African nation for a five-year period, while, contradictorily in 2009 there was a simultaneous $100-million Saudi investment to grow and export rice, wheat and barley back to itself [22].

2.2. The "food and biofuels conflict"

The recent global awareness of anthropogenic climate change and the resulting growing interest in green energy, including biofuels, have been another important motivation for land investments. Currently biofuel production is the dominant reason for land deals in countries such as Madagascar and Ethiopia, where jatropha, palm oil and sugar are major crops. According to the Global Land Report 2010 (GLP) biofuels production is an important driver for the international land investments in Africa [9]. These deals are driven in part by the international demand for renewable fuels and the shifts in energy policy among Southern African countries to fulfill their energy needs with their own natural resources [32]. This growing interest on green energy is leading investors to invest in productive land overseas which results in the opening of new land for agriculture [9]. Many see this as a strategy by the private sector to take advantage of the emerging market of green energy. For instance, countries like China wish to diversify its domestic energy sector [9] due to the increasing demand of oil and its high global prices. Hence, the growing production of biofuels has started to affect the current food production as land deals keep taking place in the international scope.

On the other hand, many Southern African countries have actively embraced the biofuels productions in their lands, as they wish to limit their dependence on future oil imports and exposure to price volatility [32] by becoming oil producers. Mozambique is a perfect example. With the goal of becoming an 'oil exporting country' on 2004 the Mozambique government urged farmers to plant jatropha - a Latin American shrub which seeds produce an oil that upon extraction can be refined to produce biodiesel - on all marginal and unused lands [33]. Although there is evidence that this crop will perform poorly under harsh agro-ecological conditions, the building of cultivation and processing facilities for the production of biodiesel derived from this plant have been initiated.

2.3. Seeking land and water

When seeking arable land overseas foreign investor's main targets are lands with access to irrigation for better potential production of food or biofuels [13]. According to the International Institute for Sustainable Development, the ultimate goal of the purchase or long-term lease of land in foreign countries is the acquisition of the water rights [34]. This practice allows major investor countries facing water scarcity to shift their domestic irrigation water to municipal water supplies [13]. We find China, India, Saudi Arabia, Kuwait, Qatar, and Bahrain among this type of investors group. African and Asian countries rich in land and water resources are the primary targets for their land investments [6]. For example, Central Africa only uses irrigation resources in two percent of its land, making an investment in this untapped water resource a very appealing proposition [35]. However, as abundant as water may seem, predictions from the Intergovernmental Panel on Climate Change (IPCC) suggest that fresh water supplies are likely to be depleted in some parts of Africa. As a result of climate change, lands will become drier, with less rainfall, affecting crop yields and making livestock farming impossible. In this possible scenario the water required to slake the investors' fields could be considerable [13]. Along this line, biofuels have been described as "one of the thirstiest products on the planet". For example, to produce one liter of biodiesel from soya (soybeans) requires 9,100 liters of water. As for the production of bioethanol from corn or sugar cane there is a requirement of as much as 4,000 liters of water for one liter of bioethanol. Still, even those biofuels considered to be optimal for arid places require large amounts of water in order to grow [24].

3. How land grabs can exacerbate climate change

Climate change as defined by the IPCC refers to "... a statistically significant variation in either the mean state of the climate or in its variability... Climate change may be due to natural internal processes or external forcing to persistent anthropogenic changes in the composition of the atmosphere or in land use." The atmospheric changes associated with this phenomenon can be observed at all spatial levels from local to regional to global. It affects average global surface temperatures and sea levels, soil moisture and local precipitation, among other variables [36]. Currently, human society practices are negatively influencing these variables and thus, exacerbating this atmospheric phenomenon. Practices such as fuel burning and deforestation for agricultural purposes can have great influence in the world's climates.

As referenced in Cotula et. al (2009), 80% of the global farmland is located in Africa and South America [37]. Most of these areas are either tropical rainforests, protected natural regions or are already used for shifting cultivation or grazing of animals [38]. However, they represent the most suitable regions for land deal investments. But the conversion of tropical forests to crop land, (mostly monocrops) come as an inevitable threat to the region's biodiversity, carbon stocks and water resources [6]. Tropical forests do not only serve as reservoirs, sinks, and sources of carbon in the world, but also provide several ecosystem services that have impacts on a region's climate. Among these services are the maintenance of elevated soil moisture and surface air humidity, reduction of sunlight penetration, weaker near-surface winds and the inhibition of anaerobic soil conditions [39]. This environmental arrangement is responsible for the rich biodiversity of tropical ecosystems [40]. However, as tropical landscapes are converted to agricultural and pasture areas, the productivity of this soil decreases as less rainfall, associated with changes in the solar radiation partitioning, is observed [39].

Many studies have demonstrated that changes in land surfaces (such as land clearing for agriculture) can influence both local and regional climates and can even have major impacts on climates in distant parts of the Earth [36]. For example, the Amazon Basin landscape is well known for having a direct influence in the flux and exchange of moisture into the atmosphere, regional convection, and hence regional rainfall. However, recent works have determined that the changes in the forest cover of the region have consequences on climates of distant places. The Sahelian drought associated with the destruction of regional vegetation [36] serves as another example of the relationship between changes in land cover and distant climates. In this sense, land deals can be considered major drivers of ecological impacts at both local and global scales. Such impacts can affect the ecosystem services that sustain human livelihoods as conversion of tropical forests to pastures takes place globally.

4. Food security under climate change

Food security defined by the United Nation's FAO is "a situation that exists when all people, at all times, have physical, social, and economic access to sufficient, safe, and nutritious food that meets their dietary needs and food preferences for an active healthy life" [40]. The definition encompasses four important dimensions of food supplies; -food availability, stability of food supplies, access to food, and utilization of food - all of which are closely linked with impacts of climate change. First, food availability refers to whether or not the agricultural productivity of a region can satisfy food demand in that region. Second, food stability is an indication of how consistently the supply meets that demand. Third, access to food literally means the ability of individuals to buy proper food resources for their dietary needs. Lastly, utilization of food references how well individuals can consume food resources without undue concern for quality and safety of food [41].

Climate change affects almost every aspect of human society and natural environment, especially production of agriculture and food in multiple ways. Since many agricultural regions in the world have already suffered from extended drought and abrupt flood induced by global climate change, weather and climate variability will possibly change conditions of land suitability and agricultural productions [41]. Although temperate regions and higher latitude zones may get benefits of agricultural productivity by increasing temperature due to climate change, negative effects such as heavy rainfall, drought, and increased evapotranspiration on other regions (e.g., rain forest, semi arid region, and Mediterranean region) may hinder food availability in general [42].

Many predictions indicate that global and regional weather fluctuations and extreme weather events are expected to increase in frequency and intensity [43]. Because of the weather fluctuations, crop yields and local food supplies will also fluctuate and thus food stability and security could be adversely affected [41]. For example, extreme weather events like typhoons, hailstorms, and droughts will bring failure of crop yields. Specifically, sub-Saharan Africa and parts of south Asia, where most of high climate variable and arable lands are located, will be exposed to the highest instability of crops and livestock production [44]. Although the FAO predicted that access to food will be getting better in the long term based on falling food prices and increasing income level [45], this prediction might not consider the effects of global climate change that can possibly deteriorate the progress of food accessibility. Thus, if the situations - food prices, amount of crop yields, and supplies - of world food markets change under certain weather events, the ability to access food would also be changed as the recent food crisis in 2007 and 2008 suggests. In addition, the IPCC recently reported that increasing temperature will increase incident of more food poisoning, specifically in temperate areas, and cause food and water-borne diseases [43]. This means that individuals will need to more cautiously select and consume their foods. Thus food utilization, the last key dimension of food security, will also be affected by climate change.

All the key dimensions of food security induced by climate change consequently affect land deals in terms of both "host" and "investor" countries. Many host countries already face food shortages and difficulty to access food within local areas where land deals take place. Since land has shifted to foreign buyers, local communities cannot utilize their immediately surrounding land to produce food [13]. Many foreign investor countries, however, may take advantages of all dimensions of food security such as food availability, stability, accessibility, and utilization. Host countries are willing to sell their land in order to take advantage of short-term economic growth opportunities, due to the large-scale nature of land acquisition by investors. [37]. This tends to increase and accelerate land grabs in developing countries overall. Therefore climate change causes food insecurity in a way that changes temperature and precipitation in the first place and then food insecurity brings us more land grab to mitigate food shortages. Those three elements - climate change, food security, and land grab - are interconnected and, unfortunately, are detriment to each other.

5. Vicious circle among climate change, food security, and land grab

In the previous sections of the chapter, each two of three elements (i.e., food security and land grab, land grab and climate change, and climate change and food security) have been investigated through literature reviews in relation to climate change, food security, bio-fuels, and land grab. Explanations gained by reviewing relationships between each two elements, however, do not efficiently reveal the causal relationship among the elements and how closely coupled they are with each other. In this section, we attempt to describe the causal relationships among them in terms of a vicious circle framework.

Causal relationships among climate change, food security, and land grab make current situation worse in Global South, where people already have been suffering from food shortage and severe weather events, and increase vulnerability to climate change. Each of three elements adversely affects people in Global South in different ways that particularly threaten their livelihood, safety, and health. As discussed in the previous sections many other factors influence each of the three phenomena. For instance, land grab did not evolved due to food insecurity alone but growing global population, green energy demand, economic growth, and political reason [9]. Climate change and food security also have many reasons other than the factors in the Figure 1, however, since climate change, food security, and land grab in the circle are closely coupled, each of three elements will be treated as main driving forces in the vicious circle framework.

Figure 1 shows a vicious circle of climate change, food security, and land grab that is proposed by this chapter based on the review of relevant studies. As noted previously, climate change is likely to affect food security by increasing extreme weather events (e.g. extended drought, frequent and severe flood, cyclones, and hailstorms) which change land suitability for food production. In addition, demands to reduce carbon dioxide and other greenhouse gases, increased by human activities, lead the international society to seek alternative energy sources, biofuels and agrofuels thus esteemed as alternative energy sources that produce less CO_2 and greenhouse gases. However, enormous amount of crops and crop fields are required to produce alternative green energy [24]. Since climate change is, again, a common driving force of increasing severe weather events and green energy demands, it degrades food security and increases demand of land grabs in general.

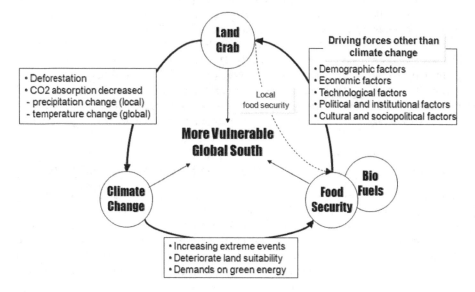

Figure 1. Vicious circle of climate change, food security, and land grab

Along with food security and biofuels, there are many other underlying driving forces that accelerate land grabs such as demographic, economic, technological, political, institutional, cultural, and sociopolitical factors [9]. For example, world population has been increasing about 34% for 2 decades by 2007 and will keep increasing until 2050 under the medium scenario projection of the United Nations [46]. This means that the average amount of land per person will keep decreasing and population disparity at the global scale will consequently increase cross-national land deals [9]. The global economy also acts on land deals when agriculture attracts as an investment opportunity. At the same time, land grab exacerbates local food insecurity; because most of the regions, where land deals take place, have already been experiencing famine for a long time, even if the regions themselves have plenty of fertile lands.

The influence of land grabs to climate change significantly increases with deforestation in tropical rainforests where protected natural areas are also located [38]. Deforestation itself, particularly in tropical rainforest, has an adverse effect on reducing carbon dioxide and greenhouse gases through the process of photosynthesis. For example, some studies show that large amount of trees cleared for palm oil crop field can actually hold up to 150 years of carbon savings, and biofuels, which are initially proposed to decrease carbon dioxide and greenhouse gases, also negatively affect climate change by increasing CO2 and greenhouse gases [24].

Analysis of causal relationship among climate change, food security, and land grab confirms existence of a vicious circle that exacerbates vulnerability of poor and small farmers to climate change, and the safety and health of the Global South. Therefore it is important that efforts should be dedicated to disconnect each element of vicious circle or, if it is possible, focused on changing vicious circle into virtuous circle, since climate change, food security, and land grab have already threatened people there.

6. Discussion

Studies have shown that global land deals have increased dramatically in recent years, especially during the food crisis of 2007 and 2008. Some countries that need to guarantee food security and biofuels production as a strategy to cope with impacts of climate change and some other factors (i.e., demographic and economic factors) increased a scale of land deals in Global South. Consequently land deals have increased possibility of the climate change impacts by increasing deforestation. Deepening climate change once again can exacerbate food security and increases biofuels demands. This implies that the relationships between land grabs, climate change, and food security make vicious circle. However, it is not easy to approach for solutions from climate change perspective to ameliorate the vicious circle, while it is relatively easy to approach solutions from land grabs and food security perspective. That is because land grabs and food security are specific issues compared to the climate change discourse - one of the most complex issues of our day. This part of the chapter thus approaches to deal with land grabs and food security issues to dismantle the vicious circle.

As illustrated in Figure 1, climate change, food security, and land grabs are connected, and each element has harmful effects to the one next to it in a predominantly counter clockwise direction. This is why we named the framework as the vicious circle and it has causal relationship among elements. Although climate change, food security and land grabs are the main subjects of our conceptual model, they do not stand alone (see in the boxes of Figure 1). Some of them affect reverse direction (e.g. land grab causes local food insecurity). However, the three elements are core sources of making vicious circle to people in Global South. It is important to dismantle and neutralize this circle, as each element is primarily responsible for the damage done to the next. There are both short-term and long-term policies or strategies that may accomplish this. For example, long-term strategies should be suggested for climate change issues. However, this chapter will not touch any policies and strategies on mitigation, adaptation, and vulnerability directly to climate change, since nature of climate change cannot shortly be improved by any efforts due to the complexity of the climate change itself. Instead, policies for land grab and food security (include biofuels) can be discussed as short-term policies.

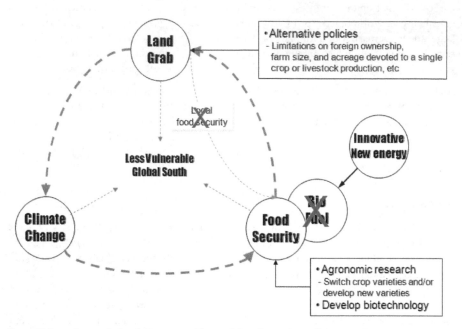

Figure 2. Loosely-coupled and disconnected by applying alternative policies over time

First, land grab can be reduced and controlled relatively in short time period, if countries targeted for land deals recognize the outcomes and tradeoffs of these deals (e.g. local food insecurity) and regulate it through alliance with countries in similar condition. In fact, some host countries are already considering alternative policies such as "strict limitations on foreign ownership of land; limitations on farm size/land ownership; limitations on the farm acreage dedicated to a single crop or livestock production (soy monoculture, for example); limitations on the total national acreage devoted to a single crop or livestock production, and; limitations on the strict controls on foreign investment in food and agriculture." [17]. Second, there might be alternatives for food security that may lead to reduce land grabs. Governments depending on imported food could help domestic farmers to increase production by agronomic research. For instance, farmers in food importing countries switch crop varieties and/or develop new varieties better able to produce in terms of volume, and develop biotechnology as well. Third, as a direct competitor to food production and another driving force of the land deals the biofuel production should be alternated by technologically innovative new energy sources [47], since production of biofuels does not really help reduce greenhouse gases, but increases land grab and worsen climate change. Countries seeking green and new alternative energy sources invest in, hydrogen fuel, wind power, solar power, and tidal power, for instance.

Innovative approaches to increase food and energy security suggested above will likely reduce land grabs. Each of the efforts suggested above will keep reducing issues along the line of vicious circle. Even though only a few short-term policies were suggested here, there might be more solutions and efforts to remove many issues in the vicious circle of climate change, food security, and land grab over time. Therefore, the short-term policies connected with long-term policies to climate change may pragmatically trigger a transformation of the vicious circle into a more virtuous circle or at least weaken the connection between each two of the elements (Figure 2).

7. Conclusion

There are many studies on the relationship between climate change and food security, food security/biofuels and land grab, and land grab and climate change. However, isolating and addressing one pair of issues at a time masks what is really going on in terms of the vicious circle which keeps each element in the circle locked into a downward spiral. The people in the Global South, especially in Africa, are thus more vulnerable to conditions out of their control. This chapter examined the relationship between climate change, food security/biofuels, and land grab as a concept of the vicious circle. To reduce problems in the vicious circle, first of all, each pairs were examined to figure out issues between pairs respectively. Second, causal relations among the elements were shown and explained as the vicious circle. Lastly, based on the examination, disconnecting solutions, in terms of policies, were briefly suggested.

Despite many other factors, this chapter has only focused on the three elements; climate change, food security, and land grab, as the vicious circle. This does not mean that policies suggested here may solve all the issues on climate change, food security, and land grab, but solve or weaken some issues of them. Furthermore, policies may trigger a disconnecting of the links in the vicious circle or transform the vicious circle into a virtuous circle with other possible policy efforts, and can be starting point of reducing vulnerability of people in Global South.

Author details

Kihwan Seo*
School of Geographical Sciences and Urban Planning, Arizona State University, USA

Natalia Rodriguez
School of Sustainability, Arizona State University, USA

Acknowledgments

The authors would like to thank the editor Netra Chhetri who gave us valuable advice and opportunity to write a chapter. We also thank Helme Castro and Evan Palmer for their careful and quick proofreading of our manuscripts.

---

* Corresponding author

8. References

[1] Damerow H (2010) International Politics, web lecture note. Available: http://faculty.ucc.edu/egh-damerow/global_south.htm. (accessed 18 June 2012).

[2] Lutz W, Samir KC. Dimensions of global population projections: what do we know about future population trends and structures? Phil. Trans. R. Soc. B 2010; 365, 2779–2791.

[3] Borlaug N, Carter J. "Food for Thought". Wall Street Journal. Oct 14, 2005 p. A10.

[4] MA (2005) Ecosystems and Human Well-being: Synthesis, Washington. DC, Island Press.

[5] Scherr SJ, McNeely JA. Biodiversity conservation and agricultural sustainability: towards a new paradigm of 'ecoagriculture' landscapes. Phil. Trans. R. Soc. B.2008; 363, 477 494.

[6] von Braun J, Meinzen-Dick R "Land grabbing" by foreign investors in developing countries: Risks and opportunities. IFPRI Policy Brief 13. International Food Policy Research Institute, Washington 2009.

[7] Abbott P, Hurt C, Tyner W."What's Driving Food Prices?", Issue Report, Farm Foundation, July 2008.

[8] Pretty J, et al. The top 100 questions of importance to the future of global agriculture. International Journal of Agricultural Sustainability 2010.
http://ucanr.org/blogs/food/blogfiles/5698.pdf. (accessed 29 November 2011).

[9] Friis C, Reenberg A. Land grab in Africa: Emerging land system drivers in a teleconnected world. GLP Report No. 1. GLP-IPO, 2010 Copenhagen.

[10] Liverman D., Kapadia K. Food Systems and the Global Environmental Change: An Overview. In: Ingram J, Ericksen P, Liverman D. (ed.) Food Security and Global Climate Change. London: Earthscan; 2010 p. 3-24.

[11] Haralambous S., Liversage H., Romano M., 2009. The growing demand for land Risks and opportunities for smallholders farmers. IFAD, 32d session of Governing Council, Rome, 17 p.

[12] Freeman, D., J. Holslag & S. Weil (2009), 'China's Foreign Farming Policy: Can Land Provide Security?', BICCS Asia Paper, vol. 3, nr 9, Brussels Institute of Contemporary China Studies, Brussels

[13]] Food and Water Europe: Global Land Grab Undermines Food Security in the Developing World. http://www.foodandwaterwatch.org/factsheet/global-land-grab/ (accessed 15 November 2011)

[14] De Schutter. United Nations. "Large-scale land acquisitions and leases: A set of core principles and measures to address the human rights challenge." June 11, 2009

[15] Hall R. Land grabbing in Southern Africa: the many faces of the investor rush, Review of African Political Economy 2011; 38:128, 193-214

[16] Cotula, L. Land deals in Africa: What is in the contracts? IIED, London 2011.

[17] Zoomers A. Globalisation and the Foreignisation of Space: Seven Processes Driving the Current Global Land Grab. Journal of Peasant Studies 2010; 37:429–447.

[18] Brown L. World on the Edge. Earth Policy Institute: London 2011.

[19] Scoones I. A Global land grab? Institute of Development Studies.(IDS)   2009. http://www.ids.ac.uk/go/news/a-global-land-grab. (accessed 16 November2011)

[20] GRAIN: Seized! The 2008 land grab for food and financial security. http://www.grain.org/briefings/?id=21. (accessed 14 November 2011)

[21] FAO: The State of Food Insecurity in the World 2008 (SOF)I: High Food Prices and Food Security- Threats and opportunities, Rome 2008, FAO.

[22] Cochrane L. Food Security or Food Sovereignty: The Case of Land Grabs. Journal of Humanitarian Assistance. http://sites.tufts.edu/jha/archives/1241. (accessed 20 October 2011)

[23] Taylor M, Bending T. Increasing commercial pressure on land: building acoordinated response. Discussion paper. Rome: International Land Coalition 2009.

[24] Burley H, Bebb A. Africa: up for grabs – the scale and impact of land grabbing for agrofuels. Friend s of the Earth Europe, Brussels, Belgium 2010. http://www.foeeurope.org/agrofuels/FoEE_Africa_up_for_grabs_2010.pdf. (accessed 14 October 2011)

[25] Ramiaramanana D. 'Impacts of land grabbing in Madagascar'. Presentation at the Regional Workshop on Commercialisation of Land and 'Land Grabbing', Southern Africa hosted by the Institute for Poverty, Land and Agrarian Studies (PLAAS), University of the Western Cape; 2010 Mar 24-25.

[26] GRAIN: Grabbing Gambela. http://www.grain.org/bulletin_board/entries/4387-grabbing-gambela#. (accessed 14 November 2011)

[27] Food First: Landmark Conference on Land Grabbing. http://www.foodfirst.org/en/Land+grabbing. (accessed 23 November 2011)

[28] Vidal J. Ethiopia at Centre of Global Farmland Rush. The Guardian [newspaper online] http://www.guardian.co.uk/world/2011/mar/21/ethiopia-centre-global-farmland-rush. (accessed 16 November 2011)

[29] Reuters: Ethiopia, UN launch food appeal for 2.8 mln people. http://af.reuters.com/article/topNews/idAFJOE7160MB20110207. (accessed 16 November 2011)

[30] Dominguez A. Why was there still malnutrition in Ethiopia in 2008? Causes and Humanitarian Accountability. Journal of Humanitarian Assistance. http://sites.tufts.edu/jha/archives/640. (accessed 24 November 2011)

[31] USAID Ethiopia – FY 2010 Implementation Plan. Available: http://www.feedthefuture.gov/documents/FTF_2010_Implementation_Plan_Ethiopia.pdf

[32] Sulle E, Nelson F. Biofuels, land access and rural livelihoods in Tanzania. London: International Institute for Environment

[33] Schut M, Slingerland M, Locke A. 'Biofuel developments in Mozambique: Update and analysis of policy, potential and reality'. Energy Policy 2010; 38(9), 5151-5165.

[34] Smaller C, Mann H. "A Thirst for Distant Lands: Foreign investment in agricultural land and water." International Institute for Sustainable Development (IISD) 2009. http://www.iisd.org/publications/pub.aspx?id=1122. (accesed 20 November 2011)

[35] FAO (2008) "Ministerial Conference on Water for Agriculture and Energy in Africa: The Challenges of Climate Change." Dec 15-17.

[36] Rosenzweig C, Tubiello F N, Goldberg R., Mills E, Bloomfield J. Increased crop damage in the US from excess precipitation under climate change, Global Environmental Change 2002; 12 197-202.

[37] Cotula L, Vermeulen S, Leonard R, Keeley J. Land grab or development opportunity? Agricultural investment and international land deals in Africa. Rome/London, UN (FAO) / (IFAD) /(IIED) 2009. http://www.iied.org/pubs/display.php?o=12561IIED

[38] Ramankutty N, Foley JA, Olejniczak. NJ. People on the land: Changes in global population and croplands during the 20th century. Ambio: A Journal of the Human Environment 2002; 31(3): 251-257.

[39] Pielke Sr, R.A, Marland G, Betts RA, Chase T.N, Eastman JL, Niles JO, Niyogi D, Running SW . The influence of land-use change and landscape dynamics on the climate system—relevance to climate change policy beyond the radiative effect of greenhouse gases. Phil. Trans. R. Soc. Lond. A 2002; 360, no. 1797 1705-1719.

[40] Food and Agriculture Organization. (2002) The State of Food Insecurity in the World 2001, Food and Agriculture Organization, Rome.

[41] Schmidhuber, J. & Tubiello. F.N. (2007) Global food security under climate change, Proceedings of the National Academy of Sciences.

[42] Rosenzweig, C., Tubiello, F. N., Goldberg, R., Mills, E. & Bloomfield, J. (2002) Increased crop damage in the US from excess precipitation under climate change, Global Environmental Change Vol.(12): 197-202.

[43] Intergovernmental Panel on Climate Change. (2007) Climate Change: Impacts, Adaptation and Vulnerability, Contribution of Working Group II to the Fourth Assessment Report of the Intergovernmental Panel on Climate Change. Cambridge Univ Press, Cambrigde, UK, in press.

[44] Bruinsma, J. (2003) World agriculture: towards 2015/2030, a FAO perspective, Earthscan, London.

[45] Food and Agriculture Organization. (2006) World agriculture: towards 2030/2050, Interim report. Food and Agriculture Organization, Rome.

[46] United Nations. (2004) World Population to 2300 - Economic & Social Affairs, United Nations, New York.

[47] Atkinson, R., Chhetri, N., Freed, J., Galiana, I., Green, C., Hayward, S., Jenkins, J., Malone, E., Nordhaus, T., Pielke, Jr. R., Prins, G., Rayner, S., Sarewitz, D., & Shellenberger, M. (2011) Climate Pragmatism: innovation, resilience and no regrets - The hartwell analysis in an american context.

Chapter 9

Climate Changes and Its Impact on Agriculture – The Case Study of Bulgaria

Nedka Ivanova and Plamen Mishev

Additional information is available at the end of the chapter

1. Introduction

The Republic of Bulgaria is situated on the eastern Balkan Peninsula with Black See on the east, Turkey and Greece on the south, Macedonia and Serbia on the west and the Danube River and Romania on the north. Bulgaria spreads on a territory of 111 thousand sq. km. with 6 NUTS2 administrative regions (Figure 1): North West Region (NWR) with 5 major towns (Vidin, Vratza, Lovech, Montana and Pleven); North Central Region (NCR) with 5 major towns (Veliko Tarnovo, Gabrovo, Razgrad, Ruse and Silistra); North East Region (NER) with 4 major towns (Varna, Dobrich, Targoviste and Shumen); South East Region (SER) with 4 major towns (Burgas, Sliven, Jambol and Stara Zagora); South Central Region (SCR) with 5 major towns (Kardgali, Pazardgik, Plovdiv, Smoljan and Haskovo) and South West Region (SWR) with 5 administrative centres (Blagoevgrad, Kjustendil, Pernik, Sofia-town and Sofia-district).

The basic climatic characteristics of Bulgaria are: temperately continental and subtropical (in the south) climate with four seasons and high variation in the temperature, precipitation and humidity among the country regions. Mountains cover 60% of the country territory as the rivers are short, low-water and unevenly allocated through the country.

Bulgaria has a moderate continental climate, with the Black Sea influencing the weather conditions in the coastal area (30-35 km along the sea shore, NER, SER). The average temperatures in the country vary between years and among the regions. The physic and geographical conditions in Bulgaria are very favourable for the development of agriculture, but there are substantial differences in climatic conditions among regions.

Due to the continental climate the summer in Bulgaria is hot and the winter – dry and cold. There are dry spells in summers in July and August. The amount of precipitation is generally low with variations among the regions. The lowest precipitation is observed in

SWR. West and northeast winds dominate and in the winter there are strong north and northeast winds. Because of the strong and steady winds the snow cover is often blown away from the flat areas and the soil gets frozen.

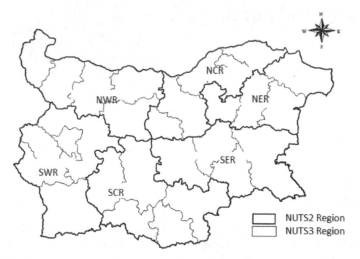

Figure 1. Administrative structure of Bulgaria

The irrigated areas in the country are about 8% of the cultivated land. Concerning water use for irrigation it should be mentioned that it is accounted about only 3% of the total water used in the country which make crop production highly dependent on climatic conditions.

In conclusion, Bulgarian agricultural production is rain-fed, crucially depends on precipitation regimes and climate changes are a very important factor for agricultural development of Bulgaria.

Agriculture plays a crucial role for the economy in Bulgaria. About 5% of GDP and 17.2% of total export of the country in 2009 were provided by agriculture. The sector is the major activity in the rural regions of the country ensuring employment and development of these regions. Over the last years crop production reached to 70% of GAO thus making agriculture highly dependent of crop output. Crop pattern and crop productivity are affected substantially by the regional climate as weather and climate factors are regarded as key factors for the crop output. Having in mind that approximately 49% of the country's territory is agricultural land and that more than 60% of it is the arable land, it is obvious that crop production plays an important role in Bulgarian economy and is crucial for the development particularly of the rural regions.

Agricultural productivity and in particular crop productivity plays an important role for the development of Bulgarian agriculture. Although technological advances such as improved seeds, cultivation methods, fertilization etc. play a major role, weather and climate can still be regarded as key factors for agricultural productivity [e.g., Anderson and Hazel, 1989;

Alexandrov and Hoogenboom, 2001; Sun et al., 2007]. Alexandrov and Hoogenboom [2000] demonstrated the importance of monthly temperature and precipitation conditions for yields of maize and winter wheat for Bulgaria, but the authors do not cover economic aspects of the observed impact of the mentioned climatic indicators. Some attempts to cover economic aspects of the climatic impact on crop yields are made under the CLAVIER project[1] as the study covers only North-East Region of the country (Mishev Pl., Ivanova N., Mochurova M., Golemanova A.,2009).

Having in mind the importance of the climatic factors for crop yields and crop production for agricultural development in Bulgaria as well as the importance of the agriculture for the national economy the main goal of this study is to evaluate the economic impacts of climate changes on Bulgarian agriculture and through then on the national economy.

2. Some national peculiarity

As mentioned above the climate in Bulgaria is temperately continental and subtropical in the south part of the country with high variation in the temperature, precipitation and humidity among the country regions. These differences in the climatic conditions reflect in different structure of land use, pattern of production and crop yields among the regions. The allocation of territory by regions, some basic climatic indicators and the share of agricultural land by regions are shown in Table 1.

indicators	Bulgaria	NWR	NCR	NER	SER	SCR	SWR
Area, '000 sq. km	111	19,1	14,8	14,4	19,7	22,3	20,7
% of UAA in total territory	49%	59%	58%	65%	50%	31%	42%
% of arable land in UAA	62%	68%	78%	81%	64%	59%	19%
% of grains and sunflower in UAA, 2009	75%	81%	83%	80%	73%	57%	47%
Climate indicators							
Average air temperature, degrees C, 2009	13,1	11,9	12,7	12,8	13,4	12,0	15,7
Annual amount of precipitation, l/m², 2009	1,0	1,4	1,1	1,0	0,9	1,1	0,7
Average humidity	56,2	57,0	57,2	61,5	56,2	51,6	54,1
Share of agriculture in GDP, 2009	4,8%	11,7%	9,2%	7,2%	5,8%	7,5%	1,5%

Source: National Statistical Institute, Statistical Yearbook different years; MAF, Agrostatistics, BANSIK different years

Table 1. Geographic Indicators

The data in the table show that the agricultural land in the country covers nearly half of the country's territory as in some regions reaches 60% - 65% of region's territory (NER, NWR,

---

[1] CLAVIER project (Climate Change and Viability: Impacts on Central and Eastern Europe): http://clavier-eu.org/

NCR). The arable land is 62% of the agricultural land but generally it is unevenly allocated among regions (between 81% in NER and 19% in SWR). The importance of the agricultural sector in for the economic development the regions varies between 1,5% of the regional GDP in SWR and 11,7% in NWR (Table 1). In all regions with exception of SWR the share of agriculture in GDP is higher than the national average which shows that the importance of the sector in these regions is even stronger for the regional development than at national level.

Grains and sunflower seeds have always been   most important crops cultivated in the country as the importance of these crops increases. Over the last decade the share of grains and sunflower seeds in total arable land increased from 74,5% in 2001 to 82,1% in 2009, which practically means that Bulgarian crop production depends highly on 4 crops only. The grains and sunflower seed are also important for the country in respect to the export. These 4 crops provide 30,6 % (in this number wheat 13%, sunflower and oil 13%) of total agricultural trade and are the main export oriented products.

Although the physic and geographical conditions in Bulgaria are very favourable for the development of agriculture, due to the substantial differences in natural and climatic conditions among regions (Table 1) the impact of climate changes would be different by regions. Due to this the regional approach for estimation of climate changes impact on crop yields have been used in the study as the results are aggregated at national level.

3. Case study framework: Brief methodological notes

The study covers three main aspects of the impact of climate on the economy:

1. Evaluation of the impact of climatic changes on crop yields;
2. Estimation of these effects in economic terms for agriculture;
3. Evaluation of the impact of changes in the sector on the overall economy.

In respect to the first aspect different methods have been developed to estimate the climate impact on crop yields. These methods can be grouped into two main groups: dynamic process-based crop models and empirical-statistical approaches [Feenstra et al., 1998; Hansen and Indeje, 2004]. For the study empirical-statistical techniques are applied to design climate-crop models in order to quantify the impacts of climate change on agricultural productivity. There are lots of publications focus on the climate factors impact on crop productivity and the statistical methods for estimating the impact of climate change on crop yields (Cline, W., 2008; Iglesias, A., L. Garrote, S. Quiroga, M. Moneo, 2009; Ciscar, Juan-Carlos, 2009; Alexandrov, V., 2008, etc.). In the studies generally multiple regression models with crop yield as dependent variable have been used. This approach has been used in the current study.

The problem of estimation the economic impact for agriculture of changes in yields caused by the climate changes is not widely considered in the literature. An approach for doing this has been developed under the CLAVIER project and this approach has been used in the study. The approach is based on constant process and no changes in land allocation thus excluding the economic factors impact on performance of the sector.

In respect to the third aspect of the study there are a lot of publication dealing with the evaluation of the impact of changes in a given sector on the economy (Johansen L., 1960; Pyatt & Round, 1985; Hertel T.W, Brockmeier M., Swaminathan P.V., 1997; Bach C.F., Frandsen S.E., Jensen H.G., 2000; Jensen H.G., Frandsen S.E., Bach C.F., 1998; Ivanova N., T. Todorov, A. Zezza, 2000 atc.). The approaches used can be classified into three main groups: input-output analysis, social accounting matrix analysis and General equilibrium analysis as input-output analysis is implicitly involved in the other two groups of analysis. For the purpose of this analysis the input-output models with multiplier analysis have been chosen.

In respect to the first aspect of the study in order to estimate the climatic factors impact on crop yields the crops to be examined should be selected first. The estimation of the climate changes impact on selected crops could be done following two possible approaches:

1. To estimate the impact of major climatic factors directly on average yields at national level for the selected crops;
2. To estimate the impact of major climatic factors on average yields of crops important for the regions and to aggregate the expected effect at national level.

The first approach is suitable to be used in case of no substantial differences in crop yields and climatic indicators among regions while the second approach could be used in case of differences in crop yields and climatic indicators among the regions. The second approach requires an additional analysis of the importance of selected crops at national level to the regions.

As seen from Table 1 the climatic factors differ substantially among the regions even on a yearly basis. Due to this the second approach is more suitable in the case of Bulgaria but in this case an additional analysis of the importance of selected crops by region should be done.. The second approach also requires more precise analysis based on differences in monthly data for temperature, precipitation and relative humidity by regions as well as the regional differences in crop yields of the selected crops. Due to this after selection of crops to be examined at national level the regional differences in yields and climatic indicators have to be analysed and on the basis of the results of this analysis to select the approach to be followed.

Following the selected approach for estimation the climate changes impact on crop yields the multiple regression crop models have to be developed for selected crops at regional level. The models use selected meteorological parameters as predictors and crop yields as the dependent variable.

To estimate the potential impact of climate changes on crop yields scenarios for the climate changes for the period 1951-2050 have to been produced. They are based on the post-processed climate simulations obtained in the VI FP project CLAVIER. For projections of climatic indicators error corrected daily data from highly resolved regional climate simulations (REMO version 5.7). Hemispheric synoptic-climatological studies were realised based on the ERA-40 re-analyses data (for the past) and the ECHAM 5 global climate model's results (for the past and the future as well)[2]. The scenario simulation (2010 – 2050) is

[2] CLAVIER project (Climate Change and Viability: Impacts on Central and Eastern Europe): http://clavier-eu.org/

based on greenhouse gas emission scenarios A1B REMO and B1 LMDZ and can be used to quantify climate change signals by comparing it to the control simulation (1951 – 2000) which is based on observed greenhouse gas concentrations. Based on these scenarios potential impact on crop yields by regions is estimated.

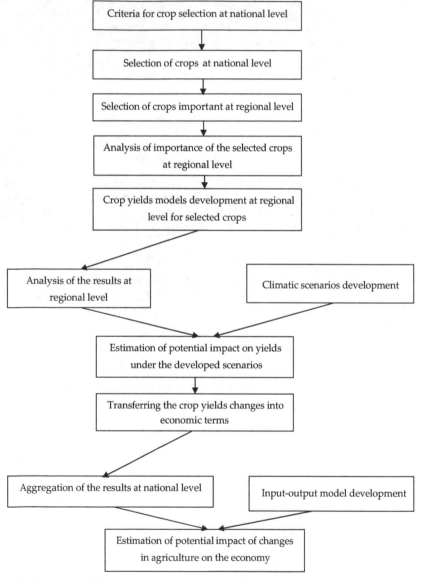

Figure 2. Methodology followed in the study

In respect to the second aspect the estimated changes in yields should be translated into economic terms. As the main economic indicator gross agricultural output (GAO) is used in the study. To avoid price changes impact the constant prices are used for estimation the yields changes in value terms. Additional assumptions used in transferring the climate changes impact on GAO are:

- no changes in land use structure;
- all other crops remain unaffected;
- value of livestock production and other activities in GAO remain constant.

Based on this assumptions changes in GAO by regions are estimated and the results are aggregated at national level and the estimated change in GAO is used as a proxy for economic impact of climate changes on agriculture.

The third aspect of the considered problem requires an input-output model at nationasl level to be developed. The model is based on make and use tables provided by the National Statistical Institute. In order to estimate direct, indirect and spillover impact of the changes in agriculture on the national economy the developed input-output model is shocked as the shock vector is constructed on the basis of change GAO. The impact of changes in agriculture on the national economy is estimated on the basis of multiplier analysis.

The methodology used in the study described above is shown on Figure 2.

4. Selection of crops to be examined

Selection of crops to be examined is based on the following criteria: share of crops in the arable land and the importance of crops in respect to the crop output. Results of the analysis of importance of crops in respect to the two criteria are shown in Table 2. As seen from the table the 4 crops mentioned above use nearly 70% of the arable land in the country and provide half of crop output. The shares of other crops, produced in Bulgaria are relative much lower than the shares of crops shown as in arable land as well as in respect to the crop output. Based on the results of the analysis the selected crops to be examined at national level are: wheat, barley, maize and sunflower.

As shown in the table the importance of the 4 crops analysed at regional level in respect to the land use and crop output is relatively high in all regions but NWR Having in mind that the share of the four crops in crop output and in arable land in SWR is relatively low in comparison with the other regions (Table 2) the SWR is excluded from the regional analysis. The same is valid for barley in NWR. Thus the analysed products by regions are the four selected products for SCR, SER, NCR and NER, and wheat, maize and sunflower for NWR. Thus the selected crops cover at least 50% of the arable land and above 40% of crop output at regional level.

Results of the analysis of crop yields at regional level are shown on Figure 3. As seen from the figure the crop yields differ quite substantially by regions as in cases of wheat, maize and sunflower the difference is quite substantial while in case of barley the yields differences are not so large. The differences in climatic factors by regions are obvious from

Table 1 although the annual data are presented in the table. The detailed analysis of the three selected climatic indicators based on the monthly data shows that the differences by the regions are even higher than on average for the year. The most substantial differences in temperature are observed in winter months when the differences reached to 90% of the country average and are the smallest in summer with deviation from the average accounted to 15%. In respect to the precipitation the most substantial differences are observed in spring and early summer as the deviation reached to 45% of the country average. The deviation in relative humidity is relatively smaller in comparison with the other two climatic indicators but during the summer the differences reached to 25% of the country average. These differences in climatic conditions as well as differences in yields are the reason for selecting the second approach for the study, i.e. analysing the impact of climate changes on yields by regions.

products	Bulgaria	NWR	NCR	NER	SER	SCR	SWR
Share in crop output							
Wheat	20%	21%	24%	23%	24%	11%	6%
Barley	5%	5%	7%	5%	9%	1%	1%
Maize	8%	12%	11%	11%	2%	3%	2%
sunflower	13%	18%	17%	16%	14%	5%	2%
Total	46%	57%	60%	55%	48%	20%	11%
Share in arable land							
Wheat	36,6%	37,7%	39,1%	37,1%	39,9%	29,9%	23,9%
Barley	7,6%	5,7%	9,1%	6,8%	11,5%	4,6%	3,6%
Maize	8,0%	10,6%	12,1%	11,6%	0,8%	3,4%	5,5%
sunflower	16,9%	22,9%	22,3%	22,9%	1,9%	13,3%	7,3%
total	69,1%	76,9%	82,6%	78,4%	54,0%	51,2%	40,4%

Source: NSI, Economic account for agriculture, 2009; MAF Agricultural Statistics Department

Table 2. Share of major crops in crop output and in arable land, 2009

5. Data and metadata used in the analysis

The source of crop yield data, historic meteorological data, prices and the national I-O table and all other economic data is the National Statistical Institute and Agro Statistics Department of MAF. The historic data used in the analysis covers the period 1961 – 2009, as for yields annual data are used, for temperature, precipitation and relative humidity average monthly data are used. For construction of I – O table Make and Use tables for 2005[3] are used.

As already mentioned for projections of climatic indicators error corrected daily data from highly resolved regional climate simulations (REMO version 5.7) are used and the scenarios run covers the period 1910 to 2050 (datasets STAT-CLIMATE-ECA-A1B and B1 LMDZ METEO REGION).

---

[3] Last available Make and Use tables, Source NSI, Revised Make and Use tables for 2005

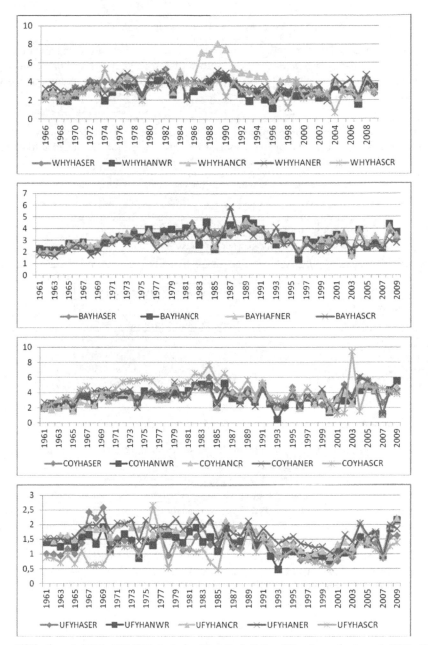

Source: NSI for the period 1961 – 2001; MAF, Agricultural Statistics bulletins, different years for the period 2002 - 2009

Figure 3. Crop yields by regions

6. Analysis of the impact of climate changes on yields

The analysis of the relation between the climatic indicators (temperature, precipitation and humidity) and the crop yields for the crops cultivated in by regions is based on the multiple regression models with yield as a dependent variable. For the purposes of this analysis the following data are used:

- Weighted average yields by regions
- Average monthly data for temperature, humidity and precipitation by regions
- Hindcast simulation data for the scenarios covering the period 1951 to 2050 (data base STAT-CLIMATE-ECA-A1B and and B1 LMDZ METEO REGION).

The mnemonic used in the figures as well as in the regression analysis is shown in Table 3.

	Data from NSI
Average monthly temperature	Tnn
Average monthly temperature changes(i.e. for the first difference of the indicator)	DTnn
Relative Humidity	RHnn
Relative Humidity changes(i.e. for the first difference of the indicator)	DRHnn
Precipitation	Rnn
Precipitation changes(i.e. for the first difference of the indicator)	DRnn
Wheat yield	WHYHA
Barley yield	BAYHA
Maize yield	COYHA
Sunflower yield	UFYHA

Note: nn is used for the month

Table 3. Mnemonics used

Generally there are three groups of factors affecting crops yields: technological development, economic factors and climate factors. In long run the first two groups of factors are associated with the trend while the third group of factors are associated with the deviation from the trend. Since the purpose of this analysis is to evaluate the impact of the third group of factors only the crop yield data are analysed more detailed. From the Figure 2 it is obvious that over the period up to 1990 there is an increasing trend in the yields, followed by a decreasing trend over the period of 90th and then with the stabilization of the economy the trend in yields became again positive for all crops analysed. Because of this in the analyses of the crop yields data either three sub-periods should be considered or a transitional dummy should be used to capture the economic factors impact on yields. For this study the second approach has been chosen.

To be able to exclude the impact of technological and economic factors the three type of trend models for yields are studied: linear trend models, logarithmic trend model and reciprocal trend model for all analysed crops by regions. In all models a transitional dummy is also used to absorb the effect of transition. It has to be mentioned that neither of the trend examined is statistically significant if the effect of transition is not taken into account. Results also show that in all cases analysed trend is not statistically significant even if the transitional dummy is included. Because of this the traditional approach for estimation of the impact of climatic factors on yields based on the two steps procedure of estimation (exclusion of trends first and them estimation of the climate impact on de-trended yields) could not be used. To solve this problem we chose to analyse the impact of climate factors change on the change in yields thus trying to exclude the impact of technological and economic development factors. This practically means that all the data (yields, temperature, relative humidity and precipitation) are transformed and the first differences of the series instead of the series alone are used in the further analysis. Further examination of the transformed yields data includes statistical properties of the data i.e. testing whether the adjusted yield series are stationary or integrated. Both tests (augmented Dickey-Fuller (ADF) tests, and the Phillips-Perron (PP) test[4]) proved that the time series of the first differences of transformed yields are stationary series at 99% confidence level, according to both, ADF tests and PP tests. Having in mind this, there was no need of further adjustments in the yields data or considering the autoregressive process (AR models) in modelling the impact of climate variables on yields.

7. Regression analysis for yields

In order to estimate the climate factors impact on transformed yields the correlation between the changes in yields and changes in climate factors has been checked for all crops and all regions analysed. The analysis of the correlation coefficients shows that in general the estimated coefficients are low (below 0,5). Never mind low correlation on the basis of the results the factors with highest correlation coefficients for any crop has been chosen. To avoid the potential multicollinearity problem that might appear in the transformed crop yields models in case of high correlation among the factors, only one of them has been chosen. The selection of factors is based on the correlation coefficients. Never mind that the number of observations is small more than 4 factors have been chosen to be tested in the regression models. Following this procedure the following factors have been chosen for the crops analysed (Table 4).

As seen from the table the change in climatic factors having impact on the change in yields of a given crop differs among the regions which confirms that the analysis should be done by regions but not at national level.

After testing various functional forms (linear, quadratic, log-linear etc.) and the significance of the variables, linear function has been chosen for modelling the change in crops yields. In the process of testing the regression models for the four crops analysed by regions combinations of the mentioned factors are used as some of them appeared to be statistically insignificant at 95% confidence level and do not improved the explained variation in change in yields or do not comply with theoretical requirements. As a consequence those factors have not been included in the models. The selected models are the ones with highest R

---

[4] Tested with a constant and a linear time trend

square that comply with the regression theory properties. Results for the best fitted models are shown in Table 5.

crop	NWR	NCR	NER	SER	SCR
	factors				
wheat	DT10(-1) DT6 DR5 DRH3 DRH2	DT7 DR3 DR4 DR5 DRH3	DT7 DT10(-1) DR3 DR4 DR5	DT7 DT5 DR3 DR4 DR6 DRH10(-1)	DT12(-1) DT5 DR3 DR7 DRH11(-1)
Barley		DT2 DT5 DR3 DRH2 DRH4	DT2 DT7 DR3 DR5 DRH2 DRH3 DRH10(-1)	DT2 DT7 DR3 DR5 DRH7	DT2 DT3 DR2 DR5 DRH6
Maize	DT5 DT8 DR4 DR6 DRH10	DT6 DR3 DR7 DRH5 DR9	DT5 DT7 DRH3 DR6 DR8 DRH10	DT6 DT11 DRH3 DRH6 DRH10	DT2 DT3 DR5 DR11 DRH10
Sunflower seeds	DT6 DT7 DR5 DR10 DRH7	DT3 DT6 DR5 DR10 DRH8	DT5 DT7 DR3 DR6 DRH5 DRH7	DT5 DT10 DT11 DR5 DRH7	DT8 DT5 DR5 DR6 DRH11

Source: Own calculations

Table 4. Climate factors with significant impact on yields

As seen from the results the explained variation in the changes in yields is relatively reliable. In some cases factors not statistically significant at 95% confidence level have been left in the model since they improve the explained variation (based on adjusted R-squared). The selected models have been tested for stability (QSUM and QSUMSQ tests) and proved to be stable.

As seen from the results the change in climatic factors explains 30% to 50% of the variation of crop yields as the less affected crop is maize (the climate factors explains between 22% and 36% of the variation in yields) and wheat is the most sensitive to the climatic changes (between 36% and 50% of the yield changes are explained by the changes in climatic factor). Results also show that climate changes affect more substantially yields in NER and SER and not so much the other regions.

crops	NER Variables and estimated coef.	NER Stat. signifi-cance (p-value)	NER R square	NCR Variables and estimated coef.	NCR Stat. signifi-cance (p-value)	NCR R square	NWR Variables and estimated coef.	NWR Stat. signifi-cance (p-value)	NWR R square	SER Variables and estimated coef.	SER Stat. signifi-cance (p-value)	SER R square	SCR Variables and estimated coef.	SCR Stat. signifi-cance (p-value)	SCR R square
wheat	0,06	0.0000		0,03	0.8204		0,05	0.609		0,04	0.5786		0,03	0.9503	
	-0,31DT7	0.0040		0,87DR3	0.0000		-0,10DT10(-1)	0.0129		-0,14DT7	0.0001		-0,68DT12(-1)	0.0003	
	-0,40DR4	0.0069	0.469	-0,24DR4	0.0363	0.407	0,03DRH3	0.0025	0.356	-0,23DR4	0.0019	0.503	1,18DR3	0.0344	0.411
	-0,23DR5	0.2200		-0,25DR5	0.0806		-0,26DR5	0.0182		0,02DRH10(-1)	0.0405		-1,70DR7	0.002	
				0,08DT3	0.1880					0,15DR6	0.2183		-0,08DRH11(-1)	0.1138	
barley	0,02	0.7863		0,02	0.8266					0,03	0.7172		0,02	0.8302	
	-0,10DT7	0.0411		0,70DR3	0.0000					-0,07DT2	0.0017		-0,05DT2	0.0434	
	0,32DR3	0.0296	0.379	-0,06DT2	0.0283	0.432				-0,24DT7	0.0004	0.453	0,08DT3	0.0385	0.279
	-0,19DR5	0.0837		-0,03DRH4	0.0337					0,21DR3	0.0891		0,13DR2	0.0852	
	0,02DRH2	0.0073								-0,04DRH7	0.0029				
maize	0,04	0.8458		0,07	0.6698		0,05	0.8008		0,05	0.722		0,04	0.8652	
	0,60DR6	0.0358		-0,29DT6	0.0064		0,77DR4	0.0039		0,16DT11	0.0055		0,27DT3	0.0201	
	0,43DR8	0.0207	0.294	0,67DR3	0.0107	0.335	-0,40DR6	0.0632	0.217	-0,29DT6	0.0351	0.365	-0,35DR11	0.0396	0.292
	-			0,27DR7	0.0717		0,09DT8	0.2104		0,07DRH3	0.0001		0,07DRH10	0.0168	
	0,07DRH10	0.0031		-0,05DRH5	0.0927					-0,06DRH6	0.0408		0,12DT2	0.0752	
sunflower	0,01	0.8395		0,01	0.8059		0,02	0.7881		0,01	0.8004		0,01	0.8121	
	-0,04DT7	0.1365		-0,05DT3	0.0049		-0,08DT6	0.0847		0,06DT11	0.0024		0,26DR6	0.0116	
	0,25DR3	0.0018	0.350	-0,12DR5	0.005	0.324	0,13DT7	0.0222	0.383	0,03DT10	0.0939	0.389	-0,01DRH11	0.0219	0.302
	0,16DR6	0.0299		0,01DRH8	0.0711		0,15DR10	0.0603		-0,19DR5	0.0083		0,03DT8	0.2275	
	-0,02DRH5	0.0175		-0,06DR10	0.0754		0,04DRH7	0.0078		0,02DRH7	0.0022				

Table 5. Estimated model results

Having in mind that 41% of wheat, 47% of barley and 43% of sunflower and 30% of maize are produced in the east part of the country (NER and SER) the obtained results stress on the fact that grains and sunflower production in the country would vary quite substantially die to the changes in climate. Furthermore, taking into account that crop production is two third of GAO, this would mean that strong variation in GAO could be expected, i.e. variations in GAO observed by now would continue.

7.1. Expected climate change

Climatic changes scenarios used in the study for projections are developed under the VI FP CLAVIER project. Climate scenarios describe the mean conditions over a longer period and hence, comparing the mean conditions in future periods (e.g., 2021 to 2050) to those in a reference period (e.g., 1961 to 1990) allows deducing the influence of climate change.

The following two scenarios and climate models are applied in the study:

1.   A1B - REMO
2.   B1 - LMDZ

These scenarios are based on the different CO_2 emissions in the future (the so called A1B and B1). The Emission Scenarios have been developed by the Intergovernmental Panel of Climate Change (IPCC).

The A1B storyline and scenario describes a future world of very rapid economic growth, global population that peaks in mid-century and declines thereafter, and the rapid introduction of new and more efficient technologies (reference to Clavier WP). Major underlying themes are convergence among regions, capacity building and increased cultural and social interactions, with a substantial reduction in regional differences in per capita income. A1B scenario is a balance across all energy sources: fossil intensive and non-fossil energy sources.

Scenario/ model	Yearly mean of the mean daily temperature /°C/	Yearly mean of the daily precipitation amount /mm/
A1B - REMO	+1.0	0.0
B1 - LMDZ	+1.8	-0.5

Source: own calculations based on CLAVIER database

Table 6. Differences in the climate parameters in the future 2021-2030 as compared to the past climate 1961-1990 in Bulgaria

The B1 storyline and scenario describes a convergent world with the same global population, that peaks in mid-century and declines thereafter, as in the A1 storyline, but with rapid change in economic structures toward a service and information economy, with reductions in material intensity and the introduction of clean and resource efficient technologies (reference to Clavier WP). The emphasis is on global solutions to economic, social and environmental sustainability, including improved equity, but without additional climate initiatives. B1 – LMDZ scenario do not provide relative humidity data, due to which humidity projections obtained from scenario A1B – REMO are used in the second scenario.

The changes expected in the future 2021-2030 as compared to the past climate 1961-1990 over the territory of Bulgaria under the two scenarios are presented in Table 6.

The changes expected in the future 2021-2030 according to scenario A1B - REMO as compared to the past climate 1961-1990 over the territory of Bulgaria under the first scenario are presented in Figure 3.

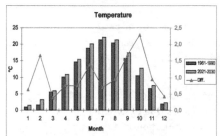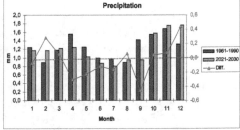

Source: own calculation based on data base STAT-CLIMATE-ECA-A1B

Figure 4. Expected climate changes over the territory of Bulgaria, A1B REMO scenario

While it is expected the mean monthly temperature to increase by 1 °C on average, the difference between the past and future climate reaches about +2 °C in autumn (September and October) and in February. There is almost no change in the mean yearly precipitation (-0.04 mm). However, a decrease in the mean monthly precipitation the can be observed during most months, especially in September, as an increase could be expected in winter.

Under scenario B1 LMDZ a moderate increase in the temperature and a decrease in the precipitation in Bulgaria is expected as compared to the A1B - REMO scenario. The most noticeable raise in the mean monthly values of the daily mean temperature is expected in spring (+2.2 °C) and in summer (+3.1 °C). A decrease in the average precipitation is projected, especially in June, July, September and October.

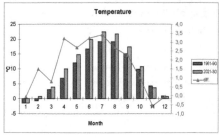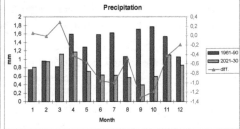

Source: own calculation based on data base B1 LMDZ METEO REGION

Figure 5. Expected climate changes over the territory of Bulgaria, B1 LMDZ scenario

8. Crop yields projections by regions

The changes in climatic factors used in the models (temperature, precipitation and relative humidity) for the historic period as well as Hindcast simulation data for the period 1950 – 2009 were tested for statistical equity. The equity tests for mean, median and variance have been performed for the month temperature data, precipitation data and relative humidity data used in the crop models. Results show that the null hypothesis is not rejected in all cases analysed and therefore there are no statistically significant differences at 95% confidence level for all climatic variables used in model and no adjustments in the data are needed.

Projections of crop yields are based on projections of changes in crop yields due to the changes in climatic factors and observed yields in 2009 (the last year in the historic period). Projected yields under scenario A1B - REMO and scenario B1 - LMDZ for the four crops analysed by regions are shown in Figure 5.

As seen from the figure the expected yields by products differ quite substantially by regions under the both scenarios. The expected variation in yields due to the changes in climate is substantial for all products and all regions. The most important is the variation in maize yields in SCR (scenario B1 – LMDZ) reaching above 85% and wheat yields (scenario A1B REMO and B1 – LMDZ) in SCR estimated at 79% and 88% followed by maize yields (scenario B1 – LMDZ) in all regions varing among the regions from 55% to 67%. The variation of yields in case of barley and sunflower is relatively smaller under both scenarios compared to maize and wheat varying among the regions between25% to 43% for barley and from 18% to 37% for sunflower .

Further analysis of the results obtained shows that the potential impact of climate factors on yields for the period 2010 – 2030 is generally positive for wheat yields in all regions under both scenarios but high deviation is expected through the years. Potential impact of climate changes on barley yields is also positive under both scenarios in all regions with exception of barley yields in SCR under scenario B1 – LMDZ where slight reduction in yields is expected. Generally the impact of climate changed on maize yields is positive much lower than the impact on wheat. For this crop slightly negative impact could be expected in SRC under scenario A1B – REMO and in NER under scenario B1 – LMDZ. At national level the impact of climate changes of sunflower yields is also positive under both scenarios, but at regional level slightly negative impact could be expected in NER and NCR under scenario B1 – LMDZ.

In estimation of the climate impact on yields towards 2025 two approaches are possible: to use the yields projected for year 2025, or to use a simple 3-year, 5-year or 10-year averages. Having in mind that climate factors projections are long run projections and are not so precise on a year by year base, a 10-year averages (from 2020 to 2029) are used as a proxy for change in yields in 2025 in the two scenarios considered.

Since change in yields in 2025 is estimated only on the basis of climate changes the change could be directly compared with yields in the last observed. Yields in 2025 are obtained on the basis of changes in yields estimated and observed yields in 2009. They are shown in Table 7. As seen from the table in 2025 the impact of climate on wheat is positive under the both scenarios showing on increase in yields between 29% (NCR under scenario B1 – LMDZ) and 86% in SCR under the same scenario. The same is valid for barley with exception of yield in SER under scenario B1 - LMDZ where a reduction in yields amounting to 6% is expected. The sunflower yields in 2025 are higher than in 2009 for all regions with exception of NER and NCR under the scenario B1 – LMDZ with reduction in yields by 1% and 2% respectively. Potential impact of climate on maize yield in 2025 is positive with exception of SCR under scenario A1B – REMO with reduction estimated at 3% and in NER under scenario B1 –LMDZ with reduction in yields by 5%.

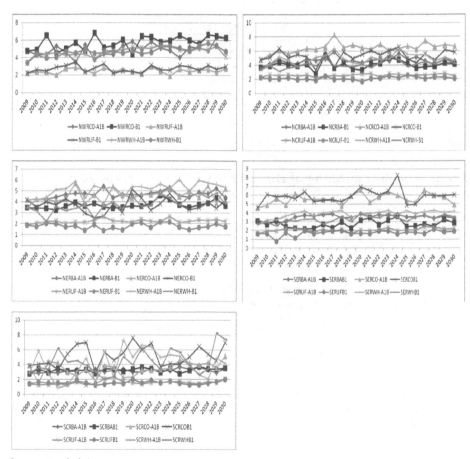

Source: own calculations

Figure 6. Projected crop yields for analysed products by regions

Yields in 2009					
	NER	NCR	NWR	SER	SCR
Wheat	3,45	3,38	3,4	2,7	2,86
Barley	3,46	3,67		3,08	2,79
Maize	4,15	4,76	5,51	4,53	3,85
Sunflower	1,85	2,21	2,15	1,62	1,39
Scenario A1B - REMO					
	NER	NCR	NWR	SER	SCR
Wheat	5,284	4,909	4,875	3,909	4,768
Barley	4,775	4,714	0,000	3,599	3,279
Maize	4,727	6,589	6,048	5,782	3,743
Sunflower	2,277	2,606	2,401	2,131	1,599
Scenario B1 - LMDZ					
	NER	NCR	NWR	SER	SCR
Wheat	4,268	4,348	5,046	3,538	5,312
Barley	3,870	4,113	0,000	2,910	3,326
Maize	3,937	5,544	6,786	6,168	5,507
Sunflower	1,835	2,173	2,691	1,790	1,610

Source: Own calculations

Table 7. Projected yields

9. The economic meaning of climate changes at sectoral and economy level

As mentioned above the economic impact of climate changes is analysed at two levels: impact on agricultural sector performance and impact on the Bulgarian economy. Following the selected methodology the physical changes in yields are transformed into value terms based on the following assumptions:

- the area cultivated is not affected by the relative change in yields
- to exclude price impact on agricultural output constant prices are used
- all other crops remain unaffected
- value of livestock production and other activities in GAO remain constant

Based on these assumptions the impact of climate on agriculture in value terms is shown in Table 8.

Table 8 shows that the estimated economic impact of climate changes on yields by regions is positive under both scenarios but differed substantially by regions. The expected changes in agricultural development under scenario A1B REMO are more favourable for NCR, NER and SER while the changes under scenario B1 MLDZ are more favourable in NCR and NWR.

As seen from Table 8 at sectoral level the expected changes in crop output and GAO due to the climate changes are positive under both scenarios as changes in climatic factors under scenario A1B REMO are more favourable to the sectorat national level then those under scenario B1 MLDZ. The estimated impact of climate changes toward 2025 under scenario A1B REMO is increase by nearly 15% of the total crop output and increase in GAO by 9,5%. The increase in total output under scenario B1 LMDZ is estimated at 10% and in GAO at 6,6%.

Change in	NER	NCR	NWR	SER	SCR	National level
Change in value of crops analysed (million leva)	162,19	153,15	116,66	97,65	55,95	585,6
change in total crop output	19%	19%	16%	15%	10%	14,9%
change in GAO	11%	11%	8%	7%	5%	9,5%
Change in value of crops analysed (million leva)	47,46	71,07	159,3	53,60	77,01	408,5
change in total crop output	6%	9%	22%	8%	14%	10,4%
change in GAO	3%	5%	10%	4%	7%	6,6%

Table 8. Economic impact of climate changes

Following the chosen methodology in order to find the effect of climate changes on the economy, obtained results for agriculture are incorporated in the input-output (I-O) model by adjusting the vector of agricultural sector. For this purpose I-O model with 20 sectors has been constructed as agriculture, forestry and food industry are considered separately, while other sectors are aggregated. Based on the constructed I-O model, gross output multipliers (type II B), income multipliers as well as employment multipliers are estimated following the commonly used methodology of multiplier analysis. This allows direct as well as indirect and induced effects caused by the change in agricultural output due to the changes in climate to be taken into account by simulating a shock in final demand. The changes in final demand are based on estimated impact of change in climate factors on GAO. In addition the multiplier analysis is used for analysis of the importance of the economic sectors for generating growth in the national economy.

Traditionally, the impact analysis within input-output models is done with the use of the backward linkages proposed by Rasmussen (1956) and Hirschman (1958) and forward linkages proposed by Augustinovics (1970). These linkages show the size of structural interdependence in an economy as well as the degree in which the enlargement of a sector can contribute directly or indirectly in the enlargement of other sectors in the model. On the basis of I-O table for year 2005, both backward and forward linkages for output, value added, income and employment for the 20 sectors are calculated (Table 9).

Table 9. Estimated Multipliers and its rang

	Chenery & Watanabe		Rasmussen & Hirschman								Elastisity				Augustinovich FL			
		rank	OBL	rank	VABL	rank	IBL	rank	EBL	rank	BOE	rank	FOE	rank	OFL	rank	VAFL	rank
Agriculture	0,517	8	1,558	7	0,504	13	1,586	9	1,750	9	0,064	8	0,003	7	1,925	6	0,558	14
Forestry	0,455	9	1,426	9	0,433	16	1,572	13	1,578	14	0,003	20	0,001	20	1,634	10	0,397	18
Mining and quarrying	0,123	20	0,745	20	0,185	20	1,454	15	1,783	8	0,004	17	0,001	17	1,837	7	0,334	19
Food and beverages	0,419	13	1,338	14	0,645	7	1,574	12	1,861	6	0,074	6	0,003	6	1,363	12	0,609	9
Tobacco industry	0,216	19	0,874	19	0,207	19	1,506	14	1,564	15	0,029	14	0,001	14	1,123	17	0,301	20
Textile; leather products	0,438	12	1,364	12	0,469	14	1,584	11	1,964	5	0,055	9	0,003	9	1,406	11	0,560	12
Chemic industry	0,633	2	1,650	5	0,353	18	2,707	1	2,619	1	0,096	2	0,003	2	2,534	1	0,996	2
Machinery & equipment	0,396	16	1,295	16	0,806	4	1,715	5	2,271	2	0,089	3	0,003	3	1,954	4	0,514	15
Furniture & secondary raw materials	0,585	5	1,729	4	0,452	15	1,452	16	1,528	16	0,017	16	0,001	16	1,336	14	0,404	17
Electrical energy, gas, water	0,581	6	1,492	8	0,829	3	1,656	7	1,617	11	0,036	12	0,002	12	1,945	5	0,807	4
Water supply services	0,396	15	1,326	15	0,630	8	1,252	19	1,492	18	0,003	18	0,001	18	1,098	18	0,681	7
Construction	0,598	3	1,784	2	1,050	1	2,030	3	2,036	4	0,081	4	0,004	4	2,433	2	0,919	3
Transport, hotels restaurants	0,595	4	1,760	3	0,950	2	2,488	2	1,842	7	0,168	1	0,008	1	1,956	3	1,481	1
Financial intermediation	0,413	14	1,349	13	0,647	6	1,595	8	1,604	13	0,077	5	0,003	5	1,792	8	0,681	6
Public administration	0,449	10	1,397	11	0,610	9	1,377	17	1,498	17	0,073	7	0,003	7	1,179	16	0,701	5
Education	0,235	18	0,987	18	0,695	5	1,162	20	1,236	20	0,037	11	0,002	11	1,022	19	0,566	11
Health services	0,446	11	1,422	10	0,363	17	1,585	10	1,400	19	0,042	10	0,002	10	1,019	20	0,496	16
Other governmental services	0,675	1	1,934	1	0,598	11	1,898	4	2,119	3	0,027	15	0,001	15	1,347	13	0,633	8
Other services	0,298	17	1,122	17	0,596	12	1,333	18	1,616	12	0,003	19	0,001	19	1,330	15	0,560	13
Trade	0,560	7	1,622	6	0,562	6	1,674	12	1,711	10	0,031	13	0,001	13	1,741	9	0,584	10

According to the estimated output backward and forward linkage coefficient (OBL & OFL) the forward linkages are generally higher than the backward linkages. The exceptions are: "Furniture and secondary row materials", "Water supply services", "Public administration", "Health services" and "Other governmental services". It is seen from the table that induced impact by the sectors is much higher than the direct impact of a change in the sector. The results also show that in total, the average of the forward linkages is higher than the total of the backward linkages (1.59 vs. 1.41).

"Other governmental services", "Construction" and "Transport, hotels & restaurants" are the sectors with the highest backward linkages in respect to the output. This implies that decreases in demand in the above three sectors, compared with all other sectors, may result in the greatest losses to the national economy. Contrary, increases in investment, export or consumption in these sectors may have the biggest potential power to augment the economy by requiring large quantities of goods and services from other sector. Since "Construction" and "Transport, hotels & restaurants" are among the most important "buyers" of agricultural inputs, potential positive climate change effect could boost the general economic development. At the same time "Agriculture" takes the 7th place (backward) and 6th place (forward) which means that the impact of changes in the sector alone will not cause strong changes in output of the economy.

Since the impact of a change in a given sector on the economy depends not only on the multiplier effect but also on the share of the sector in national economy the weighted average of both linkages were calculated (BOE, FOE). The weighs are calculated on the basis the share of each sector's input/output out of total input/output. Agriculture has the rank 8 in case of backward elasticity and rank 7 in case of forward elasticity with means that the there are sectors leading to much higher impact on the economy than agriculture and even strong impact of climate change on the sector will not cause significant impact on the economy

Looking at estimated value added backward and forward linkage coefficient (VABL & VAFL) the conclusion is again that the forward linkages are generally higher than the backward linkages, but with more exceptions than in case of output. Concerning the backward value added linkages, the first three places are taken from: "Construction" (1,05), "Transport, hotels & restaurants" (0,95) and "Electrical energy, gas, water" (0,83). As a result any external impact on the economy concerning these three sectors would cause the highest changes in value added of the economy of the country. Agriculture takes 13th (backward) and 14th (forward) places in respect to the generation of value added meaning that agriculture is not important sector in respect to the value added. But since "Agriculture" is again tightly connected with "Transport, hotels & restaurants", the impact of climate changes on agriculture might appear in the economy through this sector.

"Transport, hotels & restaurants" (2,49) and "Construction" (2,03) are again one of the most important sectors in the economy in respect to the income generation. Agriculture takes 9th place. Because of the low wages in the sector climate changes impact on the total regional economy as a whole will not be that crucial. However, in terms of social stability and source of income for the poorer parts of the population agriculture could be influential.

Regarding employment generation "Chemic industry", "Machinery and equipment" and "Other governmental services" are having the highest potential. Agriculture again is ranked at the middle that means its impact on job creation is not important in the economy but having in mind that the agriculture is a major sector in the rural areas, even not so strong impact on the national employment is important for the employment in the rural areas.

Multiplier analysis in respect to the output, value added, income and employment leads to a conclusion that the most important sectors having crucial impact on the Bulgarian economy are "Construction", "Transport, hotels and restaurants", "Chemic industry" and "Machinery and equipment". Agriculture alone does not have such a strong impact on the national economy but as mentioned above since "Construction" and "Transport, hotels & restaurants" are among the most important "buyers" of agricultural inputs, the impact of changes in the sector would be transferred to the economy via these sectors also.

9.1. Climate scenarios simulation results

To estimate the impact of the climate changes on the national economy, the simulated yields impact in value terms to GAO under both scenarios considered is incorporated into the national I – O model by adjusting the vector of agricultural sector. The simulated changes caused by the change in GAO in respect to the output, income and employment under the considered scenarios are shown Table 10. The expected magnitude of the impact of changes in GAO on the economy output is modest. The total output is expected to increase by 1% - 1,5% as the effect from scenario A1B REMO is higher than under the scenario B1 MLDZ. As seen from the table the indirect and induced impact of climate changes on agricultural output is much higher than the direct impact only (15,1% against 9,5% and 8,6% against 6,6% respectively). Results also show that in both cases the expected changes in all other sectors are less than 1%, as the highest impact is expected for sectors "Food and beverages", "Transport, hotels and restaurants" as well as "Construction". As could be expected due to the insignificant change in the economy results show no changes in the structure of the economy under scenario B1 MLDZ and an increase in the share of agriculture by 1% at the account of industry under scenario A1B REMO.

Results also show model impact on the compensation of employees due to the climate changes (Table 10). The overall changes in incomes are around 1% - 2%, as again the expected changes under scenario A1B REMO are higher. It should also be mentioned that the expected increase in income is slightly higher than the expected increase in output under both scenarios. As in the case of output the induced impact in income is higher than the direct impact on income only. As could be expected the highest increase in income is observed in agriculture, followed by the increase in "Food and beverages", "Transport, hotels and restaurants" and "Construction" sectors.

Practically the same changes are observed in respect to the employment but it should be mentioned that increase in employment in the economy is even smaller that the increase in output (around 1% under both scenarios). Increase in labour above 1% except in agriculture could be expected in "Food and beverages" and "Transport, hotels and restaurants" sectors under both scenarios and "Trade" sector under scenario A1B REMO.

	change in output		change in income		change in mployment	
	A1B REMO	B1 MLDZ	A1B REMO	B1 MLDZ	A1B REMO	B1 MLDZ
Agriculture	15,1%	8,6%	14,1%	7,9%	10,6%	7,3%
Forestry	0,1%	0,1%	0,2%	0,1%	0,2%	0,2%
Mining and quarrying	0,0%	0,0%	0,0%	0,0%	0,0%	0,0%
Food and beverages	0,8%	0,6%	2,6%	1,8%	1,7%	1,2%
Tobacco industry	0,1%	0,1%	0,1%	0,1%	0,1%	0,1%
Textile; leather products	0,1%	0,1%	0,2%	0,1%	0,2%	0,1%
Chemic industry	0,1%	0,1%	0,1%	0,1%	0,0%	0,0%
Machinery & equipment	0,0%	0,0%	0,0%	0,0%	0,0%	0,0%
Furniture & secondary raw materials	0,0%	0,0%	0,0%	0,0%	0,0%	0,0%
Electrical energy, gas, water	0,0%	0,0%	0,0%	0,0%	0,0%	0,0%
Water supply services	0,0%	0,0%	0,0%	0,0%	0,0%	0,0%
Construction	0,3%	0,2%	0,3%	0,2%	0,4%	0,3%
Transport, hotels restaurants	0,4%	0,3%	0,5%	0,4%	1,3%	1,1%
Financial intermediation	0,1%	0,1%	0,3%	0,2%	0,2%	0,2%
Public administration	0,3%	0,2%	0,4%	0,3%	0,4%	0,3%
Education	0,5%	0,4%	0,6%	0,4%	0,6%	0,4%
Health services	0,6%	0,4%	0,7%	0,5%	0,5%	0,3%
Other governmental services	0,1%	0,0%	0,1%	0,1%	0,1%	0,1%
Other services	0,0%	0,0%	0,0%	0,0%	0,0%	0,0%
Trade	0,7%	0,5%	1,0%	0,7%	1,1%	0,8%
For the economy	1,4%	1,0%	2,1%	1,5%	1,1%	0,8%

Source: Own calculations

Table 10. Impact of changes in GAO on the national economy

Considering the very insignificant impact on the Bulgarian economy it should be stressed that no other effect is taken into account except impact of climate changes on production of the 4 major for the Bulgarian agriculture crops.

10. Conclusion

This chapter tries to quantify the effects of the climate changes at two levels: sectoral level(on Agriculture) and national level (on the economy of Bulgarian) using and Input-output methodology. Additionally, some comparative analysis about the magnitude and distribution effects of the two climate scenarios was made. In this respect, the following conclusions can be derived:

- The analysed two scenarios can bring a modest contribution to the overall output increase of the national economy.
- Scenario A1B REMO provides a benchmark of the potential maximum impact of the analysed case study. If this climate situation is accomplished and crops by regions reach relevant yields, the total output of the region would be increased by 1,4%.
- Sectors with highest potential to generate output, value added, incomes and employment are: "Construction", "Transport, hotels & restaurants", "Chemic industry" and "Machinery and equipment". They may be affected by the climate changes in the agricultural sector through their linkages with the latter. This is especially important for "Transport, hotels & restaurants".
- The favourable climate effects, however, should be regarded with certain caution. There are several factors that could worsen or even completely change the optimistic view from the climate scenarios. These factors range from technological ones to global ones (financial crises, food security, trade issues). The abovementioned factors could significantly deteriorate favourable results.
- Limitations of the undertaken research have to be acknowledged, as well. First of all, it has to be taken into consideration that climate changes represent only one dimension of the potential future impacts on the national economy. From one side, even though the regions are well specialised in agricultural activities, potential shortages of agricultural goods might be solved by importing goods in order to reach market equilibrium. From the other, if the agricultural production highly exceed due to the climate change it is unlikely that it could bring significant incomes to the agricultural producers due to increased supply. When it comes to analysis of economic impact, another important issue that is not tackled in the current analysis should be borne in mind. This is the behaviour of the agricultural producers after applying the instruments of the Common agricultural policy, which might significantly guide their decision in direction of increasing or decreasing the agricultural production. Secondly, limitations of the adopted I-O methodology should be considered: no substitution among factors of production, no change in technique, constant import coefficients. However, provided that I-O table is estimated accurately, theoretically implausible assumptions of the model are in some respect overshadowed by its empirical realism and simplicity.

Author details

Nedka Ivanova and Plamen Mishev
University of National and World Economy, Sofia, Bulgaria

11. References

Alexandrov, V., 2008, Climate Change, Vulnerability and Adaptation in Agriculture: the Situation in Bulgaria, Presentation, Adagio project

Alexandrov, V.A, G. Hoogenboom, 2001: Climate variation and crop production in Georgia, USA, during the twentieth century, Climate Research, Vol. 17, pp. 33-43.

Alexandrov, V.A, G. Hoogenboom, 2000: The impact of climate variability and change on crop yield in Bulgaria, Agricultural and Forest Meteorology, Vol. 104, pp. 315-327.

Anderson, J.R, Hazell, P, 1989: Variability in Grain Yields; Implications for Agricultural Research and Policy in Developing Countries, John Hopkins University Press, Baltimore and London, 416 pp.

Bach C.F., Frandsen S.E., Jensen H.G. 2000. Agricultural and Economy-Wide Effects of European Enlargement: Modelling the Common Agricultural Policy. *Journal of Agricultural Economics*, 51, 2: 162-180.

Binswanger H.P. & Deininger K. 1997. Explaining Agricultural and Agrarian Policies in Developing Countries, *Journal of Economic Literature*, Vol: 35, pp: 1958-2005.

Haggblade S., Hammer J., Hazell P., 1991, Modelling Agricultural growth multipliers, American Journal of Agricultural Economics, 73 (2): 361 – 374

Ciscar, Juan-Carlos (ed.), 2009, Climate change impacts in Europe, Final report of the PESETA research project, Joint Research Centre, European Commission, p. 103, 109

Clavier project (Climate Change and Viability: Impacts on Central and Eastern Europe): http://clavier-eu.org

Clavier project deliverables, WP2 and WP4

Cline, W., 2008, Global warming and agriculture, Finance & Development, issue March, p. 23-27

Feenstra, J.F., I. Burton, J.B. Smith, R.S.J. Tol, 1998: Handbook on Methods for Climate Change Impact Assessment and Adaptation Strategies, United Nations Environment Programme, Institute for Environmental Studies, Version 2.0, pp. 464.

Hansen, J.W., A. Challinor, A. Ines, T. Wheeler, V. Moron, 2006: Translating climate forecasts into agricultural terms: advances and challenges, Climate Research, Vol. 33, pp. 27-41.

Hansen, J.W., M. Indeje, 2004: Linking dynamic seasonal climate forecast with crop simulation for maize yield prediction in semi-arid Kenya, Agricultural and Forest Meteorology, Vol. 125, pp. 143-157.

Hertel T.W., Brockmeier M., Swaminathan P.V. 1997. Sectoral and economy-wide analysis of integrating Central and Eastern European countries into the EU: Implications of alternative strategies. European Review of Agricultural Economics, 27: 359-386.

Iglesias, A., L. Garrote, S. Quiroga, M. Moneo, 2009, Impacts of climate change in agriculture in Europe. PESETA-Agriculture Study, JRC-IPTS, European Communities, Luxemburg, p. 32

Ivanova N., T. Todorov, A. Zezza, (2000), "The role of agricultural sector in the transition to a market economy: Bulgarian case study", в *Perspectives on Agriculture in Transition: Analytical Issues, Modeling Approaches, and Case Study Results*, (editors Witold-Roger Poganietz, Alberto Zezza etc), Wissenschaftsverlag Vauk Kiel KG, ISBN 3-8175-0323-7, c. 78-139

Jensen H.G., Frandsen S.E., Bach C.F. 1998. Agricultural and economy-wide effects of European enlargement: Modelling the Common agricultural policy. SJFI - working paper No. 11/1998. Kopenhagen, SJFI: p. 40.

Johansen L. 1960. A multi-sectoral study of economic growth. North-Holland, Amsterdam.

Ministry of Regional Development and Public Works, 2005, Northeast Planning Region –
    Regional Development Plan 2007 – 2013, Sofia
National Statistical Institute, 2008, Regions, Districts and Municipalities in Bulgaria 2006,
    Sofia
Pyatt & Round, 1985, "Social Accounting Matrices: A Basis for Planning", The World Bank.
Republic of Bulgaria, 2007a, National Strategy Plan for Rural Development 2007 – 2013
Republic of Bulgaria, 2007b, Rural Development Programme 2007-2013
Sauberei, W., P. Royston, H. Binder, 2007: Selection of important variables and
    determination of functional form for continuous predictors in multivariable model
    building, Statist. Med., Vol. 26, pp: 5512-5528.
Sun L., H. Li, M.N. Ward, D.F. Moncunill, 2007: Climate Variability and Corn Yields in
    Semiarid Ceará, Brazil, Journal of Applied Meteorology an Climatology, Vol. 46, pp.
    226-240.

Chapter 10

Climate Change and Grape Wine Quality: A GIS Approach to Analysing New Zealand Wine Regions

Subana Shanmuganathan, Ajit Narayanan and Philip Sallis

Additional information is available at the end of the chapter

1. Introduction

The influences of seasonal climate variability on the phenological dynamics of certain terrestrial communities observed mostly since the mid-20th century are seen as leading to unprecedented consequences (Richard, et al., 2009). The potential impacts of the phenomenon on the phenological development and in turn on the species composition of certain specific plant, insect, aquatic, bird and animal communities evolved in parallel over millions of years to form the existing *"make-up"* of what is referred to as the *"biodiversity"* or *"endemic species"* of these natural habitats, are depicted as significant (Peñuelas and Estiarte, 2010). Scientific research results have revealed that the recent rapid climate change effects on these systems, more specifically during the last few decades, have resulted in presently being seen *"temporal mismatch in interacting species"*. Such ecological observations are even described as early *vital signs* of imminent *"regime shifts"* in the current base climate of these regions or latitudes (Schweiger, Settele, Kudrna, & Klotz, 2008: Saino, et al., 2009). On the other hand, climatologists portray the major cause for such rapid *"climate regime shifts"* and the consequent impacts on the survival of so called co-evolved species, as *anthropogenic* (Anderson, Kelly, Ladley, Molloy, & Terry, 2011). For this reason, research relating to climate change impacts on vegetation spread over landscapes, phenological development and population dynamics of susceptible communities, in some cases even with potential threat for total extinction of *"endangered species"* under future climate change, has in recent years gained enormous momentum. In fact, this unprecedented attention has also drawn greater scrutiny and controversies at never seen before proportions in a way hindering any form of formal research on the phenomenon (Shanmuganathan & Sallis, 2010).

Interestingly, many recent studies on climate change indicate *"major shifts"* in grapevine phenology within the next few decades. The potential impacts of climate change on grape berry ripening and in turn on grape crop harvests are predicted to ultimately result in beneficial in some wine regions and detrimental outcomes in others (Jones & Davis, 2000). The implications of such dramatic shifts are expected to affect the production of wine, its taste and distinction pertaining to its style especially, in the regions that are presently well-known world over for their vintages and wine labels, grapevine growers and winemakers will be challenged in continuing with their premium quality wine production. The current climate change rate is not considered as leading to regional scale relocation of vineyards or *"alarming"* at least for the time being (Ramón, 2010). However, based on recent research on climate change impacts observed in the last few decades on viticulture and vinification, especially from the results obtained by analysing grapevine vine phenology, grape berry composition and vintage quality, it is predicted that *"shifts"* in the present *"base climate regimes"* could occur within the next decade in certain major wine regions and two such regions are; the *Mediterranean* (Jones, et al., 2005: Deloire, 2006) and Australian (Web, 2006). The potential impacts are anticipated to impose: "added pressure on increasingly scarce water supplies, additional changes in grapevine phenological timing, further disruption or alterations of balanced composition and flavour in grapes and wine, regionally-specific changes in varieties grown, necessary shifts in regional wine styles, and spatial changes in viable grape growing regions" (Jones , 2007: 3). In this context, the paper looks at the effects of seasonal climate change on the quality of vintages produced from New Zealand's wine regions spread across the north and south islands of the country at a regional scale.

Grapevine that is being described as one among the most expensive cultivated crops with significant pre-historic development even with remarkable links to human civilisation is perceived to be the most vulnerable of all crops to the recent climate change. Grapevine varieties require niche climate and environmental conditions for successful cultivation, and history proves that the climate has been one among the major factors in the rise and fall of many wine regions over centuries (Jones G. V., 2004). The current climate change is increasingly becoming yet again the ultimate *"determinant"* factor for the continued existence of some presently well-established vineyards that are still able to ripen grapes to produce the world's famous premium wine labels derived from viticulture and winemaking knowledge refined over the last century. The vine-related historical manuscript sources from the *Klosterneuburg* monastery achieves consist of the recording of vine phenological events since the 16th century, the events were originally recorded to study the climate change impact on crop and wine taste over that time (Koch, Hammerl, Maurer, Hammerl, & Pokorny, 2010). This centuries-old records, and the fairly recent wine tasting and rating systems, such as Sotheby (Stevenson, 2007) and Michael cooper (Cooper, 2008), not only reveal viticulturist/ winemaker efforts to analyse climate change impacts, but also portray the wider spatiotemporal scales at which the change impacts on viticulture and vinification have been and are being studied; the former illustrates the impacts in the world's major wine regions over the last five centuries while the latter demonstrates the vintage-to-vintage variability in labelled wines produced from a vineyard or/ and major wine regions.

Interestingly, winemakers and sommeliers continue to utilise a fascinating methodology to describe the nexuses between the vintage-to-vintage rating (in wine taste pertaining to its style) and the weather conditions that ripened the grapes. The numeric rating as well as text descriptors used to describe vintages tend to correlate wine flavours to grape berry ripening temperature as well, for example, *Chardonnay* vintages produced from grapes ripened under *"cooler"* conditions within the ideal ripening temperature range, are described as possessing *lime/lemon* meanwhile, vintages from grapes ripened under warmer conditions of the range are linked to tropical *fruit/pineapple*, in the free text format of descriptors.

The wine descriptor system was originally developed to convey wine aroma, mouth feel and after taste characteristics along with numeric ratings in a 10 or 100 point scale, in an attempt to "quantify" the variability in vintage quality (Brochet, 2001). However, scientists see this form of interpretation as a rather *"subjective way of expressing wine quality"*. Nevertheless, published work on sommelier capability to express cognitive specificity of chemical senses using distinctive wine descriptors and terms of *hedonic* (or personal likings) is seen as unique, consistent and even capable of discerning grape ripening weather conditions. Similar to the vintage label ratings, there are regional rating systems that are used to describe the overall vintage quality of major wine regions within a country or in the world. Such a chart can compiled "Wine Enthusiast" could be seen at (www.winemag.com/PDFs/Vintage_Chart_022011.pdf).

With that background on vintage rating systems, the studies that looked at analysing the effects of seasonal climate change on vintage-to-vintage rating (and vintage price change at wine auctions) at different scales are summarised in section 2. Section 3 presents an approach being investigated into analysing the seasonal climate change effects on vintages at a regional scale using an example of New Zealand wine production and regions identified as "world-famous", such as *Sauvignon Blanc* of Marlborough region.

2. Climate influences on wine vintage-to-vintage variability

The climate is considered as one of the major *"terroir"* factors when determining the quality of vintages apart from wine maker experience and capability. The studies that looked at the extent of this influence on vintages are presented in this section.

The vintage-to-vintage (inter annual) variability in local seasonal weather conditions can to a greater extent influence the grape ripening process and in turn the berry composition, such as sugar and phenols that give the specific colour, aroma and flavours to the wine. All these components when combined with winemaker talent and experience, give a unique characteristics, added finesse to the wine style produced from the winery and the end product is called the vintage i.e., 2009 *Kumeu River Mate's Vineyard Chardonnay*. Scientific research has shown that 50% of Vintage-to-vintage variability to be determined by climate and 25% from soil whereas, only 10% being attributed to factors relating to the grapevine variety or *"cultivar"* and some of such published research is discussed in this section. Meanwhile the other *"terroir"* determinants of grape wine quality are the environmental and soil related factors and they are carefully looked at associated with a site/region and these factors are given precedence when selecting a wine style for the site. Recent research efforts of *old* wine countries to unravel this old concept to improve viticulture and to further refine grape wine in style and appellation are presented here onwards.

In an interesting study by (van Leeuwen, et al., 2004), it is concluded that climate to be the major influencing factor on the berry ripening process and hence in the composition of berry components that give the unique characteristics to the end product vintage, its colour and Baume. The study was based on an analysis of all variables generally classified as *"terroir"* and *"cultiva"*, and looked at the influences of both major sets of factors simultaneously on vintage-to-vintage vine development and berry composition of non irrigated *vitis vinifera* on gravelly soil (with heavy clay subsoil and sandy soil as well as water table within the reach of roots). This wine appellation and quality study included *Merlot, Cabernet franc* and *Cabernet Sauvignon* style wines. The climate variables used for the study were maximum and minimum temperatures, degree days (base of 10°C), sunshine hours, ETo (a Reference Evapotranspiration), rainfall, and water balance for a four year period from 1996 to 2000.

Similarly, in (Grifoni, Mancini, Maracchi, Orlandini, & Zipoli, 2006) the authors argued that the climate variation to be the main influencing factor on the vintage-to-vintage variability in vintage quality and stated that the other three i.e., grape variety, rootstock and soil type, as constants. Interestingly, in the study cultivation techniques were described as human factors, as the techniques tend to be seen as responsible for long-term variability because of the long periods required by grape growers for the adoption of any modification in production methods. Monthly or multi monthly average air temperatures and cumulated precipitation for 500 haP geopotential height (over the *Mediterranean* sea) and sea surface temperatures for northern hemisphere growing season (January to October) were the major variables used in the analysis. The other variables included in the analysis were North Atlantic Oscillation and Southern Oscillation indices. The results were found to be in consistent with some previous correlation coefficients established between wine quality and May to October air temperatures. The study concluded that the high values of the climate variables to be responsible for high temperature and dry conditions during the summer months in Italy that were generally observed to be favourable for high quality wine.

In (Ashenfelter, Ashmore, & Lalonde, 1995) the authors showed how the correlations between price and quality of vintages could be modelled with an example set of French red wines using regression techniques with data on the season weather that produced the wines of respective price (in logarithm) for different vintages of a portfolio of Bordeaux Chateau wine. Variables used in the analysis were; age of vintage and weather variables, such as temperature, (during growing season i.e., April-September), rain in September and August, rain in the months preceding to the vintage i.e., October-March, average temperature in September R2 (root mean squared error) for vintages of 1952-1980 excluding 1954 and 1956, as these wines were rare, the two vintages being considered as the poorest in the decade.

Meanwhile, in (Jones, White, Cooper, & Storchmann, 2005) climate and global wine quality factors were compared using the year-to-year variability over ten years to predict the potential vintage quality for the next five decades. Citing many earlier studies the authors of this work pointed out that the analysis of the relationships between climate variables and wine prices to be based on an underlying hypothesis that beneficial climate conditions would improve the wine quality and that in the past these had in turn led to short term price hikes. They also reflected that the unavailability of consistent price data for multiple regions and with different styles over many years to be a shortcoming for any complete analysis/ study on long term effects. They also argued that the vintage ratings to be a strong determinant of the annual economic success of a wine region based on the work of (Nemani, et al., 2001) but then went on to say that the ratings could be determinants of wine quality not necessarily a predictor based on (Ashenfelter & Jones, 2000) where ratings were described to be reflective of wine somewhat in an indirect way i.e., they had the same weather factors documented to be the determinants of the same wine quality.

Frost (2001) used pairs of liking ratings and 14 wine descriptor intensities, originally collected for analysis with statistical methodologies (partial and least square regression), and created a second map with liking ratings on the y axis and the sensory descriptive data of the wines on the x axis. The model only accounted for 25% of the variation, described to be very low. However, the authors made the following observations based on the map results: 1) some of the wine descriptors used in the analysis, such as *"leather"*, and *"sour"* as exerting a *negative* effect on the preference, meaning subjects' liking scores were low for wines with these descriptors and 2) Subjects liked wines with certain descriptors, such as *"vanilla/oak"*, *"canned vegetables"* and *"green olives"* over to wines with high *"buttery"* or *"berry"*. As the model failed to explain the driving factors for the remaining 75% of the variation, the majority of the sample, the authors cautioned the readers of their results. Please note that the second approach was not with free text on wine taste instead used data obtained from a trained panel to rate the intensity of each of 14 wines used in the study.

3. New Zealand wine industry

New Zealand's (NZ) wine industry continues to grow rapidly in total grapevine cultivation area and in the production of premium wine catering to both domestic and export markets. The extremely diverse climate and environmental conditions combined with incredible enology skills enable NZ wineries to achieve fine quality wine with some unique flavours in a substantially wide range of appellations catering to global markets meeting considerably high standards. This rapid growth has in recent times led to increased interest in scientifically understanding the link between the country's climate conditions, site specific attributes, berry component formation and the overall impact on the ultimate end product wine and its quality (vintage). The section gives a summary of major New Zealand wine regions, varieties cultivated and ecological niche (described in terms of climate, environmental, soil and topographic factors) favourable for grapevine growth and different *"clutiva"* or varieties, before discussing the approach being investigated for modelling the correlations between the seasonal variability in weather conditions and vintage rating using data on related variables made available for this study.

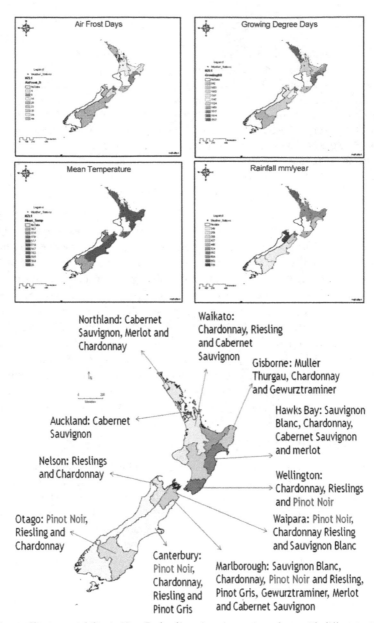

Figure 1. a-e: Climate variability in New Zealand's major wine regions along with different wine styles produced from the regions. For example, *Chardonnay* varieties are grown in regions of the North and South islands but not *Pinot Noir* Varieties as the latter needs extreme climate conditions i.e., cold for dormancy and high temperatures for berry ripening.

The research presented here is conducted using regional vintage ratings of white and red wine styles provided by Michael Cooper (Cooper, 2008) and climate variability extracted from National Institute of Water and Atmospheric Research (NIWA) meteorological web portal (http://cliflo.niwa.co.nz). The weather and wine quality data sets are collectively analysed using computational and formal statistical methods for map based interpretation of climate change effects on NZ wine regions, as is typical in a geographic information system (GIS) that manipulates and analyses geospatial data for informed decision making about, in this case, wine quality in relation to location and climate. The advantage of using GIS in addition to standard data analysis is the use of map-based visualisations that express spatial relationships in the data. The results of the approach experimented show interesting patterns between New Zealand's white and red wine regional vintage ratings, and certain temperature variables found to be significant by ANOVA test results across the country in the geographical context.

New Zealand's major wine regions, climate regimes and varieties

New Zealand is one among the *New* world wine producing countries. Its growth in fine wine production has been rather unprecedented especially over the last two decades. The country's grapevine cultivation area increased from 4,880 hectares in 1990 to 31,002 projected for 2010, with an estimated wine production of 200 Million litres (in 2008) from 582 wineries. The 2008 production consisted of 50 for domestic and 87.8 Million litres for export markets. The value of New Zealand grape wine export for 2008 was 773.9 Million dollars, 76.1 % of the export (by volume for June 2008) comprised of *Sauvignon Blanc*, the main export destinations being 34.6 % to UK, 27.6% to Australia and 21.5% to USA.

The ideal grapevine clones (wine styles) grown in New Zealand wine regions over the last ten years based on base climate are presented in figures (1a-d). The figures show the variability in air frost days/ year, growing degree days (GDD), annual mean temperature and monthly rainfall in the major wine regions of New Zealand. It is noteworthy of mention that *Chardonnay* varieties are grown in all the regions of New Zealand's North and South islands but not the *Pinot Noir* Varieties as the latter needs extreme climate conditions i.e., cold during grapevine dormancy and high temperatures for berry ripening.

4. The methodology for modelling wine quality using regional vintage ratings

In this research, available monthly weather variables and a complete set of regional vintage ratings (while and red wine styles) for the ten major wine regions of New Zealand are analysed using statistical and data mining methods to produce models that best fit the data and are outlined. The methodology adopted consists of the following steps:

1. Regional scale white and red vintage ratings of New Zealand's ten wine regions as presented in Michael Cooper's wine atlas (1st and 2nd editions) from 1993 to 2006 are compiled in to one data file. The ten well-known NZ wine regions included are: Auckland 1, Canterbury 2, Gisborne 3, Hawks Bay 4, Marlborough 5, Nelson 6, Northland 7, Waikato 8, Wairarapa 9, Central Otago 10.

2.   The original regional vintage ratings 2-7 for white and red wine styles are reclassified into binary 1(2-5) and 2(6&7) for ANOVA analysis with available weather data. The weather data extracted from NIWA is compiled to match each of the NZ regional vintage rating relating to white and red wines. Hence, each regional vintage rating is combined with a set of climate variables covering its grapevine growing season (12 months prior to harvest). The 14 monthly weather variables used in the ANOVA tests over the growing season (May – April) are listed below:
   a.   Rainfall
   b.   Mean Air Temperature
   c.   Extreme Maximum Air Temperature
   d.   Mean 20cc Earth Temperature
   e.   Mean 20cc Earth Temperature
   f.   Mean Vapour pressure
   g.   Growing degree days (GDD)
   h.   Days of Snow
   i.   Low Maximum Air Temperature
   j.   Standard (std) Day mean Temperature
   k.   Low Daily Mean Temperature
   l.   High (hi) Daily Mean Temperature
   m.   Mean 9 am Relative Humidity (RH)
   n.   Mean 9 am Temperature
3.   The most influencing weather variables (significant at $p\text{-}value$=0.05) selected from one way ANOVA tests are further analysed using rule and decision tree based data mining techniques (C5 decision tree of Clementaine) to establish the correlations between white and red original vintage ratings (2-7) and the selected significant climate variables.

Despite the inconsistencies in the weather data set, rules created using this regional vintage quality and weather data show interesting patterns in the seasonal effects of weather on the quality of white and red vintages at the regional level across New Zealand and are discussed in the next section.

5. Results: New Zealand wine regional vintage quality and climate change

The vital factors relating to seasonal climate conditions that contribute to higher/lower vintage ratings for white and red wine styles at the regional scale in wine quality identified by ANOVA tests are listed Table 1 a-c. Table 1 a: consists of variables for white (left) and red wine (right), b: shows the data distribution of significant white vintage variables and c: shows the data distribution of significant red vintage variables. The C5 rules generated for white and red vintage regional ratings using the climate variables (presented in Table 1a) are listed in Tables 2 and 3 respectively.

The ANOVA results indicate a few variables i.e., December, February and March rainfall, as well as February and March mean 9 am relative humidity, as common deterministic variables for both, white and red regional vintage ratings in New Zealand. The possible interpretation for this could be as follows:

wine	variable	F	sig	wine	variable	F	sig
white	Dec rainfall	9.113	0.003	red	Dec rainfall	5.381	0.022
	Feb rainfall	4.061	0.046		Feb rainfall	6.960	0.009
	March rainfall	11.906	0.001		March rainfall	19.581	0
	May extreme Max air T	6.473	0.013		April rainfall	6.127	0.014
	Sep extreme Max air T	12.233	0.001		July mean air T	4.527	0.035
	Dec extreme Max air T	5.792	0.019		Aug low Max air T	6.719	0.011
	Mar extreme Max air T	4.470	0.038		Feb mean 9am RH	6.038	0.015
	April extreme Max air T	6.750	0.011		March mean 9am RH	12.803	---
	Feb mean 20cc Earth T	4.744	0.032				
	March mean 20cc Earth T	4.020	0.048				
	May std daily mean T	3.971	0.048				
	Sep high daily mean T	7.938	0.006				
	Feb mean 9am RH	4.965	0.027				
	March mean 9am RH	13.710	---				
	April mean 9am RH	7.479	0.007				

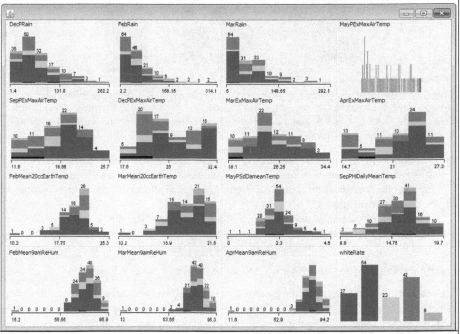

a: ANOVA results showing the monthly climate variables that affect the regional wine quality in NZ. Only the variables that have *P-value*<0.05 are listed and used in the subsequent artificial intelligent rule development tests using a C5 algorithm. T: Temperature Max: Maximum RH: Relative Humidity
b: Data distribution of monthly climate variables (based on ANOVA test results) that affect the regional white vintage quality in NZ. For each of the variables in Table 1 a (left), the distribution in terms of vintage rating value (7, 6, 5, 4, 3, as given by the histogram colours at the bottom right hand corner) against frequency (*y-axis*) and value (*x-axis*).

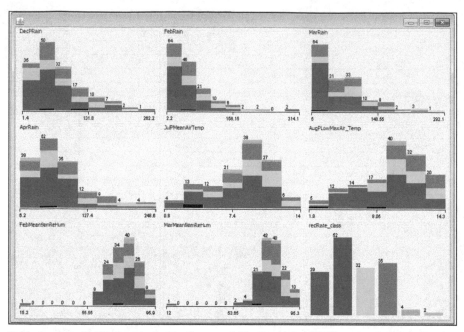

Table 1. c: Data distribution of monthly climate variables (based on ANOVA test results) that affect the regional red wine quality in NZ. For each of the variables in Table 1 a (right), the distribution in terms of wine rating value (7, 6, 5, 4, 3, 2) as given by the histogram colours at the bottom right hand corner) against frequency (*y-axis*) and value (*x-axis*).

1. Monthly rainfall totals in December, February and March: heavy rains in December affect the pollination in flowers and in turn reduce berry formation or fruitfulness meanwhile, increased rainfall in February and March affects the berry ripening process and reduce fruitfulness.
2. February and March monthly mean 9 am relative humidity: this is the berry ripening period in northern and southern regions in New Zealand. Higher relative humidity and rainfall increase the possibilities of increased fungal spread affecting the subsequent crop.

Seasonal climate effects on white and red wine regional vintage ratings

For **white** vintage ratings, February monthly rain (condition 1 in Table 2) seen as the major deterministic factor, i.e., increase in rainfall (>15.8 mm) when met with other conditions leads to higher vintage ratings (6.7) in all NZ wine regions. Meanwhile, March mean 9 am relative humidity (condition 1 in Table 3) is seen as the major deterministic factor for **red** vintage ratings for all NZ regions, however its effects are mixed, negative in the eastern and southern parts of the North Island (Gisborne (3) and Hawke's Bay (4)) and in some northern parts (Marlborough (5) and Nelson (6)) of the South Island except for Wairarapa (9) in southern North Island, and in rest of the New Zealand regions seems to be causing a positive effect on **red** vintage ratings.

Seasonal climate effects on white wine regional vintage ratings

region	rate	rule No	Condition 1	Condtion 2	Condition 3	Condition 4		
	6	1/10	Feb rain <= 18.5					
Auckland 1	4	1/7	Feb rain > 15.8	Feb mean 9am RH <=85.5	Sep hi dmean temp <= 15.2			
	5	1/11	Feb rain > 15.8	Feb mean 9am RH >85.5				
	6	2/10	Feb rain >15.8	Feb mean 9am RH <=85.5	Sep hi dmean temp > 15.2	Mar Ex max air temp <= 24.8		
	7	1/6	Feb rain >15.8	Feb mean 9am RH <= 85.5	Sep hi dmean temp > 15.2	Mar Ex max air temp > 24.8		
Caterbury 2	4	2/7	Feb rain > 15.9	Sep hi dmean temp <= 14.2	Sep Ex max air temp > 20.9			
	5	2/11	Feb rain > 15.8	Sep hi dmean temp <= 14.2	Sep Ex max air temp <= 20.10			
	6	3/10	Feb rain > 15.8	Sep hi dmean temp > 14.2	Mar mean 9am RH > 67.3			
	7	2/6	Feb rain > 15.8	Sep hi dmean temp > 14.2	Mar mean 9am RH <= 67.3			
Gisborne 3	4	3/7	Feb rain > 15.8	Mar mean 9am RH >76.9	Feb rain > 37.6			
	5	3/11	Feb rain > 15.8	Mar mean 9am RH >76.10	Feb rain <= 37.6			
	6	4/10	Feb rain > 15.8	Mar mean 9am RH <= 76.10	Mar mean 9am RH > 73.2			
	7	3/6	Feb rain > 15.8	Mar mean 9am RH <= 76.10	Mar mean 9am RH <= 73.2			
Hawks Bay 4	4	4/7	Feb rain > 15.8	Apr mean 9am RH > 77.3	Sep hi dmean temp <= 17			
	5	4/11	Feb rain > 15.8	Apr mean 9am RH > 77.3	Sep hi dmean temp > 15.1			
	6	5/10	Feb rain > 15.8	Apr mean 9am RH <= 77.3	Sep hi dmean temp > 17			
	7	4/6	Feb rain > 15.8	Apr mean 9am RH > 77.3	Sep hi dmean temp <= 15.1			
Marlborough 5	3	1/3	Feb rain > 15.8	Apr ex max air temp <= 22.8				
	5	5/11	Feb rain > 15.8	Apr Ex max air temp > 22.8	Mar Ex max air temp <= 26	Sep Ex max air temp > 15.8		
	6	6/10	Feb rain > 15.8	Apr Ex max air temp > 22.8	Mar Ex max air temp <= 26.7			
	7	5/6	Feb rain > 15.8	Apr Ex max air temp > 22.8	Mar Ex max air temp > 26.7	Sep Ex max air temp > 15.8		
Nelson 6	5	6/11	Feb rain > 15.8	Mar mean 20cc Earth temp <= 18.6				
	6	7/10	Feb rain > 15.8	Mar mean 20cc Earth temp > 18.6				
Northland 7	3	2/3	Feb rain > 15.8	Mar mean 20cc Earth temp > 19	Dec rain <= 123.6	Mar rain > 46.6	Maysd dmean temp <=1.4	
	3	3/3	Feb rain > 15.8	Mar mean 20cc Earth temp > 19	Dec rain <= 123.6	Mar rain > 46.6	Maysd dmean temp > 1.4	Dec rain > 79.4
	4	5/7	Feb rain > 15.8	Mar mean 20cc Earth temp > 19	Dec rain < 123.6	Mar rain > 46.6	Maysd dmean temp > 1.4	Dec rain <= 79.4
	5	7/11	Feb rain > 15.8	Mar mean 20cc Earth temp > 19	Dec rain <= 123.6	Mar rain <= 46.6		
	5	8/11	Feb rain > 15.8	Mar mean 20cc Earth temp > 19	Dec rain > 123.6			
	6	8/10	Feb rain > 15.8	Mar mean 20cc Earth temp <= 19.5				
Waikato 8	4	6/7	Feb rain > 15.8	Sep hi dmean temp > 15.3				
	5	9/11	Feb rain > 15.8	Sep hi dmean temp <= 15.3				
Wairarapa 9	5	10/11	Feb rain > 15.8	Feb rain > 57.2	Feb mean 9am RH <=83.4			
	6	9/10	Feb rain > 15.8	Feb mean 9 am RH > 83.4				
	7	6/6	Feb rain > 15.8	Feb rain <= 57.2				
Otago 10	4	7/7	Feb rain > 15.8	Apr mean 9 am RH > 77				
	5	11/11	Feb rain > 15.8	Mar mean 9am RH <=77	Apr Ex max air temp <= 16.3			
	6	10/10	Feb rain > 15.8	Mar mean 9am RH <=77	Apr Ex max air temp > 16.3			

Figure 2. Climate variables determinant (identified through ANOVA tests) for NZ **white** grape wine, regional vintage rating and rules created using a C5 algorithm.
rain: rainfall, RH: relative humidity, hi: high, temp: temperature, Ex: extreme, max: maximum, d: daily.

Based on Tables 1 and 2, September extreme maximum and daily high temperatures (see conditions 2 and 3 in Table 2) and are found to be the significant determinants of white vintage regional rating in New Zealand, the reason for this could be that this is the month in which buds burst occurs in southern hemisphere, an important phenological event of grapevine growth cycle. During this time buds begin to swell and burst into leaves and inflorescence, finally forming a shoot. For Auckland, Waikato, Hawke's Bay (of North Island) and Canterbury (from upper South Island) September high temperature seems to be one of the deterministic factors in the regional vintage rating of white wine styles. For example, for Canterbury (2), along with February total rainfall > 15.8 and <=18.5 mm, September high daily temperature >14.2 °C and March mean 9 am relative humidity <= 67.3 produced the highest ratings of 4-7. For the same region, September high daily temperature <14.2°C brought the rating down to 5/4, the monthly extreme maximum temperature >20.9 °C led to 4 and greater i.e., <=20.10 °C to 5. Otago region does not seem to be affected by the September extreme temperature. For the other regions data was not included in the analysis due to unavailability.

In Table 2, monthly rainfall totals in March and December (conditions 2, 3 and 4) are also seen as major deterministic factors for white wine regional vintage rating for Northland (7), Gisborne (3) and Wairarapa (9) regions, all of them from northern New Zealand, interestingly, it is more so in the extreme north, regions that are more exposed to the oceans.

In the Marlborough region in addition to the February monthly rain (condition 1 in Table 2), monthly extreme maximum temperatures of April > 22.8 °C, March >26.7 °C and September 15.8 °C (conditions 2-4 in Table 2) are seem to be the deterministic factors in the vintage ratings of white wine styles, higher temperatures in all three months experienced in this region seem to be producing the highest rating of 4-7.

Seasonal climate effects on red wine regional vintage rating

Based on Table 3, August low maximum (max) air temperate (condition 3 Table 3) in Auckland (1) and Hawke's Bay (4) seem to be the final deterministic factor, low temperatures leading to higher (6 and 7 respectively) confirm the fact that low temperatures during dormancy is vital for ultimate quality of the berry in the following season (growth cycle). This indicates that increase in August low maximum temperatures in these regions could lead to decrease in red vintage ratings.

Except for Auckland, all the other regions have either or all of December, February and March monthly total rainfall as deterministic factors (conditions 2-4 of Table 3). Interestingly, for all three months, higher rainfall leads to lower ratings, this implies that in general, any increase in rainfall could affect the red vintage rating in New Zealand. Again, in Canterbury (2), Hawk's Bay (4) and Wairarapa (9) (of the North Island) lesser monthly rainfall in December (condition 2 in Table 3) seems to lead to higher ratings, December being the time during which pollination takes place and heavy rains could severely hider the opening of flowers (cap fall) affecting pollination and in turn fruitfulness for the season.

The major deterministic factor for red wine regional rating is March mean 9 am relative humidity (RH) (condition 1 of Table 3). This is the time berry ripening occurs and higher percentages of relative humidity could create favourable conditions for fungal infections that severely hamper the crop unless fungicides are sprayed to avoid extensive damage to berry bunches and leaves. Therefore, in general during this period, low moisture and relative humidity conditions are preferred and this is evident from this rule. But based on condition 1-4 of Table 3, moderately high March relative humidity (72.8 - 80.6 %) has increased the red vintage ratings in Auckland (1) Canterbury (2), Northland (7) and Waikato (8). However, in Wairarapa (9) > 86.9 % along with December rain <=102.6 mm has led to the highest rating 7. On the hand, in Otago (10) increase in March relative humidity and February rainfall (>57.3 mm) are seen to be contributing to the higher rating of 5-6. The reason for this could be that the Otago (10) wine region is in the most southern part of New Zealand, where grape ripening is delayed with late harvest in June hence the region is benefited by any increase in both of these conditions.

region	rate	rule No	Condition 1	Condtion 2	Condition 3	Condition 4
Auckland 1	5	2/11	Mar mean 9am RH > 72.8	Mar mean 9am RH <= 81.9	Aug low max air temp > 12.7	
	5	3/11	Mar mean 9am RH > 72.8	Mar mean 9am RH > 81.9		
	6	4/12	Mar mean 9am RH > 72.8	Mar mean 9am RH <= 81.9	Aug low max air temp <= 12.7	Mar mean 9am RH > 80.6
	7	1/9	Mar mean 9am RH <= 72.8			
	7	7/9	Mar mean 9am RH > 72.8	Mar mean 9am RH <= 81.9	Aug low max air temp <= 12.7	Mar mean 9am RH <= 80.6
Caterbury 2	4	1/7	Mar mean 9am RH <= 72.8	FebRain > 62		
	4	2/7	Mar mean 9am RH > 72.8	DecPRain > 83.6		
	6	1/12	Mar mean 9am RH <= 72.8	FebRain <= 62	FebRain <= 40.8	
	6	5/12	Mar mean 9am RH > 72.8	DecPRain <= 83.6	FebRain > 18.6	
	7	2/9	Mar mean 9am RH <= 72.8	FebRain <= 62	FebRain > 40.8	
	7	8/9	Mar mean 9am RH > 72.8	DecPRain <= 83.6	FebRain <= 18.6	
Gisborne 3	4	3/7	Mar mean 9am RH > 72.8	JulPMeanAirTemp > 8.9	DecPRain <= 54.8	
	5	4/11	Mar mean 9am RH > 72.8	JulPMeanAirTemp > 8.9	DecPRain > 54.8	
	6	6/12	Mar mean 9am RH > 72.8	JulPMeanAirTemp <= 8.9		
	7	3/9	Mar mean 9am RH <= 72.8			
Hawks Bay 4	4	4/7	Mar mean 9am RH > 72.8	FebRain > 47.8		
	5	5/11	Mar mean 9am RH > 72.8	FebRain <= 47.8	Aug low max air temp > 10.1	
	6	2/12	Mar mean 9am RH <= 72.8	DecPRain > 38.4		
	6	7/12	Mar mean 9am RH > 72.8	FebRain <= 47.8	Aug low max air temp <= 10.1	
	7	4/9	Mar mean 9am RH <= 72.8	DecPRain <= 38.4		
Marlborough 5	4	5/7	Mar mean 9am RH > 72.8	MarRain > 68		
	5	6/11	Mar mean 9am RH > 72.8	MarRain <= 68		
	7	5/9	Mar mean 9am RH <= 72.8			
Nelson 6	4	6/7	Mar mean 9am RH > 72.8	MarRain > 113		
	5	7/11	Mar mean 9am RH > 72.8	Aug low max air temp > 10		
	6	8/12	Mar mean 9am RH > 72.8	Aug low max air temp <= 10		
	7	6/9	Mar mean 9am RH <= 72.8			
Northland 7	4	7/7	Mar mean 9am RH > 72.8	MarRain > 33.7	Mar mean 9am RH <= 88.1	MarRain > 62.2
	5	8/11	Mar mean 9am RH > 72.8	MarRain <= 113	MarRain > 84.8	
	6	9/12	Mar mean 9am RH > 72.8	MarRain <= 113	MarRain <= 84.8	
Waikato 8	2	1/1	Mar mean 9am RH > 72.8	MarRain <= 33.7		
	3	1/2	Mar mean 9am RH <= 72.8			
	5	9/11	Mar mean 9am RH > 72.8	MarRain > 33.7	Mar mean 9am RH <= 88.1	MarRain <= 62.2
Wairarapa 9	3	2/2	Mar mean 9am RH > 72.8	DecPRain > 102.6		
	5	10/11	Mar mean 9am RH > 72.8	DecPRain <= 102.6	Mar mean 9am RH <= 80	
	6	10/12	Mar mean 9am RH > 72.8	MarRain > 33.7	Mar mean 9am RH > 88.1	
	6	11/12	Mar mean 9am RH > 72.8	DecPRain <= 102.6	Mar mean 9am RH > 80	Mar mean 9am RH <= 86.9
	7	9/9	Mar mean 9am RH > 72.8	DecPRain <= 102.6	Mar mean 9am RH > 80	Mar mean 9am RH > 86.9
Otago 10	5	1/11	Mar mean 9am RH <= 72.8	FebRain > 36.6		
	5	11/11	Mar mean 9am RH > 72.8	FebRain <= 57.2		
	6	3/12	Mar mean 9am RH <= 72.8	FebRain <= 36.6		
	6	12/12	Mar mean 9am RH > 72.8	FebRain > 57.2		

Table 2. Climate variables found as significant deterministic factors (through ANOVA tests) for NZ **red** grape wine, regional rating and rules created using a C5 algorithm.

6. Conclusion

A few critical studies on recent climate change effects on the phenological development and ecological dynamics of some natural habitats described in the introduction, reported on the current and potential detrimental outcome on the biodiversity of the particular habitats. On the extreme, the effects are witnessed to be causing *"temporal mismatch in interacting species"* that have co-evolved over millions of years. As far as viticulture and wineries are concerned even though *"regime shifts"* in base viticulture climate have not been reported yet, signs are that this could happen in two popular wine regions in the very near future and they are: 1) within the next two decades in certain major wine producing regions of Australia, and 2) with an over 1 °C increase in average temperature in some of the world famous *Mediterranean* wine regions.

The studies from literature relating to climate change effects on viticulture, wine production and quality discussed in section 2 have used conventional statistical methods and data spanning at least three decades. However, in *New* wine countries consistent data is not available for such time span hence it is not possible to study the climate change effects on viticulture and wine production, especially using conventional methodologies. In addition, New Zealand's wine regions have never been mapped for grape wine growing/ zoning/ marketing purposes. In this context, the paper presented an approach using statistical (ANOVA) and data mining (rule and tree based), with results interpreted in a geographical context. With this approach, it is possible to establish the key climate variables and the conditions that are vital to successful grapevine growing and the production of premium quality wines across the New Zealand at a regional scale.

The research discussed in the paper showed how disparate multi-sourced data could be transformed into useful knowledge with a GIS approach. The research results showed how inconsistent the climate change effects have been in the past within New Zealand and also gave some insight into potential future climate effects on both red and white vintage ratings in the country's major wine regions that have become world famous for their premium fine wine labels produced with some unique flavours.

The daily extreme temperatures (August low high, September daily high) and monthly rainfall effects on the regional vintage ratings of **white** and **red** wines produced from the north and south islands of the country clearly showed the varying climate effects across the country. In the extreme north and south increase in temperatures seems to influence the regional vintage ratings favourably in **white** wine varieties but not in the **red** vintages. The only temperature variable that showed significant impact on red vintages is August low maximum temperature in Auckland (1) and Hawke's Bay (4 in the North Island possibly affecting grapevine dormancy. Increased February rainfall (>15.8 mm) in all NZ regions along with other conditions (mainly relating to December and March rain and September high daily mean temperature) seems to lead to increase in **white** wine vintage ratings across the county.

Interestingly, monthly rainfall and relative humidity of certain months have been the significant climate variables for **red** vintages in New Zealand's wine regions. Increased rainfall in December, February and March seems to affect the regional vintage ratings of **red** wine except for Otago where >57.2 mm February rainfall combined with increased March mean 9 am relative humidity > 72.8 %, is seen to be leading to an increase in **red** vintage rating i.e., 6. The reason for Otago's positive response i.e., leading to a higher rating, to increases in these conditions could be that it is in the most southern part of New Zealand where grape ripening is delayed with late harvest is June. But March relative humidity in general seems to have negative effects on red vintages produced from the wine regions in eastern and southern parts of the North Island as well as in upper South island.

Author details

Subana Shanmuganathan, Ajit Narayanan and Philip Sallis

Auckland University of Technology, Auckland, New Zealand

Acknowledgement

The authores wish to thank Dr Ana Jagui Perez Kuroki for help with ArcGIS mapping.

7. References

Anderson, S. H., Kelly, D., Ladley, J. J., Molloy, S., & Terry, J. (2011). *Cascading Effects of Bird Functional Extinction Reduce Pollination and Plant Density.* Published Online February 3 2011 Science 25 February 2011:Vol. 331 no. 6020 pp. 1068-1071 DOI: 10.1126/science.1199092.

Ashenfelter, O., & Jones, G. V. (2000). *The demand for expert opinion: BordeauxWine.* VDQS Annual Meeting, d'Ajaccio, Corsica, France. October, 1998. Report published in Cahiers Scientifique from the Observatoire des Conjonctures Vinicoles Europeenes, Faculte des Sciences.

Ashenfelter, O., Ashmore, D., & Lalonde, R. (1995). Bordeaux Wine Vintage Quality and the Weather. *Chance 1995 vol 8 No. 41995: 7-14.*

Cooper, M. (2008). *Wine Atlas of New Zealand (2nd Ed).* New Zealand: Hodder Moa.

Deloire, A. (2006). Climate Trends In A Specific Mediterranean Viticultural Area Between 1950 And 2006: Climate and viticulture in the South of France. [Internet]. Version 4. Knol. 2008 Aug 11. Available from: http://knol.google.com/k/alain-deloire/climate-tre.

Frost, M. B. (2001). A Preliminary study of the effect of knowledge and sensory expertise on liking for red wines. *American Journal of Enology and Viticulture. 2002 vol. 53(4) :275-284.*

Grifoni, D., Mancini, M., Maracchi, G., Orlandini, S., & Zipoli, G. (2006). Analysis of Italian Wine Quality Using Freely Available Meteorological Information,. *Am. J. Enol. Vitic. 2006 57:3:339-346,* Maracchi, G., Orlandini, S. and Zipoli, G.

Jones, G. V. (2004). Making Wine in a Changing Climate. Vol. 50, No. 7: 22-27 www.sou.edu/envirostudies/gjones_docs/Jones%20Geotimes.pdf.

Jones, G. V. (2007). Climate Change: Observations, Projections, and General Implications for Viticulture and Wine Production. In E. Essick, P. Griffin, B. Keefer, S. Miller, & K. Storchmann, *Economics Department working paper No. 7.* Economics Department working paper No. 7 Whitman College, ISSN. 1933-8147 pp3-17 (https://dspace.lasrworks.org/bitstream/handle/10349/593/WP_07.pdf?sequence=1).

Jones, G. V., & Davis, R. E. (2000). Climate Influences on Grapevine Phenology, Grape Composition, and Wine Production and Quality for Bordeaux, France. *Am. J. Enol. Vitic. 51:3:249-261 (2000).*

Jones, G. V., Duchene, E., Tomasi, D., Yuste, J., Braslavksa, O., Schultz, H., Guimberteau, G. (2005). Changes in European Winegrape Phenology and Relationships with Climate. GESCO 2005. August 2005.

Jones, G. V., White, M. A., Cooper, O. R., & Storchmann, K. (2005). Climate and Global Wine Quality. *Climatic Change by Springer vol. 73:319–343.*

Koch, E., Hammerl, C., Maurer, C., Hammerl, T., & Pokorny, E. (2010). BACCHUS historical phenological and early temperature records from Eastern Austria, Burgundy and the Swiss Plateau. *Phenology 2010: Climate Change Impacts and Adaptation*. Oral Presentations – Climate Change Joly Theatre – Monday, 17.30: Dept. of Botany, Trinity College Dublin, College Green, Dublin 2.

Nemani, R. R., White, M. A., Cayan, D. R., Jones, G. V., Running, S. W., & Coughlan, C. (2001). Asymmetric climatic warming improves California vintages. *Clim. Res. Vol 19:25–34*.

Peñuelas, J., Filella, I., & Estiarte, M. (2010). *This Rutishauser (2010)Phenology in local, regional and global ecology, Oral Presentations - Plenary, Phenology 2010: Climate Change Impacts and Adaptation Trinity College Dublin, Ireland 14 - 17 June 2010*. Retrieved from http://www.tcd.ie/Botany/phenology/assets/docs/Abstract%20booklet.pdf

Ramón, M. d. (2010). Climate change associated effects on grape and wine quality and production,. *Food Research International, Volume 43, Issue 7, August 2010, Pages 1844-1855*.

Richard, P. B., Hiroyoshii, H., & Abraham, J. M.-R. (2009). The impact of climate change on cherry trees and other species in Japan. *Biological Conservation 142 (2009) 1943–1949, Biological Conservation 142 (2009) 1943–1949*.

Richard, P. B., Ibáñez, I., Higuchi, H., Lee, S. D., Miller-Rushing, A. J., Wilson, A. M., & Silander Jr., J. A. (2009). Spatial and interspecific variability in phenological responses to warming temperatures. *Biological Conservation, Vol. 142, Issue 11, November 2009, 2569-2577*.

Saino, N., Rubolini, D., Lehikoinen, E., Sokolov, L. V., Bonisoli-Alquati, A., Roberto, A., Møller, A. P. (2009). Climate change effects on migration phenology may mismatch brood parasitic cuckoos and their hosts. *Biol. Lett. (2009) 5, 539–541,* http://dx.doi.org/10.1098/rsbl.2009.0312 or via http://rsbl.royalsocietypublishing.org.

Schweiger, O., Settele, J., Kudrna, O., & Klotz, S. (2008). Climate Change Can Cause Spatial Mismatch Of Trophically Interacting Species. *Ecology, 89(12), 2008, 3472–3479 [doi:10.1890/07-1748.1]*.

Shanmuganathan, S., & Sallis, P. (2010). Web Mining for Modelling Climate Effects on Wine Quality, Climate Change and Variability,. In S. Simard. ISBN: 978-953-307-144-2, Sciyo pp389-407,.

Shanmuganathan, S., Sallis, P., & Narayanan. (2009). Unsupervised artificial neural nets for modelling the effects of climate change on New Zealand grape wines. In *In B. Anderssen et al. (eds) /18th IMACS World Congress - MODSIM09 International Congress Congress on Modelling and Simulation/, 13-17 July 2009, Cairns, Australia. ISBN: 978-0-9758400-7-8*. (pp. 803-809.).

Shanmuganathan, S., Sallis, P., & Narayanan, A. (2010). Modelling the seasonal climate effects on grapevine yield at different spatial and unconventional temporal scales. In *Fifth Biennial Meeting International Environmental Modelling and Software Society (iEMSs) 2010 International Congress on Environmental Modelling and Software Modelling for Environment's Sake, Ottawa, Canada, July 5-8 http://www.iemss.org/iemss2010/Volume2*.

Stevenson, T. (2007). *The Sotheby's Wine Encyclopedia*. London: Dorling Kindersley.

van Leeuwen, C., Friant, P., Choné, X., Tregoat, O., Koundouras, S., & Dubourdieu, D. (2004). Influence of Climate, Soil, and Cultivar on Terroir. *Am. J. Enol. Vitic. 2004 55:207-217*.

Web, L. B. (2006). The impact of projected greenhouse gas-induced climate change on the Australian wine industry, PhD thesis,. School of Agriculture and Food Systems University of Melbourne pp 277.

Chapter 11

Characterization of the Events of the Dry Spell in a Basin Northern Tunisia

Majid Mathlouthi and Fethi Lebdi

Additional information is available at the end of the chapter

1. Introduction

One form of drought is the interruption of the rainy season by a so called dry spell. Dry spell can be defined as a sequence of dry days including days with less than a threshold value of rainfall. The analysis of the historical occurrence of droughts and its probability of recurrence is important. This information is extremely useful for planning and design applications in agriculture and environment and many other sectors. Drought is perceived as a two-dimensional phenomenon (intensity and duration) which is integrated on a spatial basis (regional drought). The two basic dimensions (intensity and duration) require that bivariate frequency analysis is required for linking return periods to both dimensions levels. This approach should be considered as an intermediate step towards a more comprehensive approach, which is related to anticipate damages of specific sectors.

One criticism, however, is that the first-order Markov chain has a relatively short memory that may limit the model's ability to reproduce adequately long dry and/or wet spells, as well as interannual variability [1, 2, 3, 4, 5]. Higher-order Markov chains [6, 7] often improve these inadequacies; however, they require an estimation of more parameters thereby placing stricter demands on the amount of input data. Moreover, the estimation of more parameters results in a higher uncertainty in these parameters and subsequently, the model itself. An alternative to the Markov chain process is to use the wet-dry spell model. This is known as an alternating renewal model, that is, to simulate wet and dry spells separately by fitting their durations to an appropriate probability distribution such as the negative binominal or geometric distribution [2, 6, 8, 9, 10, 11, 12], or empirical distribution [13]. The characteristics of multi-day wet and dry spells is often important for investigating likely scenarios for agricultural water requirements, reservoir operation for analyses of antecedent moisture conditions [8, 14], and runoff generation in a watershed. The main objective of this paper is the event-based analysis of the dry spells based on daily records and the influence of a climatic evolution for identifying dry events under Mediterranean climatic conditions. Previous literature on the statistics of dry spells based on daily records is limited. Studies have primarily dealt with the length of dry and wet spells [10, 15, 16, 17, 18, 19, 20, 21].

2. Data

In Tunisia, regardless of the origin of disturbances, the situations that give rise to copious rainfall remain dependent of important polar meridian flows, especially at altitude. Yet located on the southern margin of the Mediterranean, Tunisia is, wide inter-annual, that very irregularly affected by these flows. In years when the advections of polar air to the south are deep and frequent correspond to rainy periods in the country. In years when the cold air flows are limited in frequency and extension, constitute on the other hand, periods of low rainfall or drought.

So, given the character of atmospheric circulation, rainfall in Tunisia can only be unevenly distributed across seasonal and highly variable inter-annual scale. These same characters of atmospheric circulation are, with the assistance of geographic factors, in the origin of important regional contrasts rainfall.

The case study used as a base for the approach and the methodology of this research is the downstream basin of the Ichkeul Lake, situated in northern Tunisia (Figure 1). This area has a surface of approximately 1500 square kilometres. The wet season extends from September to April, although the beginning or the end of this wet season can move several weeks. Average seasonal precipitation is 600 mm, varies from 450 to 700 mm following the location and altitude. Mean evaporation (from a free water surface) is around 1490 mm yearly. The average annual temperature is around 17.5°C. The rainfalls are recorded in Ghézala-dam rain gauge, from 1968 to 2010. The mean is 701.3 mm; the coefficient of variation is 0.23. In this 42-year period the wettest year was recorded in 1995-1996, with 1104.3 mm, and the driest year in 1987-1988, with 408.6 mm. For the monthly rainfall, the mean is 56.8 mm; and the coefficient of variation is 0.67. The wettest month during the period from 1968 to 2010 was February 1996, with 341.1 mm, and the driest month, during the rainy season, was October 2001, with 0.5 mm. On average, for this period, the wettest month was December, with 105.5 mm of rainfall, and the driest month was May, with 26.8 mm of rainfall. The wet seasons are separated by dry seasons from almost four months. The rainfall events seem to be grouped over several wet days, separated by dry periods from variable duration. However, rainfall events lasting only one day can be observed. Daily precipitation series, available for a sufficiently long period, of five rain gauges located in this region have been analyzed.

3. Method

In the wet-dry spell approach, the time-axis is split up into intervals called *wet periods* and *dry periods* (Figure 2). A rainfall *event* is an interval in which it rains continuously (it is an uninterrupted sequence of wet periods). The definition of event is associated with a rainfall

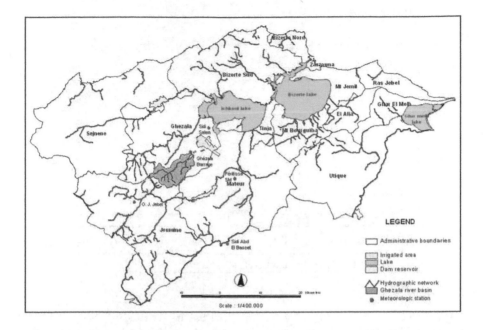

Figure 1. Location of the study area, Tunisia

threshold value which defines *wet*. The limit 4 mm day^{-1} has been selected because it corresponds to the average daily evapotranspiration in the area. This amount of water corresponds approximately to the expected daily evaporation rate, thus marking the lowest physical limit for considering rainfall that may produce utilizable surface water resources during the rainy season which lasts from September to April [8, 11]. In this approach, the process of rainfall occurrences is specified by the probability laws of the length of the wet periods (storm duration), and the length of the dry periods (time between storms or inter-event time).

Several distributions have been used for the length of the wet and dry periods; the exponential distribution, the discrete negative binomial distribution "see [11]". For the wet period length, the Weibull distribution has also been used to model short time-increment rainfall occurrences [22]. Other studies used different probability distributions for the length of wet and dry periods [8, 10, 23].

The varying duration of the events requires that the cumulative rainfall amounts corresponding to each event should be conditioned by the duration of the event. The identification and fitting of conditional probability distributions to rainfall amounts may be problem, especially in the case of short records and for events with extreme (long) durations.

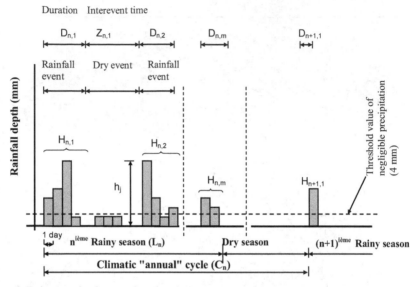

Figure 2. Definitions for the event based analysis

A rainfall event m in a given rainy season n will be characterized by its duration $D_{n,m}$, symbolizing the number of subsequent rainy days, and by the total accumulated rainfall depth of $H_{n,m}$ of $D_{n,m}$ rainy days in mm [8].

$$H_{n,m} = \sum_{j=1}^{D_{n,m}} h_j, \quad n = 1, 2, \dots N \quad \text{and} \quad m = 1, 2, \dots M_n \tag{1}$$

Where N is the total number of observed rainy season and M_n the number of events/rainy season n; h_j stands for the daily rainfall totals in mm. Note that $h_j > 0$ and that for at least one $h_j > 4$ mm. In order to define the temporal position of an event within the rainy season, a time parameter is needed. In study of [24], this time parameter is usually the interarrival time. In this investigation, the interarrival time is replaced by the interevent time or dry event $Z_{n,m}$ (Figure 2). Dry event represents the number of days without rainfall between two subsequent rainfall events. The beginning of the first rainfall event in autumn, in September, marks the beginning of the rainy season, while the end of the last rainfall event in spring, in April, marks the term of the rainy season. Thus, a wet season, with variable length, must start with a rainfall event and end with a rainfall event. Thus, the dry season lasts approximately four months. The length of the rainy season is defined as the time span between the start of the first and the end of the last event of the given season; while the "annual" climatic cycle is determined as the time lapsed between the onsets of two subsequent rainy seasons (Figure 2). The climatic cycle define the position of the first rainfall event within the rainy season. However, the length of the year is fixed at 365 days, and it is taken as a constant [8].

4. Results

Duration of dry events

If the series of successive precipitations do not form independent events, the waiting time follows a gamma distribution with two parameters instead of an exponential distribution [24]. Consequently, if the time is discretized in days, the distribution of time separating two events is represented by the negative binomial distribution [11, 25] which is the equivalent discrete distribution of the gamma distribution:

$$f(n) = \frac{(r + n - 1)}{n!(r - 1)!} . p^r . q^n \tag{2}$$

where $n = 0, 1, 2, \ldots$ and r and p are estimated by:

$$r = \frac{m.p}{1-p}; \quad p = \frac{m}{\sigma_n^2}; \quad q = 1 - p; \quad m = \bar{n} - 1 \tag{3}$$

It is necessary to subtract 1 to n because the negative binomial distribution starts with $n = 0$ whereas time separating two events lasts, per definition, at least 1 day ($n = 1$). Consequently $f(n=0)$ is the probability of a dry event lasting 1 day.

As a regression analysis shown, the length of the dry event (time lapsed between rainfall events) or interevent time can be assumed to be independent from all other characteristics of the rainfall event. Thus the distribution of the dry event follows an unconditional probability distribution function. The dry event duration can only assume integer values. As the example Figure 3 reveals, the shortest interruption (one day) is the most frequent one. Almost one fourth of the observed interevent times are only one day long. Nevertheless, the observed range is much longer than that of the rainfall event duration (table 1). Dry periods up to 30 or even days may be recorded (table 2), even though the probability of such extreme length occurring in the middle of the rainy season is small. The very fact that the mean length fluctuates between 7 and 9 days and the high standard deviation are both serious warnings about the unreliability of assuming an evenly distributed precipitation during the rainy season. The univariate negative binomial pdf has been found as best fitted to describe the distribution of the dry event (Figure 3).

Rain gauge	Observation period	No. of data n	Longest observed in duration days	Arithmetic mean (day)	Standard deviation	Coefficient of variation
Ghézala-dam	1968/2010	895	56	7.39	8.00	1.08
O. J. Jebel Antra	1961/2010+	984	55	7.56	8.03	1.06
Sidi Salem	1959/2010	1078	64	7.58	8.22	1.08
Frétissa	1982/2010	532	60	8.96	9.09	1.01
Sidi Abd el Basset	1968/2010++	569	81	9.18	9.49	1.03

+Not observed for 3 years; ++not observed for 14 years.

Table 1. Parameters of the distribution of the dry event duration in Ichkeul basin

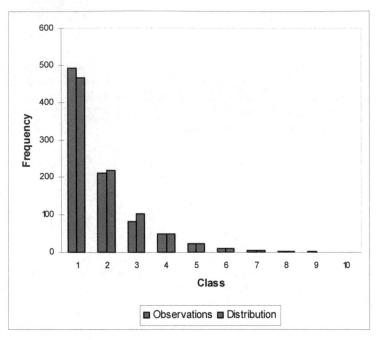

Figure 3. Distribution of dry event duration, O. J. Jebel Antra rain gauge

Rain gauge	Observation period	No. of data	Maximum no. of seasonal dry events observed	Arithmetic mean of no. of dry event per rainy season	Standard deviation of the no. of dry event per rainy season	Duration of the maximum seasonal dry event (day)	Arithmetic mean of the maximum seasonal dry event (day)	Standard deviation (day)
Ghézala-dam	1968/2010	895	30	21.31	4.39	56	30.66	9.53
O. J. Jebel Antra	1961/2010[+]	984	30	21.39	4.30	55	30.45	9.73
Sidi Salem	1959/2010	1078	34	21.15	4.24	64	30.82	10.80
Frétissa	1982/2010	532	35	20.07	4.82	60	35.37	10.83
Sidi Abd el Basset	1968/2010[++]	569	31	20.32	5.13	81	34.53	12.66

[+] Not observed for 3 years; [++] not observed for 14 years.

Table 2. Comparison of the variability of the number of events/season and that of the maximum seasonal dry event

For planning purposes, the longest dry spells associated with different return periods are of fundamental importance. These values were obtained by modelling this process by GEV (General Extreme Value distributions) distributions which shows the best fit (Figure 4 and 5). Table 3 shows the estimated duration of extreme dry events obtained. From table 3 a few rainy seasons characterized by a favourable distribution of rainfall can hide the statistics fact that for a statistical recurrence period of one year, may it produce at least one of more than 20 days (rain gauges of Frétissa and Sidi Abdel Basset). This can be justified by low altitude and unfavourable exposure to rain for these stations (opposite the prevailing wind north-west) and a low annual rainfall in the two rain gauges. For the median the values obtained are critical, the duration of the extreme dry event is almost or more 30 days for all rain gauges; about 4 decades in the Frétissa rain gauge (35 days). For a hundred-year recurrence period, a longest dry spells up to 71.5 days may be registered at Sidi Abdel Basset rain gauge of low average annual rainfall (450 mm).

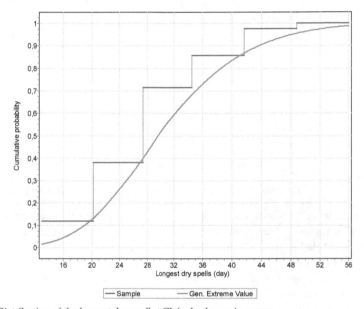

Figure 4. Distribution of the longest dry spell at Ghézala-dam rain gauge

Probability of exceedance	Return period	Ghézala-dam	O. J. Jebel Antra	Sidi Salem	Frétissa	Sidi Abd el Basset
99	1.01	11.9	13.5	11.4	15.3	16.5
95	1.05	16.3	16.8	15.8	20.2	20.3
90	1.11	18.8	18.9	18.3	23.1	22.7
80	1.25	22.2	21.6	21.7	26.8	25.9
50	2	29.4	28.1	29.2	35	33.5
20	5	37.9	36.8	38.1	44.8	43.7
10	10	42.8	42.6	43.4	50.5	50.4
4	25	48.2	49.8	49.5	57.1	58.9
2	50	51.9	55.1	53.6	61.5	65.2
1	100	55.1	60.4	57.4	65.6	71.5
0.5	200	58.1	65.6	60.8	69.3	77.8
0.1	1000	63.8	77.7	67.9	76.8	92.2

Table 3. Estimates of extreme dry event durations

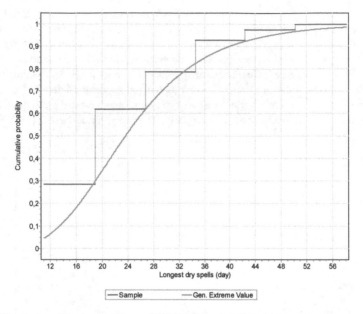

Figure 5. Distribution of the longest dry spell (middle season) at Ghézala-dam rain gauge

To analyze the severity of extreme dry events, the central part of the rainy season for the period from December to March was chosen. Dry events occurring in the core of the rainy season were identified as those ending within the timespan of December - March. Thus any dry event resulting from a rain of start or end of rainy season is not counted. It is important to examine the occurrence of these longest dry spells, during the central part of the rainy season and the whole season, for different return periods. The exceedance probability Pe (N), that an extreme long dry event would occur at least once within a given statistical recurrence period of T years must be equal to the reciprocal value of the product λT:

$$Pe(N) = \frac{1}{\lambda T} \qquad (4)$$

where λ denotes the expected number of dry events/year (season). λT specifies the expected number of trials needed to observe at least once the extreme duration of N days associated with the return period of T years. The length of the extreme dry spell N can then be obtained from the cumulative negative binomial pdf:

$$p = 1 - Pe(N) = \sum_{n=1}^{N} f(n) \qquad (5)$$

Figure 6 shows the negative binomial distribution fitted to the station Ghézala dam. Table 4 shows the estimated duration of dry events N obtained by this method. Significant differences for a small number of observations between the results obtained using the method of event-based analysis and the extreme seasonal value approach are due to the conceptual difference between these methods. It appears that few rainy seasons characterized by a favorable distribution of rainfall can hide the statistics fact that on 21 or 22 dry events it is likely to produce at least one of more than 24 days.

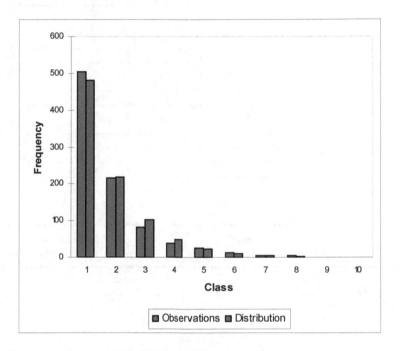

Figure 6. Distribution of the time lapsed between rainfall events (inter-event time), Ghézala Dam rain gauge

Duration of rainfall events

By focusing on the dry spell event, the duration of the rainfall event $D_{n,m}$ will in fact be identified as inter-event time. This change of roles fits the original Poisson model better. Since rainfall events are shorter, their duration follows the geometrical pdf, as theoretically required. The analysis show that approximately, 50 % of the events indeed last at most one day, the persistence of uninterrupted sequences of rainy days sometimes lasting beyond two weeks (the maximum observed duration is 17 days) (table 5). However the frequency of such long-duration events decreases rapidly with increasing duration. The empirical and fitted geometric pdf of event duration at the Ghézala-dam rain gauge are displayed in Figure 7.

Probability of exceedance	Statistical recurrence period	GEV distribution fitted to the seasonal extreme values		Model series of non-extreme values Negative binomial pdf	
		Maximum number of dry days in the core of the rainy season	Extreme number of dry days in the rainy season	Duration of the seasonal extreme event	Expected value of the sample size to be considered
99	1.01	6	11	23	21.52
95	1.05	10	16	24	22.37
90	1.11	12	18	25	23.65
80	1.25	16	22	26	26.63
50	2	24	29	29	42.62
20	5	34	38	36	106.55
10	10	40	43	42	213.1
4	25	48	48	49	532.75
2	50	53	52	55	1065.5
1	100	58	56	56	2131
0.5	200	63	59	58	4262
0.1	1000	74	65	65	21310

Table 4. Estimates of extreme dry event durations at Ghézala-dam rain gauge

Rain gauge	Observation period	No. of data n	Longest observed in duration days	Arithmetic mean (day)	Standard deviation	Coefficient of variation
Ghézala-dam	1968/2010	937	13	2.89	1.98	0.68
O. J. Jebel Antra	1961/2010+	1029	16	2.62	1.85	0.70
Sidi Salem	1959/2010	1129	17	2.53	1.83	0.72
Frétissa	1982/2010	569	11	2.32	1.63	0.70
Sidi Abd el Basset	1968/2010++	597	10	1.89	1.21	0.64

+ not observed for 3 years; ++ not observed for 14 years.

Table 5. Parameters of the distribution of the rainfall event duration in Ichkeul basin

Number of events per rainy season

By taking account of the assumption of the sequential independence of the rainfall events, as formulated above, the Poisson density function should adequately describe the distribution of the number of events per season:

$$f(N,\lambda) = \frac{e^{-\lambda}\lambda^N}{N!}, \quad N = 0, 1, 2, \ldots \tag{4}$$

Where N described the number of events during a rainy season. The parameter is the average number of events per rainy season.

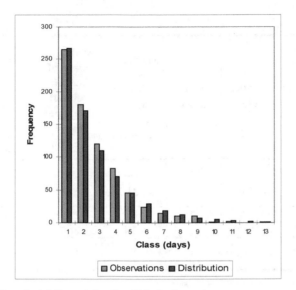

Figure 7. Distribution of rainfall event duration, Ghézala-dam rain gauge

Figure 8 shows the fitted Poisson probability density function. Table 6 summarizes the parameters of the pdfs for all rain gauges. The arithmetic mean is 22.31 and the standard deviation 4.39 for Ghézala-dam rain gauge. The goodness-of-fit has been assessed by the Kolmogorov-Smirnov test at the 95% significance level. The arithmetic mean appears to provide a stable estimate of the parameter λ of the Poisson pdf, since this statistic is also the maximum likelihood estimator; it is used to estimate λ, in preference to the sample variance, which shows more substantial fluctuations [23].

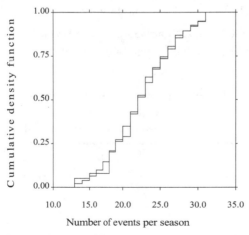

Figure 8. Distribution of the number of rainfall events per season, Ghézala-dam rain gauge

Rain gauge	Observation period	No. of data n	Maximum no. of seasonal rainfall events observed	Arithmetic mean of no. of rainfall event per rainy season	Standard deviation of the no. of rainfall event per rainy season	Variance of the no. of rainfall event per rainy season
Ghézala-dam	1968/2010	42	31	22.31	4.39	19.34
O. J. Jebel Antra	1961/2010[+]	46	31	22.39	4.30	18.51
Sidi Salem	1959/2010	51	35	22.15	4.24	18.01
Frétissa	1982/2010	27	36	21.07	4.82	23.30
Sidi Abd el Basset	1968/2010[++]	28	32	21.32	5.13	26.37

[+] not observed for 3 years; [++] not observed for 14 years.

Table 6. Parameters of the distribution of the number of rainfall event per season in Ichkeul basin

Length of the climatic cycle

The phenomenon of a rainy season followed by a dry season constitutes an annual cycle. As shown in table 7, the expected value of the cycle confirms the annual characteristic of this phenomenon, and the low coefficient of variation indicates the stability of this expected value. Nevertheless, a negative skewness can be consistently observed. The log Pearson type III pdf provides a good fit to distribution of the length of the climatic cycle. Figure 9 presents an example of empirical versus fitted theoretical pdf.

It is found that a climatic cycle with an above average length is generally followed by a shorter one, thus preventing any long lasting shift of the rainy season.

Rain gauge	No. of data points	Arithmetic mean (day)	Standard deviation	Coefficient of variation	Coefficient of skewness
Ghézala-dam	42	364.6	15.25	0.042	0.355
Oued J. Jebel Antra	45	364.1	18.34	0.050	0.196
Sidi Salem	50	365.6	22.87	0.063	-0.007
Frétissa SM	26	368.2	23.49	0.064	-0.375
Sidi Abd el Basset	27	364.7	17.94	0.049	-0,306

Table 7. Statistical parameters for the length of the climatic cycle in Ichkeul basin

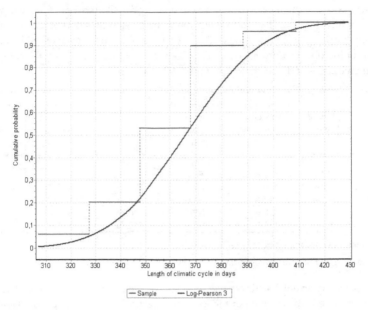

Figure 9. Distribution of the length of the climatic cycle, Sidi Salem rain gauge

Critical of the influence of a climatic evolution

In order to carry out a criticism of the influence of a climatic evolution three cases are considered for the Ghézala-dam rain gauge: i) the complete daily precipitation series, ii) series divided into two sub-series, iii) series divided into three sub-series. Table 8 shows the calculated statistics for Ghézala-dam rain gauge.

From the data available at the Ghézala-dam rain gauge, it is possible to bring to the force the non-stationarity of the rainfall time-series for the period from 1968 to 2010. It is possible to isolate the periods of rainfall anomaly (strongly rainy periods or drought). One try to evaluate the behaviour of some statistical methods largely employed in research of climatic variability.

Simulated variability is that of a brutal change of the average. The procedures concerned are the Pettitt and Buishand test [26] (see Figure 10 and 11). Since the p-value calculated (0.374 for the Pettitt test and 0.652 for Buishand test) are greater than the significance level alpha = 0.05, we can not reject the null hypothesis H0 stating that the data are homogeneous. The risk of rejecting the null hypothesis H0 even tough is true is 37.37% for Pettitt test and 65.19% for Buishand test. The results show that the null assumption (absence of rupture) is accepted at 99 % significance level around the p-value. The threshold value 4 mm day^{-1} was selected. Table 9 summarizes the statistical characteristics of dry events for Ghézala-dam rain gauge in this hypothesis. It indicates that for the second period (89-2010), the maximum number of observed seasonal dry events is identical to the complete time series, whereas that of the first period (68-89), it is slightly inferior. The arithmetic mean of maximum seasonal dry events differs respectively during the two periods. The larger the standard deviation is high; east the dispersion of the observations of a variable. Consequently compared with the complete time series, the time series of the second period (89-2010) is dispersed.

Period of observation	No. of data	Minimum annual precipitation (mm)	Maximum annual precipitation (mm)	Arithmetic mean of annual precipitation (mm)	Standard deviation of annual precipitation	Coefficient of variation of annual precipitation
1968/2010[+]	42	408.6	1104.3	701.39	168.03	0.23
1968/1989[*]	21	408.6	960.7	671.90	128.82	0.19
1989/2010[*]	21	420.3	1104.3	730.88	198.65	0.27
1968/1982[§]	14	514.4	903.6	671.31	107.68	0.16
1982/1996[§]	14	408.6	1104.3	720.17	218.49	0.30
1996/2010[§]	14	461.9	940.8	712.70	167.99	0.23

[+] complete time series, [*] two sub-series, [§] three sub-series.

Table 8. Statistics of the annual precipitation at the Ghézala-dam rain gauge

The maximum dry event (56 days) was occurred during the second period. Similarly, the maximum arithmetic mean value of the longest duration is during this period. However, the coefficient of variation, which represents the dispersion of the observations, is practically identical to the complete time series. The maximum number of observed seasonal dry events is recorded during the second period (82-96) and the third period (96-2010). For the later period, the arithmetic mean is approximately equal to that of the complete time series. A different standard deviation shows dispersion more distinguished during this period (96-2010). The maximum dry event occurred during the period (82-96). The average duration of the longest dry event is more remarkable during the last two periods (82-96) and (96-2010). A standard deviation, during the period (82-96), is much more significant (12.82) explaining a more significant dispersion of this series (C.V. = 0.40 while C.V. = 0.31 for the complete time series).

Period of observation	No. of data	Maximum no. of seasonal dry events observed	Arithmetic mean of no. of dry event per rainy season	Standard deviation of the no. of dry event per rainy season	Duration of the maximum seasonal dry event (day)	Arithmetic mean of the maximum seasonal dry event (day)	Standard deviation (day)
1968/2010[+]	42	30	21.31	4.39	56	30.66	9.53
1968/1989[*]	21	28	22.04	4.08	46	28.04	8.34
1989/2010[*]	21	30	20.57	4.67	56	33.28	10.11
1968/1982[§]	14	27	23.14	2.71	40	28.28	5.83
1982/1996[§]	14	30	19.78	5.60	56	31.85	12.82
1996/2010[§]	14	30	21.00	4.00	47	31.85	8.87

[+] complete series, [*] two sub-series, [§] three sub-series.

Table 9. Comparison of the variability of the number of events/season and that of the maximum seasonal event, Ghézala-dam rain gauge.

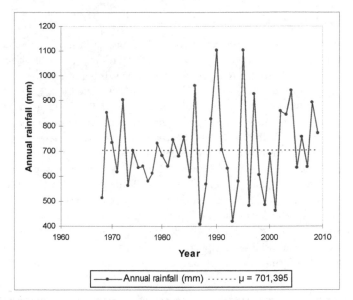

Figure 10. Test of Pettitt for the annual precipitation, Ghézala-dam rain gauge

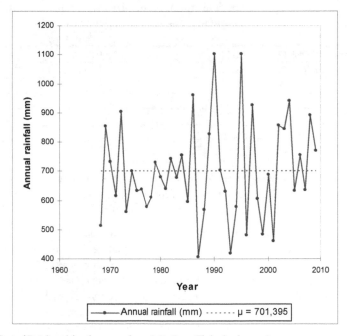

Figure 11. Test of Buishand for the annual precipitation, Ghézala-dam rain gauge

5. Discussion and conclusions

This case study, using rainfall records of the Ichkeul basin, illustrates the independency between the durations of wet and dry events. It is shown that dry spells occur randomly during the rainy season. In this region dry spells can well be described by the negative binomial pdf. The procedure defines the inter-event time as being the dry event period. For the rainfall event duration, the theoretical requirement of the fitted geometric pdf are satisfied (Figure 7). It has to be pointed out that the event-based definition of the rainy season does not exactly fit the theoretical condition. Rainy seasons have variable lengths, as they are a stochastic function of the events themselves. For planning purposes, the longest dry spells associated with the various statistical recurrence periods are derived on the basis of the fitted GEV distributions. Event-based analysis is also useful to check the spatial properties of dry events. Event-based analysis, even if it is carried out on the basis of few years of observation, can rely on large number of data points (table 2). While the expected number of events/season is still derived from very few data, this estimate is more reliable than the approximate expected length of the longest seasonal dry spell, since this variability of the former is usually less than that of the latter, for the same data sets (table 10).

Rain gauge	No. of data	Coefficient of variation of the number of dry events-season (Cv_1)	Coefficient of variation of the longest seasonal dry event (Cv_2)	Cv_2/Cv_1
Ghézala-dam	42	0.20	0.31	1.55
Oued J. Jebel Antra	46	0.20	0.31	1.55
Sidi Salem	51	0.20	0.35	1.75
Frétissa	27	0.24	0.30	1.25
Sidi Abd el Basset	28	0.25	0.36	1.44

Table 10. Coefficient of variation of the number of dry events-season and that of the longest seasonal dry spell.

The procedure adopted allowed us to study the influence of climatic variation and the threshold value of negligible precipitation in identifying dry events. Another application of the event-based analysis is the study of the effects of climate change.

This study, which relied on daily data, describes characteristics of dry spell and their extreme cases. These features are important when estimating the drought risks. Although the region selected for study is not semi arid, data were not always available at the required level of detail. As a result, the detail of the studied events depended on the data collection stations involved. This lack of data is partially compensated by the daily series. In general, this study aimed to define droughts under specific conditions in terms of dry event. We also synthesised information for use in models of climatic risk, infrastructure damage mitigation and environmental management.

6. Adaptive measure in reponse to longest dry spells in the basin

During 1987 – 1989 and 1993 – 1995 drought was characterized by two subsequent dry years. The drought management system has been applied. It was based on reactive decisions. The system is based in three phases: (i) Before drought (preparedness and early warning); (ii) Drought management (mitigation when drought is upon) and (iii) Subsequent drought (when drought is over). During each of the three phases different measures are applied. The Drought Management Phase is characterized by the execution of the planning programs of drought mitigation. Depending on the type, intensity and duration of the drought event, different scenarios are adopted [27].

The year is considered dry when the precipitation deficit is beyond 50% of the mean historically established value. The probability to have a dry year is 7 to 23% in the region. From such situation results a substantial shortage in the available water resources, a production falling and range shortage, and some problems related to the domestic water supply appear. Livestock sickness could be observed, because of diet change and unbalanced nutrition regime. In order to attenuate those problems a mitigation program is executed. For livestock safeguard, the identification of the animal nutriments stocks is established and an importation planning, if necessary, is fitted in order to gap the deficiency. Prevision of vaccination campaigning on the livestock against the sickness related to drought. To satisfy the domestic water demand, in urban as well as in rural areas, a program of aquifers uses is adopted, the use of the surface water resources is avoided or minimized. Particular attention is given to water transportation until the rural drought sensitive regions. Establishing reservoirs water management plan regarding the evolution of the climatic conditions. The eventuality of a second dry year is specially taken in account. Intensification of the preparedness operations related to the next year up going (short loans, soil tillage, seeds distribution …). When drought is over, several measures are taken. This is the intensification of the vulgarization program related to the soil tillage and farming practices in order to maximize the valorisation of the precipitation coming during the subsequent drought wet year. Available water resources evaluation (reservoirs and aquifers) is made. Reconstitution of the aquifers water reserves by the artificial recharging is followed. Like other action initiated, evaluation of the mitigation program efficiency and estimation of their cost.

In the following, the hydrological drought in this basin, including the water resources reservoirs management under drought and the water supply conditions, is presented. It is important to underline those water resources reservoirs management involving the application of water supply rationing depends on water storage at end April. Rainfall deficit during the period between the beginning of September and end April is an important index indicating the water resources availability situation. This period is the most determining of the hydrological drought and its impact on the reservoirs storages.

During 1987 – 1989 drought was severe in the whole country. During this period the water catchments in the reservoirs has been less than 50% of dams' capacity. During 1993 – 1995 drought was similar to the former event described above. Rainfall deficit was ranged between 33 to 56% and was around 35% on the national scale. During this drought, data on the water catchments recorded in dams located in this region that were under exploitation in this date are presented in table 11.

Dams	Use starting date	Minimal water input (Mm³)	Year
Joumine	1983 – 1984	17.9	1993 – 1994
Ghézala	1984 – 1985	0.5	1993 – 1994

Table 11. Minimal water input recorded in dams under exploitation located in this region

The decision tools used for water management, during 1992 – 1995, were based on the measurement instrumentation spread in the water supply locations, information and data transfer to the Central Direction for decision taking, the filed databases, specialized software and simulation models related to the water management optimization. The drinking water supply (for domestic ...) has been ensured without any restriction during the successive years 1992 – 1993, 1993 – 1994 and 1994 - 1995.

The agricultural water demand was satisfied during 1992 – 1993 and 1993 – 1994. A restriction plan was prepared for 1994 – 1995. This plan was applied in March 1995 and adapted in July 1995 regarding the water resources situation. For the irrigated areas, the restriction has been about 50% (with normal year as reference). This restriction ranged from 19.5 to 27.5% referring to a dry conditions year (1993 – 1994). Therefore during this year (1993 – 1994), in spite of the dry conditions all agricultural demand was satisfied. Farmers adopted some "self modifications" in their farming systems to adapt them to drought situation [27].

From all the previous drought management events, actions related to agricultural production (crop systems and livestock care) were well monitored. It is from the practical decisions taken during 1992 – 1995 drought events that tools decisions linked to water reservoir management were tested.

Author details

Majid Mathlouthi
Research Laboratory in Science and Technology of Water in INAT, Tunis, Tunisia

Fethi Lebdi
National Agronomic Institute of Tunisia (INAT), University of Carthage, Tunis, Tunisia

7. References

[1] Wallis T. W. R, Griffiths J. F (1995) An assessment of the weather generator (WXGEN) used in the erosion/productivity impact calculator (EPIC). Agric. Meteorol. 73: 115–133.

[2] Semenov M. A, Brooks R. J, Barrow E. M, Richardson C. W (1998) Comparison of WGEN and LARS-WG stochastic weather generators for diverse climates. Clim. Res. 10: 95-107.

[3] Cahill A. T (2003) Significance of AIC differences for precipitation intensity distributions. Advances in Water Resources 26(4): 457-464.

[4] Hui W, Xuebin Z, Elaine M. B (2005) Stochastic Modelling of Daily Precipitation for Canada. Atmosphere-Ocean 43 (1): 23-32.

[5] Muller A (2006) Comportement asymptotique de la distribution des pluies extrêmes en France. Thèse de l'Université de Montpellier II, Sciences et Techniques de Languedoc.

[6] Wilks D. S (1999) Interannual variability and extreme-value characteristics of several stochastic daily precipitation modes. Agric. Meteorol. 93: 153-169.

[7] Hayhoe H. N (2000) Improvements of stochastic weather data generators for diverse climates. Clim. Res. 14: 75-87.

[8] Mathlouthi M, Lebdi F (2008a) Event in the case of a single reservoir: the Ghèzala dam in Northern Tunisia. Stochastic Environ. Res. and Risk Assessment 22 : 513–528.

[9] Mathlouthi M, Lebdi F (2008b) Evaluation de la fiabilité de gestion d'un barrage réservoir pour des événements secs. Hydrol. Sci. J. 53(6): 1194–1207.

[10] Mathlouthi M, Lebdi F (2009a) Analyse statistique des séquences sèches dans un bassin du nord de la Tunisie. Hydrol. Sci. J. 54(3): 442-455.

[11] Mathlouthi M (2009) Optimisation des règles de gestion des barrages réservoirs pour des évènements extrêmes de sècheresse. Thèse de Doctorat, Institut National Agronomique de Tunisie, Tunis, Tunisie. 162 p. Available: http://www.birsa.agrinet.tn

[12] Mathlouthi M, Lebdi F (2010) Caractérisation des événements secs dans un bassin du Nord de la Tunisie. In: Eric Servat, Siegfried Demuth, Alain Dezetter & Trevor Daniell, editors. Global Change: Facing Risks and Threats to Water Resources (Proc. of the Sixth World FRIEND Conference, Fez, Morocco, October 2010). IAHS Publ. 340, 2010. IAHS Press, Wallingford, UK. pp. 86–94.

[13] Rajagopalan B, Lall U (1999) A k-nearest-neighbor simulator for daily precipitation and other weather variables. Water Resour. Res. 35: 3089-3101.

[14] Mathlouthi M, Lebdi F (2009b) Emploi de la série chronologique des événements secs générés dans l'optimisation de la gestion des barrages. Hydrol. Sci. J. 54(5): 841-851.

[15] Longley R.W (1953) The length of dry and wet periods. Q J R Meteorol Soc 79: 520.

[16] Williams C.B (1952) Sequences of wet and of dry days considered in relation to the logarithmic series. Q J R Meteorol Soc 78(335): 91–96. doi:10.1002/qj.49707833514.

[17] Feyerherm A.M, Bark L.D (1967) Goodness of fit of a Markov chain model for sequences of wet and dry days. J Appl Meteorol 6: 770–773.

[18] Caskey J.E (1963) A Markov chain model for the probability of precipitation occurrence in intervals of various length. Mon Weather Rev 91: 298–301.

[19] Gringorten II (1971) Modelling conditional probability. J Appl Meteorol 10: 646–657.

[20] Naumann G, Vargas W, Minetti J.L (2008) Dry spells in the La Plata Basin. Monitoring and trend stability. Drought implication. Meteorologica 33: 61–85.

[21] Vargas W.M, Naumann G, Minetti J.L (2011) Dry spells in the River Plata Basin: an approximation of the diagnosis of droughts using daily data. Theor. Appl. Climatol. (2011) 104: 159–173. DOI 10.1007/s00704-010-0335-2.

[22] Eagleson P. S (1978) Climate, soil and vegetation 2: The distribution of annual precipitation derived from observed storm sequences. Water Resources Research 14(5): 713-721.

[23] Bogardi J. J, Duckstein L, Rumambo O. H (1988) Practical generation of synthetic rainfall event time serie in semi-arid climatic zone. Journal of Hydrology 103: 357-373.

[24] Fogel M. M, Duckstein L (1982) Stochastic precipitation modelling for evaluating non-point source pollution. In: Singh V, editors. Statistical Analysis of Rainfall and Runoff (Proc. Int. Symp. on Rainfall–Runoff Modelling 1981). Water Resources Publications. Littleton, Colorado, USA. pp. 119–136.

[25] Mathlouthi M, Lebdi F (2008) Characterization of dry spell events in a basin in the North of Tunisia. Proceedings of the First International Conference on Drought management: Scientific and technological innovations. Zaragoza, Spain, June 12-14, 2008. In: Lopez-Francos A, editors. Drought management: Scientific and technological innovations (Options Méditerranéennes , Series A: Séminaires Méditerranéens 80). CIHEAM Pub.2008. pp. 43-48

[26] Lubès-Niel H, Masson J. M, Paturel J. E, Servat E (1998) Variabilité climatique et statistique. Etude par simulation de la puissance et de la robustesse de quelques tests utilisés pour vérifier l'homogénéité de chroniques. Revue des Sciences de l'Eau 11(3): 383-408.

[27] Louati M.H, Bergaoui M, Lebdi F, Mathlouthi M., El Euchi L, Mellouli H.J (2007) Application of the Drought Management Guidelines in Tunisia. In: Iglesias A, Moneo M, Lopez-Francos A, editors. Drought Management Guidelines Technical Annex. (Options Méditerranéennes Serie B n. 58). CIHEAM Pub.2007. pp. 417-467

Climate Change and Human Health

Chapter 12

Impact of Climate Change on Zoonotic Diseases in Latin America

Alfonso J. Rodríguez-Morales and Carlos A. Delgado-López

Additional information is available at the end of the chapter

1. Introduction

The United Nations Framework Convention on Climate Change (UNFCC), located in Bonn, Germany, defines climate change as "a change of climate which is attributed directly or indirectly to human activity that alters the composition of global atmosphere and which is in addition to natural climate variability observed over comparable time periods". Although many other definitions can be found and have been stated by many authors and research groups, the important key message, is that global climate change poses a serious threat to the World, which can generate social upheaval, population displacement, economic hardships, and environmental degradation, among many other relevant consequences. In order to reach a green as less anthropogenically impacted World, according to the new ecological trends in the society, mitigation of global climate change should be a priority for the society and its governments (Rodriguez-Morales, 2011).

As has been previously stated, climate change is already a widely known problem to multiple disciplines (Rodriguez-Morales, 2005). Although its origins can converge in a complex of multiple interacting phenomena, for some disciplines, such as biological and medical sciences, their consequences are more studied and highlighted for their current and further implications. Even more, its impacts and squeals are cause of concern at a global level (Rodriguez-Morales et al, 2010). This growing threat represents in the XXI century a significant challenge for the humankind. Its effects even include many aspects that have been not studied by the society at different levels.

Right now, there is now no serious scientific debate: human actions are changing the world's climate, and are set to do so at an increasing rate in coming decades (McMichael et al, 2012; Rahmstorf, 2010; Meinshausen et al, 2009). Urgent action is now required to reduce emissions of carbon dioxide (the dominant long acting greenhouse gas); if global temperature rises are not to exceed 2°C—the International Energy Authority warns that "the door to 2°C is closing"

(International Energy Agency, 2011). Indeed, emissions must be hugely curtailed within just two decades, and then zero net emissions achieved by later this century, assisted by increased biosequestration of carbon dioxide from the atmosphere (Friedlingstein et al, 2011). However, emissions continue to rise, having increased by 49% since 1990 and by an accelerated annual rate of 5.9% in 2010 (McMichael et al, 2012; Peters et al, 2012).

In the context of the multiple impacts that climate change can pose in the World and the society, growing evidence include direct and indirect influences on human health. Good health of the population depends in large magnitude of the delicate balance or interaction between ecological, physical and socioeconomical systems of the biospheres (WHO, 2003). This is one of the many spheres that have been recently highlighted by multiple research reports in regard to the importance of climatic change for global public health (Martens et al, 1997, Rodriguez-Morales et al, 2010, Rodriguez-Morales, 2011). The role that climate change may play in altering human health, particularly in the emergence and spread of diseases, is an evolving area of research, that is even being so more complex and specialized year to year (Hambling et al, 2011). It is important to understand this relationship because it will compound the already significant burden of diseases on national economies and public health, although many times limited for some of them, such as zoonotic diseases. Authorities need to be able to assess, anticipate, and monitor human health vulnerability to climate change, in order to plan for, or implement action to avoid these eventualities (Hambling et al, 2011).

In some regions of the World numerous populations will be displaced by the increase of the level of the sea or will be seriously affected by droughts and famines, decrease of suitable lands for the agriculture, increase of the food-borne diseases, water-borne diseases, vector-borne diseases as well increase of premature deaths and diseases related to the air pollution (Mills, 2009; PAHO, 2008; United Nations, 2006; Diaz, 2006).

In this context infectious diseases have been highlighted, however in the scope of them, impact of climate change on zoonotic diseases has been largely neglected. Although these emerging conditions are very prone to increase due to shifts in the distribution and behaviour of vectors and animal species, which indicates that biologic systems are already adapting to ecological variations, more research is still necessary to fully understand their interacting roles as well how control them.

Zoonotic infections are in general defined as infections transmitted from animal to man (and less frequently vice versa), either directly (through contact or contact with animal products) or indirectly (through an intermediate vector as an arthropod or an insect) (Pappas, 2011). Although the burden of zoonotic infections worldwide is major, both in terms of immediate and long-term morbidity and mortality (Christou, 2011; Akritidis, 2011) and in terms of emergence/reemergence and socioeconomical, ecological, and political correlations (Cascio et al, 2011), scientific and public health interest and funding for these diseases remain relatively minor (Pappas et al, 2012).

Regard the distribution of vectors and pathogens, such as those of malaria and dengue, is clear that climate plays an important role even as a determinant persistence and occurrence at certain places, particularly vulnerable to changes (e.g. in zones of ecological transition in the environmental conditions or ecotones). However, more information is required and necessary for zoonoses, especially those that are important in different regions in the World, such as Leishmaniasis, Chagas disease, Toxocariasis, Brucellosis in Latin America (Rodriguez-Morales et al, 2010).

In this neotropical region recent contributions in the field have demonstrated strong links between climate variability, climate change and emerging and reemerging infectious diseases that represent public health issues for the region. These and other zoonotic diseases represent a significant burden of disease, geographically extended in the region from Mexico in the north to Argentina in the south, including conditions ranging from deserts in the subtropical areas to highlands that are rapidly changing and allowing, for their new climate conditions, the adaptation of vectors and reservoirs that usually were not present there in the past. These varied ecological scenarios have also suffered the impacts of climate change in the socioeconomical systems, such as agriculture and fishing as a consequence of the phases of the El Niño Southern Oscillation (ENSO) phenomena, but also in very specific health conditions such as infectious, tropical and zoonotic diseases such Leishmaniasis, Chagas disease, Toxocariasis, Brucellosis, among others.

Trying to understand these complex relationships between biological and ecological systems in the context of climate change, and its final consequence on human health, different statistical analysis, most of them based on linear regressions, have linked extreme climatic anomalies with significant alterations in the epidemiological patterns of diseases, sometimes coupled directly and indirectly on time and space. Additionally to statistic techniques, geographical information systems (GIS) and remote sensing (spatial epidemiology) have also supported these observations and are currently helping in the developing of systems for prediction and forecasting of such diseases based on climate variability and climate change, as has been previously reported (Rodriguez-Morales et al, 2010; Rodriguez-Morales, 2011) and already in use for tropical diseases such as malaria (Le Sueur et al, 1997; Rodriguez-Morales, 2005; Beck et al, 2000).

In this chapter, an updated review on the evidences about the impact of climate change on zoonotic diseases in Latin America is outlined.

2. General epidemiological and ecological aspects of zoonotic diseases

Emerging zoonotic diseases have increased in importance in human and animal health during the last 10 years, each emerging from an unsuspected quarter and causing severe problems (Benitez et al, 2008; Brown, 2004). Many new emerging and re-emerging diseases are caused by pathogens (bacteria, viruses, parasites) with an animal origin (from different classes and species) and given ecological and temporal conveying. Then, effective surveillance, prevention, and control of zoonotic diseases pose a significant challenge for many countries (Meslin et al, 2000) but are of utmost importance. Particularly in developing countries data record on these diseases is neglected and many times not even done, representing a major barrier to know and understand the epidemiological situation and burden of such diseases. Additionally, given the ecological scenarios where zoonotic diseases arise, different ecoepidemiological aspects need to be incorporated in the analyses, but these, unfortunately, are still not considered by many health agencies, particularly in developing countries.

The burden of zoonotic infections worldwide exceeds involves more than sheer morbidity and mortality, which has been recently analysed for different zoonotic agents by multiple authors (Pappas, 2011; Christou, 2011; Akritidis, 2011). The effect of zoonoses on various parameters of human life can be quantified, e.g. by estimating the economic impact of zoonotic epidemics, which, for the period between 1995 and 2008, exceeded 120 billion dollars in the World (Budke et al, 2006; Cascio et al, 2011). As will be shown later, such epidemiological and social conditions can be directly or indirectly affected by the climatic change (PAHO, 2003; PAHO, 2008; United Nations Development Programme, 2008; United Nations, 2006).

Just to mention some recent figures about the burden of zoonosis in Latin America, in Venezuela, after known outbreaks of acute orally-transmitted Chagas disease beginning in December 2007 (Rodriguez-Morales, 2008), a zoonosis that was considered by many authors as gone (Ache & Matos, 2001), although many reports indicated that never happens (Añez et al, 2004), was again highlighted by scientist and public health authorities, including more care about the report. Then, according to the last National Mortality Report (2009, published in November 2011) (Ministerio del Poder Popular para la Salud, 2011), this zoonosis, also known as American trypanosomiasis (International Code of Diseases, ICD-10, B57), caused the death of 700 persons (a mortality rate of 2.6 deaths/100,000 pop.). In Colombia, according to the National Institute of Health during year 2011, 57,236 exposures to rabies were reported (for a rate of 124.3 events/100,000 pop.) (Instituto Nacional de Salud, 2012). In Perú, the Ministry of Health reported 330 cases of yellow fever, between years 2003 and 2009, for an incidence rate ranging between 15 to 120 cases/100,000 pop (Ministerio de Salud, 2010).

In the last 20 years significant evidences have been generated from multiple science fields demonstrating how the climate change affects directly and indirectly disease vectors (particularly mosquitoes) (Diaz, 2006; Parry et al, 2007), but also animal reservoirs of zoonotic diseases (Benitez et al, 2008; Cardenas et al, 2006; Cardenas et al, 2008). Climate change can accelerate biological development and increase vectors population available to transmit pathogens and diseases as a consequence of its changes on the environment, altitude, cold and heat, and water reservoirs and particularly wetlands (Rodriguez-Morales et al, 2010).

Understanding zoonotic infections as a multifactorial issue is critical, predominantly for preventing their expansion, in terms of geographical and social prevalence. Factors associated with this (either de novo or resurfacing) expansion can roughly be categorized as factors related to the pathogens and factors related to human behaviour. These factors are not independent: modifications of human behaviour result in modifications of pathogen ecology and life cycle in more than one pathway (Cascio et al, 2011).

With a more spread and greater population of vectors and reservoirs, disease risk spectrum as a consequence of more time of exposition, is increasing (Rodriguez-Morales et al, 2010; Gubler et al, 1981; Sukri et al, 2003; Rifakis et al, 2005; Halstead, 2006).

3. Climate change and zoonotic diseases in Latin America

3.1. Zoonotic diseases endemic in Latin America

Previously described, it is well established that climate is an important determinant of the distribution of vectors and pathogens (Rodriguez-Morales et al, 2010). This has been extensively described for some tropical non-zoonotic diseases, such as those of malaria (vectorized by *Anopheles spp.* and caused by *Plasmodium falciparum, P. vivax, P. ovale, P. knowlesii*) and dengue (vectorized by *Ae. aegypti* and caused by Dengue viruses) (Rodriguez-Morales, 2009; Herrera-Martinez & Rodriguez-Morales, 2010; Zambrano et al, 2012).. Although not still accepted, recent evidences imply that malaria would be also a zoonotic disease (Lee et al, 2011). In the case of zoonotic diseases, leishmaniasis (vectorized by sandflies *Phlebotomus spp.* [in the Old World] *and Lutzomyia spp.* [in the New World] and caused by *Leishmania spp.*), should be probably the most studied zoonotic disease regard the impacts of climate change and variability in Latin America (Rodriguez-Morales, 2005; Rodriguez-Morales et al, 2010), with available evidences from different countries in the region. However, other zoonoses, such as Chagas disease, with a high burden in many countries, that accounts for close to 20 million people living infected in the region (Von et al, 2007), has been largely neglected regard studies assessing the impact of climate change variability on its epidemiology (Araújo et al, 2009).

Tropical areas of Latin America have been suitable for zoonotic diseases for many years; these are endemic, and climate change now is triggering its increase, persistence, even re-emergence in non-previous endemic areas or in areas where them were eliminated, eradicated or controlled (Benitez et al, 2008). In Table 1 are summarized selected zoonotic diseases that have been reported or are considered endemic in Latin American countries. Some of them have been studied regard the climate variability and climate change impact on their epidemiological patterns (Table 1).

3.2. Evidences regard climate change and its potential effect on disease: Cutaneous and visceral leishmaniasis

In this regard, evidences from Latin America have cumulated useful qualitative and quantitative information that indicates how climate variability and change influenced particularly tropical diseases (McMichael et al, 2003; Arria et al, 2005). In Northeastern Colombia the impacts of El Niño Southern Oscillation climatic fluctuations during 1985–2002 in the occurrence of cases of leishmaniasis in two northeastern provinces of the country (North Santander and Santander) were reported. During that period, it was identified that during El Niño, cases of leishmaniasis increased up to 15.7% in disease incidence in North Santander and 7.74% in Santander, whereas during La Niña phases, leishmaniasis cases decreased 12.3% in Santander and 6.8% in North Santander. When mean annual leishmaniasis cases were compared between La Niña and El Niño years, it was found significant differences for North Santander (p=0.0482) but not for Santander (p=0.0525) (Cárdenas et al, 2006). For the same study period in southern provinces effects of climate variability and change were also studied regard leishmaniasis incidence. In this study 11 southern departments of Colombia were analyzed: Amazonas, Caquetá, Cauca, Huila, Meta, Nariño, Putumayo, Tolima, Valle, Vaupes and Vichada. Climatic data were obtained by satellite and epidemiologic data were obtained from the Health Ministry. National Oceanographic and Atmospheric Administration (NOAA) climatic classification and SOI (Southern Oscillation Index)/ONI (Oceanic Niño Index) indexes were used as indicators of global climate variability. Yearly variation comparisons and median trend deviations were made for disease incidence and climatic variability. During this period there was considerable climatic variability, with a strong El Niño for six years and a strong La Niña for eight. During this period, 19,212 cases of leishmaniasis were registered, for a mean of 4,757 cases/year. Disease in the whole region increased (mean of 4.98%) during the El Niño years in comparison to the La Niña years, but there were differences between departments with increases during El Niño (Meta 6.95%, Vaupes 4.84%), but the rest showed an increase during La Niña (between 1.61% and 64.41%). Differences were significant in Valle (p=0.0092), Putumayo (p=0.0001), Cauca (p=0.0313), and for the whole region (p=0.0023), but not in the rest of the departments (Cárdenas et al, 2008). These informations show how climatic changes influence the occurrence of leishmaniasis in northeastern and southern Colombia.

Similar results have been described in Venezuela. Between 1994 and 2003, an study in 2,212 cases of cutaneous leishmaniasis cases also linked climate variability to disease incidence in an endemic area of the country, Sucre state. During that period, three important El Niño phases were observed: 1994-1995, 1997-1998, and 2001-2003, being the one in 1997-1998 the most relevant one, which was followed by a chilly and rainy season in 1999 (La Niña). During 1999-2000, 360 cutaneous leishmaniasis cases were recorded in Sucre, with an important variability within a year, and a 66.7% increase in cutaneous leishmaniasis cases (F=10.06, p=0.0051) associated with the presence of a weak La Niña phenomenon (not too cold and rainy). Models showed that with higher Southern Oscillation Index (SOI) values, there was a reduced incidence of cutaneous leishmaniasis (r^2=0.3308; p=0.0504). The increase with respect to the average trend in rain was associated with increases in trends for cutaneous leishmaniasis in the period from 1994 to 2003 (p=0.0358) (Cabaniel et al, 2005).

Although not described in such detail, in Suriname cutaneous leishmaniasis is a seasonal disease. The rainy seasons are from November to January and from May to July. In a recent study (2008), most patients with disease were registered during the short dry season in March (35%) (van der Meide et al, 2008). In Brazil studies made on leishmaniasis vector have characterized spatial distribution of them. In Mato Grosso, the vector sandfly *Lu. whitmani* s.l. have been positively correlated with deforestation rates and negatively correlated with the Brazilian index of gross net production (IGNP), a primary indicator of socio-economic development. Authors found that favourable habitats occur in municipalities with weaker economic development which confirms that vector occurrence is linked to precarious living conditions, found either in rural settlements of the Brazilian government's agrarian reform program, or in municipalities with intense migratory flows of people from lower social levels (Zeilhofer et al, 2008). In Colombia, another entomological study in 5,079 sandflies collected (*Lu. spinicrassa* represented 95.2% of them) have linked population densities to climate. The climatic period where the collection of vectors was done corresponded to a dry season of El Niño (highest Oscillation Niño Index in the last 2006 trimester). In general, the main components analyses evidenced a significant inverse relation between *Lu. spinicrassa* abundance and the relative humidity (p<0.05), as well also with the rainfall (p<0.05), but not for the average temperature (p>0.05) (Galvis et al, 2009). In Costa Rica and Bolivia, recent studies have also linked social and climate changes with cutaneous leishmaniasis (Chaves and Pascual, 2006; Gomez et al, 2006).

Disease	Present in Latin America	Studies have shown climate variability impact on disease
Anthrax	Yes	Yes, but not in Latin America (Joyner et al, 2010)
Babesiosis	Yes (Montenegro-James, 1992)	Yes, but not in Latin America (Hoch et al, 2012)
Balantidiasis	Yes	No
Barmah Forest virus	No	Yes, but not in Latin America (Naish et al, 2009)
Bartonellosis	Yes	Yes (Chinga-Alayo et al, 2004)
Bilharzia or schistosomiasis	Yes	Yes, but not in Latin America (Mas-Coma et al, 2009)
Bolivian hemorrhagic fever	Yes	No
Borreliosis (Lyme disease and others)	Yes (Santos et al, 2011)	No
Bovine tuberculosis	Yes	No
Brucellosis	Yes	No
Campylobacteriosis	Yes	Yes, but not in Latin America (Kovats et al, 2005)
Chagas disease	Yes	No
Cholera	Yes	Yes (Lama et al, 2004)
Cowpox	Yes (Megid et al, 2008)	No
Creutzfeldt-Jakob disease (vCJD)	Yes	No
Cryptosporidiosis	Yes	Yes, but not in Latin America (Britton et al, 2010)
Cutaneous larva migrans	Yes	No
Eastern equine encephalitis virus	Yes	No
Echinococcosis	Yes	No

Disease	Present in Latin America	Studies have shown climate variability impact on disease
Erysipeloid	Yes	No
Fasciolosis	Yes	No
Giardiasis	Yes	Yes (Hermida et al, 1990)
H1N1 influenza	Yes	No
Hantavirus	Yes	Yes (Magrin et al, 2007)
Leishmaniasis	Yes	Yes (see section 3.2)
Leptospirosis	Yes	Yes, but not in Latin America (Codeço et al, 2005)
Listeriosis	Yes	No
Lymphocytic choriomeningitis virus infection	No	No
Paragonimiasis	Yes	No
Rabies	Yes	Yes (Rifakis et al, 2006)
Rift Valley fever	No	Yes (Hightower et al, 2012)
Rotavirus diarrhea	Yes	Yes, but not in Latin America (Hashizume et al, 2008)
Salmonellosis	Yes	No
Sparganosis	Yes (Moulinier et al, 1982)	No
Streptococcus suis infection	Yes (Costa et al, 1995)	No
Toxocariasis	Yes (Delgado & Rodriguez-Morales, 2009)	No
Toxoplasmosis	Yes	No
Trichinosis	Yes	No
Tularemia	Yes (Machado-Ferreira et al, 2009)	No
Venezuelan equine encephalitis virus	Yes	No
Venezuelan hemorrhagic fever	Yes	No
West Nile virus	Yes (Bosch et al, 2007)	Yes, but not in Latin America (Ruiz et al, 2010)
Western equine encephalitis virus	Yes (Ruiz-Gomez & Espinosa, 1981)	Yes, but not in Latin America (Sellers & Maarouf, 1988)
Yellow fever	Yes	Yes (Vasconcelos et al, 2001; Rodriguez-Morales et al, 2004).
Yersiniosis	Yes	Yes, but not in Latin America (Ari et al, 2010)

Table 1. Selected zoonoses present in Latin America and which of them have been studied regard the impact of climate variability and climate change on disease epidemiology.

In the case of visceral leishmaniasis, also studies in Latin America have linked its incidence to climate. Prolonged droughts in semi-arid north-eastern Brazil have provoked rural-urban migration of subsistence farmers, and a re-emergence of visceral leishmaniasis (Confalonieri, 2003). A significant increase in visceral leishmaniasis in Bahia State (Brazil) after the El Niño years of 1989 and 1995 has also been reported (Franke et al, 2002).

Besides these reports, there are no more recent studies and in other countries in Latin America, indicating the impact and importance of climate change on the epidemiology of leishmaniasis.

3.3. Evidences regard climate change and its potential effect on disease: Other zoonotic parasitic diseases

Other zoonotic parasitic diseases, such as schistosomiasis have been linked to climate variability (Kelly-Hope & Thomson, 2008), although no specifically in Latin America (Mas-Coma et al, 2009). Closely to leishmaniasis, other parasitic zoonotic disease extensively present in the region is Chagas disease, however, although probably will be influenced by climate change, there are no specific studies addressing this important aspect. Zoonotic parasite diseases that have been studied regard the impact of climate change in their epidemiology, now include (Table 1): babesiosis (Hoch et al, 2012) (influences not yet studied for Latin America), cryptosporidiosis (Britton et al, 2010) (influences not yet studied for Latin America), giardiasis (Hermida et al, 1990) (Table 1).

Although many others, according to their ecology and related environmental aspects would be suitable and susceptible to be affected by climate change and variability, evidences have been not yet generated (Table 1).

3.4. Evidences regard climate change and its potential effect on disease: Viral zoonotic diseases

Some viral zoonotic diseases in Latin America have been linked to climate variability and climate change. This is the case of rabies, hantavirus and yellow fever, among others such as H1N1 influenza and other viral haemorrhagic fevers that probably are susceptible to climate change impact. An study between 2002-2004 linked rabies occurrence in Venezuela to climate variability. Rabies in Venezuela has been important in last years, affecting dogs, cats, and human, among other animals, being a reportable disease. In Zulia state, it is considered a major public health concern. Recently, a considerable increase in the incidence of rabies has been occurring, involving many epidemiological but also ecoepidemiological and social factors. These factors were analyzed in 416 rabies cases recorded during the study period. Incidence has been increasingly significantly, affecting mainly dogs (88.94%). Given this epidemiology it was associated ecoepidemiological and social factors with rabies incidence in the most affected state, Zulia. This area has varied environmental conditions. It is composed mostly of lowlands bordered in the west by mountain system and in the south by the Andes. The mean temperature is 27.8ºC, and mean yearly rainfall is 750 mm.

Climatologically, year 2002 corresponded with El Niño (drought), middle 2003 evolved to a Neutral period, and 2004 corresponded to La Niña (rainy); this change may have affected many diseases, including rabies. Ecological analysis showed that most cases occurred in lowland area of the state and during rainy season (p<0.05) (Rifakis et al, 2006). For hantaviruses, outbreaks of hantavirus pulmonary syndrome have been reported for Argentina, Bolivia, Chile, Paraguay, Panama and Brazil after prolonged droughts (Williams et al., 1997; Magrin et al, 2007), probably due to the intense rainfall and flooding following the droughts, which increases food availability for peri-domestic (living both indoors and outdoors) rodents (Magrin et al, 2007). In Brazil and Venezuela, yellow fever outbreaks have been linked to climate variability (Vasconcelos et al, 2001; Rodriguez-Morales et al, 2004) (Table 1).

3.5. Evidences regard climate change and its potential effect on disease: Bacterial zoonotic diseases

Bacterial infections have been also linked to an increase related to climate variability, climate change and global warming (Table 1). Zoonotic bacteria, such as *Leptospira* has been also linked to climate variability. Flooding produces outbreaks of leptospirosis in Brazil, particularly in densely populated areas without adequate drainage (Kupek et al, 2000). In 1998, increased rainfall and flooding after hurricane Mitch in Nicaragua, Honduras, and Guatemala caused a leptospirosis outbreak, and an increased number of cases of malaria, dengue fever, and cholera (Costello et al, 2009). In Perú, one of the non-zoonotic forms of bartonellosis, Carrion's disease (*Bartonella bacilliformis*) has been linked also to climate variability (Huarcaya et al, 2004). *Vibrio cholerae* is another zoonotic bacterial pathogen in which its incidence has been linked to climate variability. As ocean temperatures rise with global warming and more intense El Niños, cholera outbreaks might increase as a result of more plankton blooms providing nutrients for *Vibrio cholerae*. Studies in Peru, Ecuador, Colombia, Mexico and Venezuela have evidenced these relationships (Patz et al, 2005; Farfan et al, 2006; Chavez et al, 2005; Franco et al, 1997; Lama et al, 2004).

As shown in table 1, many other zoonotic bacterial agents would be affected by climate change, but evidences have been not yet generated in Latin America regard them, such as: anthrax (Joyner et al, 2010), borreliosis (lyme disease and others), bovine tuberculosis, brucellosis, campylobacteriosis (Kovats et al, 2005), listeriosis, *Mycobacterium marinum* infection, salmonellosis, *Streptococcus suis* infection, tularemia and yersiniosis (Ari et al, 2010).

3.6. Evidences regard climate change and its potential effect on disease: Other zoonoses

For veterinary public health, climate may be associated with seasonal occurrence of diseases in animals rather than with spatial propagation (Table 1). This is the case for pathogens or parasitic diseases, such as fascioliasis, in areas with higher temperatures, when seasonality is extended as a consequence of the increased survival of the parasite outside the host or, conversely, shortened by increased summer dryness that decreases their numbers. For other pathogens, such as parasites that spend part of their life cycle as free stages outside the host, temperature and humidity may affect the duration of survival. Climate change could modify the rate of development of parasites, increasing in some cases the number of generations and then extending the temporal and geographical distribution. New World screwworm is frequently found in South America, with infestations increasing in spring and summer and decreasing in autumn and winter (Rodriguez-Morales, 2006; Paris et al, 2008). West Nile Virus is a disease in which both long-distance bird migration and insect population dynamics (*Culex*) are driven by climate conditions. Vesicular stomatitis (VS) affects horses, cattle and pigs and is caused by various vesiculoviruses of the family Rhabdoviridae. Seasonal variation is observed in the occurrence of VS: it disappears at the end of the rainy season in tropical areas and at the time of the first frosts in temperate zones (Pinto et al, 2008).

3.7. Climate change and zoonotic diseases: Public health perspectives

Given the substantial burden of already studied zoonotic disease associated to climate change in developing tropical countries, such as most of Latin America, it is of utmost relevance to incorporate climate changes into public health thinking, including not only at health authorities and systems, but also in the whole public health education and faculties. However, as shown in Table 1, many diseases are still neglected regard the study of climate change impact on them, and deserve immediate analyses in order to establish their magnitude and relevance.

Although many studies still may have some limitations such as a lack of incorporation of other meteorological factors into the analysis (temperature, rainfall, sun radiation, transpiration or evotranspiration, relative humidity, vegetation indexes [Normalized Difference Vegetation Index, NDVI and Enhanced Vegetation Index, EVI] among others) (Cárdenas et al, 2006), as well more deep analyses, it has been suggested that such findings are relevant from a public health perspective to better understand the ecoepidemiology of different communicable diseases (Rodriguez-Morales, 2005; Rodriguez-Morales et al, 2010).

However, further research is needed in this region and in other endemic areas to develop monitoring systems that will assist in predicting the impact of climate changes in the incidence of tropical diseases in endemic areas with various biological and social conditions.

4. Conclusions

Given the substantial burden of zoonotic disease associated to climate change in developing tropical countries, such as most of Latin America, it is of utmost relevance to incorporate climate changes into public health thinking and prevention. Although many studies still may have some limitations such as a lack of incorporation of other meteorological factors into the analysis, it has been suggested that such findings are relevant from a public health perspective to better understand the ecoepidemiology of different diseases.

Global warming is an ecological emergency, but its implications for human disease caused by zoonotic infectious agents remains understudied. Animals' migration may be affected by global temperature alterations, they could seek novel migratory routes that may also transfer a novel zoonosis to a previously non-endemic area, as has been previously reported particularly in birds (Cascio et al, 2011).

Further research is needed in the World, and particularly in Latin America and specially in endemic areas and countries for each specific zoonotic disease to develop monitoring systems that will assist in predicting the impact of climate changes in the incidence of zoonotic diseases in endemic areas with various biological and social conditions.

Author details

Alfonso J. Rodríguez-Morales
Research Group Infection and Immunity and Department of Community Medicine, Faculty of Health Sciences, Universidad Tecnológica de Pereira (UTP), Office of Scientific Research, Cooperativa de Entidades de Salud de Risaralda (COODESURIS), Working Group on Zoonoses, International Society for Chemotherapy (ISC) and Committee on Zoonoses and Haemorrhagic Fevers, Asociación Colombiana de Infectología (ACIN), Pereira, Risaralda, Colombia
Instituto Experimental José Witremundo Torrealba, Universidad de Los Andes, Trujillo, Venezuela

Carlos A. Delgado-López
Faculty of Sciences for Health, Universidad de Caldas, Manizales, Caldas, Colombia

5. References

Aché, A. & Matos, A.J. (2001). Interrupting Chagas disease transmission in Venezuela. *Revista do Instituto de Medicina Tropical de Sao Paulo*, Vol.43, No.1, (January 2001), 37-43.

Añez, N., Crisante, G. & Rojas, A. (2004) Update on Chagas disease in Venezuela--a review. *Memorias do Instituto Oswaldo Cruz* Vol.99, No.8, (Dec 2004), 781-787.

Akritidis, N. (2011). Parasitic, fungal and prion zoonoses: an expanding universe of candidates for human disease. *Clinical Microbiology and Infection*, Vol.17, No.3, (January 2011), 331–335.

Alfaro, W. & Rivera, L. (2008). *Cambio Climático en Mesoamérica: Temas para la creación de capacidades y la reducción de la vulnerabilidad.* The International Development Research Centre (IDRC) y Department for International Development (DFID-UK), London.

Araújo, C.A., Waniek, P.J., Jansen, A.M. (2009). An overview of Chagas disease and the role of triatomines on its distribution in Brazil. *Vector borne and zoonotic diseases*, Vol.9, No.3, (June 2009) 227-234, ISSN 1530-3667

Ari TB, Gershunov A, Tristan R, Cazelles B, Gage K, Stenseth NC. Interannual variability of human plague occurrence in the Western United States explained by tropical and North Pacific Ocean climate variability. Am J Trop Med Hyg. 2010 Sep;83(3):624-32.

Arria, M.; Rodríguez-Morales, A.J. & Franco-Paredes, C. (2005). Ecoepidemiología de las Enfermedades Tropicales en Países de la Cuenca Amazónica. *Revista Peruana de Medicina Experimental y Salud Publica*, Vol.22, No.3, (July 2005) 236-240, ISSN 1726-4634

Barrera, R., Delgado, N., Jimenez M. & Valero S. (2002). Eco-epidemiological factors associated with hyperendemic dengue hemorrhagic fever in Maracay city, Venezuela. *Dengue Bulletin*, Vol.26, No.1, (December 2002) 84–95, ISBN 9290222565

Beck, L.R., Lobitz, B.M. & Wood, B.L. (2000). Remote sensing and human health: new sensors and new opportunities. *Emerging Infectious Diseases*, Vol.6, No.3, (2000) 217-227, ISSN 1080-6040

Benítez, J.A., Rodríguez-Morales, A.J., Sojo, M., Lobo, H., Villegas, C., Oviedo, L. & Brown, E. (2004). Descripción de un Brote Epidémico de Malaria de Altura en un área originalmente sin Malaria del Estado Trujillo, Venezuela. *Boletín de Malariología y Salud Ambiental*, Vol.44, No.2, (August 2004) 93-100, ISSN 1690-4648

Benítez, J.A., Sierra, C. & Rodríguez-Morales, A.J. (2005). Macroclimatic Variations and Ascaridiasis Incidence in Venezuela. *American Journal of Tropical Medicine & Hygiene*, Vol.73, No.(6 Suppl), (November 2005) 96, ISSN 0002-9637

Benítez JA, Rodriguez-Morales AJ, Vivas P, Plaz J. (2008). Burden of zoonotic diseases in Venezuela during 2004 and 2005. Ann N Y Acad Sci. 2008 Dec;1149:315-7.

Bosch I, Herrera F, Navarro JC, Lentino M, Dupuis A, Maffei J, Jones M, Fernández E, Pérez N, Pérez-Emán J, Guimarães AE, Barrera R, Valero N, Ruiz J, Velásquez G, Martinez J, Comach G, Komar N, Spielman A, Kramer L. West Nile virus, Venezuela. *Emerg Infect Dis*. 2007 Apr;13(4):651-3.

Botto, C., Escalona, E., Vivas-Martinez, S., Behm, V., Delgado, L. & Coronel, P. (2005). Geographical patterns of onchocerciasis in southern Venezuela: relationships between environment and infection prevalence. *Parassitologia*, Vol.47, No.1, (March 2005) 145–150, ISSN 0048-2951

Boutayeb, A. & Boutayeb, S. (2005). The burden of non communicable diseases in developing countries. *International Journal for Equity in Health*, Vol4., No., (2005), ISSN 1475-9276

Britton E, Hales S, Venugopal K, Baker MG. The impact of climate variability and change on cryptosporidiosis and giardiasis rates in New Zealand. J Water Health. 2010 Sep;8(3):561-71.

Brown, C. (2004). Emerging zoonoses and pathogens of public health significance–an overview. *Rev. Sci. Tech.* 23: 435–442.

Budke CM, Deplazes P, Torgerson PR. (2006). Global socioeconomic impact of cystic echinococcosis. *Emerg Infect Dis* 2006; 12: 296–303.

Cabaniel, G., Rada, L., Blanco, J.J., Rodríguez-Morales, A.J. & Escalera, J.P. (2005). Impacto de Los Eventos de El Niño Southern Oscillation (ENSO) sobre la Leishmaniosis Cutánea en Sucre, Venezuela, a través del Uso de Información Satelital, 1994 - 2003. *Revista Peruana de Medicina Experimental y Salud Pública*, Vol.22, No.1, (January 2005) 32-38, ISSN 1726-4634

Cárdenas, R., Sandoval, C.M., Rodriguez-Morales, A.J. & Franco-Paredes, C. (2006). Impact of Climate Variability in the Occurrence of Leishmaniasis in Northeastern Colombia. *American Journal of Tropical Medicine & Hygiene*, Vol.75, No.2, (August 2006) 273-277, ISSN 0002-9637

Cárdenas, R., Sandoval, C.M., Rodriguez-Morales, A.J. & Vivas, P. (2008). Zoonoses and Climate Variability: the example of Leishmaniasis in Southern Departments of Colombia. *Annals of the New York Academy of Sciences*, Vol.1149, No.1, (January 2008) 326-330, ISSN 0077-8923

Cascio, A., Bosilkovski, M., Rodriguez-Morales, A.J., Pappas, G. (2011). The Socio-Ecology of Zoonotic Infections. *Clin Microbiol Infect* Vol.17, No.3, (Mar 2011) 336-342.

Céspedes, V.M. (2007). Los desastres, la información y el Centro Latinoamericano de Medicina de Desastres. *ACIMED*, Vol.16, No.2, (2007) 0-0, ISSN 1024-9435

Chaves, L. F. & Pascual, M. (2006). Climate cycles and forecasts of cutaneous leishmaniasis, a nonstationary vector-borne disease. *Plos Medicine*, Vol.3, No.8, (August 2006) e295, ISSN 1549-1277

Chavez, M.R.C., Sedas, V.P., Borunda, E.O. & Reynoso, F.L. (2005). Influence of water temperature and salinity on seasonal occurrences of *Vibrio cholerae* and enteric bacteria in oyster producing areas of Veracruz, Mexico. *Marine Pollution Bulletin*, Vol.50, No.12, (December 2005) 1641–1648, ISSN 0025-326X

Chinga-Alayo E, Huarcaya E, Nasarre C, del Aguila R, Llanos-Cuentas A. (2004). The influence of climate on the epidemiology of bartonellosis in Ancash, Peru. Trans R Soc Trop Med Hyg. 2004 Feb;98(2):116-24.

Christou, L. (2011). The global burden of bacterial and viral zoonotic infections," *Clinical Microbiology and Infection*, Vol.17, No.3, (January 2011), 326–330

Codeço CT, Lele S, Pascual M, Bouma M, Ko AI. A stochastic model for ecological systems with strong nonlinear response to environmental drivers: application to two water-borne diseases. J R Soc Interface. 2008 Feb 6;5(19):247-52.

Confalonieri, U. (2003). Variabilidade climática, vulnerabilidade social e saúde no Brasil. *Terra Livre*, Vol.1, No.20, (January 2003) 193-204, ISSN 0102-8030.

Costa AT, Lobato FC, Abreu VL, Assis RA, Reis R, Uzal FA. Serotyping and evaluation of the virulence in mice of Streptococcus suis strains isolated from diseased pigs. Rev Inst Med Trop Sao Paulo. 2005 Mar-Apr;47(2):113-5.

Costello, A., Abbas, M., Allen, A., Ball, S., Bell, S., Bellamy, R., Friel, S., Groce, N., Johnson, A., Kett, M., Lee, M., Levy, C., Maslin, M., McCoy, D., McGuire, B., Montgomery, H., Napier, D., Pagel, C., Patel, J., de Oliveira, J.A., Redclift, N., Rees, H., Rogger, D., Scott, J., Stephenson, J., Twigg, J., Wolff, J. & Patterson, C. (2009). Managing the health effects of climate change Lancet and University College London Institute for Global Health Commission. *Lancet*, Vol.373, No.9676, (May 2009) 1693-1733, ISSN 0140-6736

Delgado, O. & Rodríguez-Morales AJ. Aspectos clínico-epidemiológicos de la toxocariasis: una enfermedad desatendida en Venezuela y América Latina. *Boletín de Malariología y Salud Ambiental* 2009 Ene/Jul; 49(1):1-33.

Depradine, C. & Lovell, E. (2004). Climatological variables and the incidence of Dengue fever in Barbados. *International Journal of Environmental Health Research*, Vol.14, No.6, (December 2004) 429–441, ISSN 0960-3123

Diaz, J.H. (2006). Global climate changes, natural disasters, and travel health risks. *Journal of Travel Medicine*, Vol.13, No.6, (November 2006), 361-72, ISSN 1195-1982

Ebi, K.L. & Paulson, J.A. (2007). Climate change and children. *Pediatrics Clinics of North America*, Vol.54, No.2, (April 2007) 213-226, ISSN 0031-3955

Ebi, K. Climate change and health risks: assessing and responding to them through 'adaptive management'. Health Aff (Millwood). 2011 May 30;924-30.

Farfan, R., Gomez, C., Escalera, J.P., Guerrero, L., Aragundy, J., Solano, E., Benitez, J.A., Rodriguez-Morales, A.J. & Franco-Paredes C. (2006). Climate Variability and Cholera in the Americas. *International Journal of Infectious Diseases*, Vol.10, No.Suppl 1, (June 2006) S12-S13, ISSN 1201-9712

Fernando, J., Brunstein J., Fernando, J. & Jankilevich, S.S. (2001). *Disyuntivas para el diseño de políticas de mitigación de la contaminación atmosférica global y local. El caso de la Ciudad de Buenos Aires. Documento de Trabajo N° 69*, Universidad de Belgrano, Buenos Aires.

Franke, C.R., Ziller, M., Staubach, C. & Latif, M. (2002). Impacts of the El Niño/Southern Oscillation on visceral leishmaniasis, Brazil. *Emerging Infectious Diseases*, Vol.8, No., (September 2002) 914-917, ISSN 1080-6040

Friedlingstein, P., Solomon, S., Plattner, G.-K., Knutti, R., Ciais, P., Raupach, M.R. (2011) Long-term climate implications of twenty-first century options for carbon dioxide emission mitigation. *Nature Climate Change*, Vol.1, 457-461.

Galvis-Ovallos, F., Espinosa, Y., Gutiérrez-Marín, R., Fernández, N., Rodriguez-Morales, A.J. & Sandoval, C. (2009). Climate variability and *Lutzomyia spinicrassa* abundance in an area of cutaneous leishmaniasis transmission in Norte de Santander, Colombia. *International Journal of Antimicrobial Agents*, Vol.34, No.Suppl 2, (July 2009) S4, ISSN 0924-8579

Gomez, C., Rodríguez-Morales, A.J. & Franco-Paredes, C. (2006). Impact of Climate Variability in the Occurrence of Leishmaniasis in Bolivia. *American Journal of Tropical Medicine & Hygiene*, Vol.75, No.(5 Suppl), (November 2006) 42, ISSN 0002-9637

Gubler, D.J., Suharyono, W., Lubis, I., Eram, S. & Gunarso, S. (1981). Epidemic dengue 3 in central Java, associated with low viremia in man. *American Journal of Tropical Medicine & Hygiene*, Vol.30, No.5, (September 1981) 1094-1099, ISSN 0002-9637

Gubler DJ, Reiter P, Ebi KL, Yap W, Nasci R, Patz JA. (2001). Climate variability and change in the United States: potential impacts on vector- and rodent-borne diseases. *Environ Health Perspect*. 2001 May;109 Suppl 2:223-33.

Hambling, T., Weinstein, P., Slaney, D. (2011). A review of frameworks for developing environmental health indicators for climate change and health. *Int J Environ Res Public Health* Vol.8, No.7, (July 2011) 2854-2875.

Halstead, S.B. (2006). Dengue in the Americas and Southeast Asia: do they differ? *Revista Panamericana de Salud Publica*, Vol.20, No.6, (December 2006) 407-415, ISSN 1020-4989

Hashizume M, Armstrong B, Wagatsuma Y, Faruque AS, Hayashi T, Sack DA. Rotavirus infections and climate variability in Dhaka, Bangladesh: a time-series analysis. Epidemiol Infect. 2008 Sep;136(9):1281-9.

Hermida RC, Ayala DE, Arróyave RJ. Circannual incidence of *Giardia lamblia* in Mexico. Chronobiol Int. 1990;7(4):329-40.

Herrera-Martinez, A.D. & Rodriguez-Morales, A.J. (2010). Potential Influence of Climate Variability on Dengue Incidence Registered in a Western Pediatric Hospital of Venezuela. *Tropical Biomedicine*, Vol.27, No.2, (August 2010) 280-286, ISSN

Hightower A, Kinkade C, Nguku PM, Anyangu A, Mutonga D, Omolo J, Njenga MK, Feikin DR, Schnabel D, Ombok M, Breiman RF. Relationship of climate, geography, and geology to the incidence of Rift Valley fever in Kenya during the 2006-2007 outbreak. Am J Trop Med Hyg. 2012 Feb;86(2):373-80.

Hoch T, Goebel J, Agoulon A, Malandrin L. (2012). Modelling bovine babesiosis: A tool to simulate scenarios for pathogen spread and to test control measures for the disease. *Prev Vet Med*. 2012 Feb 15.

Huarcaya, E., Chinga, E., Chávez, J.M., Chauca, J., Llanos, A., Maguiña, C., Pachas, P. & Gotuzzo, E. (2004). Influencia del fenómeno de El Niño en la epidemiología de la bartonelosis humana en los departamentos de Ancash y Cusco entre 1996 y 1999. *Revista Médica Herediana*, Vol.15, No., (2004) 4-10, ISSN 1018-130X

Hurtado-Diaz, M., Riojas-Rodriguez, H., Rothenberg, S., Gomez-Dantes, H. & Cifuentes, E. (2007). Impact of climate variability on the incidence of dengue in Mexico. *Tropical Medicine & International Health*, Vol.12, No.11, (October 2007) 1327-1337, ISSN 1360-2276

Instituto Nacional de Salud (2012). *Exposición Rábica a semana 52*. Instituto Nacional de Salud (January 2012).

International Energy Agency. World Energy Outlook November 2011. www.worldenergyoutlook.org/docs/weo2011/executive_summary.pdf.

Joyner, T.A., Lukhnova, L., Pazilov, Y., Temiralyeva, G., Hugh-Jones, M.E., Aikimbayev, A., Blackburn, J.K. (2010). Modeling the potential distribution of Bacillus anthracis under multiple climate change scenarios for Kazakhstan. *PLoS One*. 2010 Mar 9;5(3):e9596.

Kelly-Hope, L. & Thomson, M.C. (2008). Climate and Infectious Diseases (Chapter 3), In: *Seasonal Forecasts, Climatic Change and Human Health*, Thomson, M.C., Garcia-Herrera, R. & Beniston, M. (Ed), 31-70, Springer Science, ISBN 978-1-4020-6876-8, New York.

Koehlmoos TP, Anwar S, Cravioto A.Global health. Chronic diseases and other emergent issues in global health. Infect Dis Clin North Am. 2011 Sep 25; 623-38, ix.

Kovats, S. & Haines, A. (1995). The potential health impacts of climate change: an overview. *Medicine and War*, Vol.11, No.4, (October 1995), 168-78, ISSN 0748-8009

Kovats RS, Edwards SJ, Charron D, Cowden J, D'Souza RM, Ebi KL, Gauci C, Gerner-Smidt P, Hajat S, Hales S, Hernández Pezzi G, Kriz B, Kutsar K, McKeown P, Mellou K, Menne B, O'Brien S, van Pelt W, Schmid H. Climate variability and campylobacter infection: an international study. *Int J Biometeorol.* 2005 Mar;49(4):207-14.

Kupek, E., de Sousa Santos Faversani, M.C. & de Souza Philippi, J.M. (2000). The relationship between rainfall and human leptospirosis in Florianópolis, Brazil, 1991–1996. *Brazilian Journal of Infectious Diseases*, Vol.4, No.3, (June 2000) 131-134, ISSN 1413-8670

Lama, J.R., Seas, C.R., Leon-Barua, R., Gotuzzo, E. & Sack, R.B. (2004). Environmental temperature, cholera, and acute diarrhoea in adults in Lima, Peru. *Journal of Health, Population and Nutrition*, Vol.22, No.4, (December 2004) 399–403, ISSN 1606-0997

Lapola, D.M., Oyama, M.D., Nobre, C.A. & Sampaio, G. (2008). A new world natural vegetation map for global change studies. *Anais da Academia Brasileira de Ciências*, Vol.80, No.2, (June 2008) 397-408, ISSN 0001-3765

Le Sueur, D., Binka, F., Lengeler, C., De Savigny, D., Snow, B., Teuscher, T., Toure, Y. (1997). An atlas of malaria in Africa. *African Health*, Vol.19, No.2, (January 1997), 23-24.

Lee KS, Divis PC, Zakaria SK, Matusop A, Julin RA, Conway DJ, Cox-Singh J, Singh B. *Plasmodium knowlesi*: reservoir hosts and tracking the emergence in humans and macaques. *PLoS Pathog.* 2011 Apr;7(4):e1002015. Epub 2011 Apr 7.

Liverman, D. (2009). *Suffering the Science. Climate change, people, and poverty*, Oxfam, ISBN XXXX, Boston

Machado-Ferreira E, Piesman J, Zeidner NS, Soares CA. Francisella-like endosymbiont DNA and Francisella tularensis virulence-related genes in Brazilian ticks (Acari: Ixodidae). J Med Entomol. 2009 Mar;46(2):369-74.

Magrin, G., Gay García, C., Cruz Choque, D., Giménez, J.C., Moreno, A.R., Nagy, G.J., Nobre, C. & Villamizar, A. (2007). Latin America, In: *Climate Change 2007: Impacts, Adaptation and Vulnerability. Contribution of Working Group II to the Fourth Assessment Report of the Intergovernmental Panel on Climate Change*, Parry, M.L., Canziani, O.F., Palutikof, J.P., van der Linden, P.J. & Hanson, C.E., (Ed), 581-615, Cambridge University Press, ISBN 978 0521 88009-1, Cambridge, UK,.

Martens, W.J., Slooff, R. & Jackson, E.K. (1997). Climate change, human health, and sustainable development. *Bulletin of the World Health Organization*, Vol.75, No.6, (1997) 583-588, ISSN 0042-9686

Mathers, C.D., Loncar, D. (2006) Projections of global mortality and burden of disease from 2002 to 2030. *PLoS Medicine*, Vol.3, No.11, (2006) e442, ISSN

Mas-Coma S, Valero MA, Bargues MD. Climate change effects on trematodiases, with emphasis on zoonotic fascioliasis and schistosomiasis. Vet Parasitol. 2009 Aug 26;163(4):264-80.

McMichael, A.J., Campbell-Lendrum, D.H., Corvalan, C.F., Ebi, K.L., Scheraga, J.D. & Woodwards, A. (2003). *Climate change and human health. Risk and responses*, World Health Organization, ISBN 92-4-156248-X, Geneva.

McMichael T, Montgomery H, Costello A. (2012). Health risks, present and future, from global climate change. *BMJ*, Vol.344, (March 2012), e1359.

Megid J, Appolinário CM, Langoni H, Pituco EM, Okuda LH. Vaccinia virus in humans and cattle in southwest region of Sao Paulo state, Brazil. Am J Trop Med Hyg. 2008 Nov;79(5):647-51.

Meinshausen, M., Meinshausen, N., Hare, W., Raper, S.C.B., Frieler, K., Knutti, R., et al. (2009) Greenhouse-gas emission targets for limiting global warming to 2°C. *Nature*, Vol.458, 1158-1162.

Meslin, F.X., K. Stohr & D. Heymann. (2000). Public health implications of emerging zoonoses. *Rev. Sci. Tech.* 19: 310–317.

Mills, D.M. (2009). Climate change, extreme weather events, and us health impacts: what can we say? *Journal of Occupational & Environmental Medicine*, Vol.51, No.1, (January 2009) 26-32, ISSN 1076-2752

Ministerio de Salud. (2010). *Perú: Incidencia de Fiebre Amarilla*. Ministerio de Salud de Perú, (Diciembre 2010).

Ministerio del Poder Popular para la Salud. (2011). *Anuario de Mortalidad 2009*. Ministerio del Poder Popular para la Salud de Venezuela, (November 2011).

Montenegro-James, S. (1992). Prevalence and control of babesiosis in the Americas. Mem Inst Oswaldo Cruz. 1992;87 Suppl 3:27-36.

Moulinier R, Martinez E, Torres J, Noya O, de Noya BA, Reyes O. Human proliferative sparganosis in Venezuela: report of a case. Am J Trop Med Hyg. 1982 Mar;31(2):358-63.

Naish S, Hu W, Nicholls N, Mackenzie JS, Dale P, McMichael AJ, Tong S. Socio-environmental predictors of Barmah forest virus transmission in coastal areas, Queensland, Australia. Trop Med Int Health. 2009 Feb;14(2):247-56

OPS. (2001). *Desigualdades en el acceso, uso y gasto con el agua potable en América Latina y el Caribe*, OPS, Washington, D.C.

Ortega García, J.A. (2007). El pediatra ante el cambio climático: desafíos y oportunidades. *Boletín de la Sociedad de Pediatría de Asturias, Cantabria, Castilla y León*, Vol.47, No.202, (January 2007) 331-343, ISSN 0214-2597

PAHO. (1988). *Hippocrates. Airs, waters, places. Pag. 18 Part I Historical development. The challenger of epidemiology. Issues and selected readings*, PAHO, Washington, D.C.

PAHO. (2003). *Protecting New Health Facilities from Natural Disasters: Guidelines for the Promotion of Disaster Mitigation*, PAHO, ISBN 92 75 124841, Washington, D.C.

PAHO. (2007). *Health in the Americas 2007. Volume I. Regional. Scientific and Thecnical Publication No. 622*, PAHO, Washington, D.C.

PAHO. (2008). Climate Change and Disaster Programs in the Health Sector. *Disasters: Preparedness and Mitigation in the Americas*, Vol.110, No.1, (October 2008) 1, 11, ISSN 1564-0701

Pappas, G. (2011). Of mice and men: defining, categorizing and understanding the significance of zoonotic infections. *Clinical Microbiology and Infection*, Vol.17, No.3, (January 2011), 321–325

Pappas, G., Cascio, A. & Rodriguez-Morales AJ. The immunology of zoonotic infections. *Clin Dev Immunol* 2012; 2012: 208508

Paris, L.A., Viscarret, M., Uban, C., Vargas, J. & Rodríguez-Morales, A.J. (2008). Pin-site myiasis: a rare complication of a treated open fracture of tibia. *Surgical Infections*, Vol.9, No.3, (June 2008) 403-406, ISSN 1096-2964

Parry, M.L., Canziani, O.F., Palutikof, J.P., van der Linden, P.J. & Hanson, C.E. (2007). *Climate Change 2007: Impacts, Adaptation and Vulnerability. Contribution of Working Group II to the Fourth Assessment Report of the Intergovernmental Panel on Climate Change*, Cambridge University Press, ISBN 9780521705974, Cambridge, United Kingdom and New York, NY, USA.

Patz, J.A., Campbell-Lendrum, D., Holloway T. & Foley, J.A. (2005). Impact of regional climate change on human health. *Nature*, Vol.438, No.7066, (November 2005) 310–317, ISSN 0028-0836

Peters, G.P., Marland, G., Le Quéré, C., Boden, T., Canadell, J.G., Raupach, M.R. (2012). Rapid growth in CO2 emissions after the 2008-2009 global financial crisis. *Nature Climate Change*, Vol.2, 2-4

Peterson, A.T., Martinez-Campos, C., Nakazawa, Y. & Martinez-Meyer, E. (2005). Time-specific ecological niche modeling predicts spatial dynamics of vector insects and human dengue cases. *Transactions of the Royal Society of Tropical Medicine and Hygiene*, Vol.99, No.9, (September 2005) 647-655, ISSN 0035-9203

Pinto, J., Bonacic, C., Hamilton-West, C., Romero, J. & Lubroth J. (2008). Climate change and animal diseases in South America. *Revue scientifique et technique (International Office of Epizootics)*, Vol.27, No.2, (August 2008) 599-613, ISSN 0253-1933

Poveda, G.J., Rojas, W., Quiñones, M.L., Vélez, I.D., Mantilla, R.I., Ruiz, D., Zuluaga, J.S. & Rua, G.L. (2001). Coupling between annual and ENSO theme scales in the malaria climate association in Colombia. *Environmental Health Perspectives*, Vol.109, No., (May 2001) 489-493, ISSN 0091-6765

Prüss-Üstün, A. & Corvalán, C. (1988). *Preventing disease through healthy environments*, WHO, Geneva.

Ramal, C., Vásquez, J., Magallanes, J. & Carey, C. (2009). Variabilidad climática y transmisión de malaria en Loreto, Perú: 1995-2007. *Revista Peruana de Medicina Experimental y Salud Pública*, Vol.26, No.1, (January 2009) 9-14, ISSN 1726-4634

Rahmstorf, S. (2010) A new view on sea level rise. *Nature Reports Climate Change*, Vol.4, 44-45.

Rifakis, P., Gonçalves, N., Omaña, W., Manso, M., Espidel, A., Intingaro, A., Hernández, O. & Rodríguez-Morales, A.J. (2005). Asociación entre las Variaciones Climáticas y los Casos de Dengue en un Hospital de Caracas, Venezuela, 1998-2004. *Revista Peruana de Medicina Experimental y Salud Pública*, Vol.22, No.3, (July 2005) 183-190, ISSN 1726-4634

Rifakis, P.M., Benitez, J.A., Rodriguez-Morales, A.J., Dickson, S.M. & De-La-Paz-Pineda, J. (2006). Ecoepidemiological and social factors related to rabies incidence in Venezuela during 2002-2004. *International Journal of Biomedical Science*, Vol.2, No.1, (February 2006) 3-7, ISSN 1550-9702

Rodríguez-Morales, A.J., Barbella, R.A., Cabaniel, G., Gutiérrez, G. & Blanco J.J. (2004). Influence of Climatic Variations on Yellow Fever Outbreaks In Venezuela, 2002-2003, *Proceedings of 20th Clinical Virology Symposium and Annual Meeting Pan American Society for Clinical Virology*, pp. TM12, ISBN 0000-0000, Clearwater Beach, Florida, USA, april 2004, Pan American Society for Clinical Virology, Clearwater Beach, Florida, USA

Rodríguez-Morales, A.J. (2005). Ecoepidemiología y Epidemiología Satelital: Nuevas Herramientas en el Manejo de Problemas en Salud Pública. *Revista Peruana de Medicina Experimental y Salud Pública*, Vol.22, No.1, (January 2005) 54-63, ISSN 1726-4634

Rodríguez-Morales, A.J. (2006). Enfermedades Olvidadas: Miasis. *Revista Peruana de Medicina Experimental y Salud Pública*, Vol.23, No.2, (April 2006) 143-144, ISSN 1726-4634

Rodriguez-Morales, A.J., Rodríguez, C. & Meijomil P. (2006). Climate Variability Influence and Seasonal Patterns of Gram-positive Cocci Infections in Western Caracas, 1992–2001. *International Journal of Infectious Diseases*, Vol.10, No.Suppl 1, (June 2006) S13-S14, ISSN 1201-9712

Rodríguez-Morales, A.J. (2008). Impacto potencial para la salud pública latinoamericana del lanzamiento y puesta en órbita del satélite VENESAT-1. *Revista Peruana de Medicina Experimental y Salud Pública*, Vol.25, No.4, (October 2008) 444-445, ISSN 1726-4634

Rodriguez-Morales, A.J. (2008b). Chagas disease: an emerging food-borne entity?. *Journal of Infection in Deveveloping Countries*, Vol.2, No.2, (April 2008), 149-150.

Rodríguez-Morales, A.J. (2009). Cambio climático y salud humana: enfermedades transmisibles y América Latina. *Revista Peruana de Medicina Experimental y Salud Pública*, Vol.26, No.2, (April 2009) 268-269, ISSN 1726-4634

Rodríguez-Morales, A.J., Echezuria, L., Risquez, A. (2010). Impact of Climate Change on Health and Disease in Latin America (Chapter 24). In: Simar S (Editor). Climate Change and Variability. Sciyo, Croatia, 463-486, ISBN 9789533071442

Rodríguez-Morales, A.J. (2011). Climate Change. In: Ogunseitan O (General Editor). Green Health – An A-toZ Guide [Encyclopedia]. Robbins P (Series Editor). The SAGE Reference Series on Green Society Toward a Sustainable Future. SAGE Publications, California, USA, 111-115, ISBN 9781412996884

Rohr JR, Dobson AP, Johnson PT, Kilpatrick AM, Paull SH, Raffel TR, Ruiz-Moreno D, Thomas MB. Frontiers in climate change-disease research. Trends Ecol Evol. 2011 Jun;26(6):270-7.

Ruiz MO, Chaves LF, Hamer GL, Sun T, Brown WM, Walker ED, Haramis L, Goldberg TL, Kitron UD. Local impact of temperature and precipitation on West Nile virus infection in *Culex* species mosquitoes in northeast Illinois, USA. Parasit Vectors. 2010 Mar 19;3(1):19.

Ruiz-Gómez J, Espinosa EL. Serum epidemiology of eastern, western and Venezuelan equine encephalitides in Mexico. Arch Invest Med (Mex). 1981;12(3):395-419.

Santos M, Ribeiro-Rodrigues R, Talhari C, Ferreira LC, Zelger B, Talhari S. Presence of Borrelia burgdorferi "Sensu Lato" in patients with morphea from the Amazonic region in Brazil. Int J Dermatol. 2011 Nov;50(11):1373-8.

Schreiber, K. V. (2001). An investigation of relationships between climate and dengue using a water budgeting technique. *International Journal of Biometeorology*, Vol.45, No.2, (July 2001) 81–89, ISSN 0020-7128

Sellers RF, Maarouf AR. Impact of climate on western equine encephalitis in Manitoba, Minnesota and North Dakota, 1980-1983. Epidemiol Infect. 1988 Dec;101(3):511-35.

Sukri, N.C., Laras, K., Wandra, T., Didi, S., Larasati, R.P., Rachdyatmaka, J.R., Osok, S., Tjia, P., Saragih, J.M., Hartati, S., Listyaningsih, E., Porter, K.R., Beckett, C.G., Prawira, I.S., Punjabi, N., Suparmanto, S.A., Beecham, H.J., Bangs, M.J. & Corwin, A.L. (2003). Transmission of epidemic dengue hemorrhagic fever in easternmost Indonesia. *American Journal of Tropical Medicine & Hygiene*, Vol.68, No.5, (May 2003) 529-535, ISSN 0002-9637

Tavares S. (2005). *La bioética, el agua y el saneamiento*, Editorial Disinlimed, Caracas.

Thong, H.Y. & Maibach, H.I. (2008). Global warming and its dermatologic implications. *International Journal of Dermatology*, Vol.47, No.5, (May 2008) 522-524, ISSN 0011-9059.

Thomson, M.C., Garcia-Herrera, R. & Beniston, M. (2008). *Seasonal Forecasts, Climatic Change and Human Health*, Springer Science, ISBN 978-1-4020-6876-8, New York.

United Nations. (2006). *Global Survey of Early Warning Systems*, United Nations, ISBN 9789027725523, New York.

United Nations Development Programme. (2008). Fighting climate change: Human solidarity in a divided world. New York: Oxford University Press; 2006.

van der Meide, W.F., Jensema, A.J., Akrum, R.A.E., Sabajo, L.O.A., Lai, A., Fat, R.F.M., Lambregts, L., Schallig, H.D.F.H., van der Paardt, M. & Faber, W.R. (2006). Epidemiology of Cutaneous Leishmaniasis in Suriname: A Study Performed in 2006. *American Journal of Tropical Medicine & Hygiene*, Vol.79, No.2, (February 2006) 192-197, ISSN 0002-9637

Vasconcelos, P.F., Costa, Z.G., Travassos Da Rosa, E.S., Luna, E., Rodrigues, S.G., Barros, V.L., Dias, J.P., Monteiro, H.A., Oliva, O.F., Vasconcelos, H.B., Oliveira, R.C., Sousa, M.R., Barbosa Da Silva, J., Cruz, A.C., Martins, E. C. & Travassos Da Rosa, J.F. (2001). Epidemic of jungle yellow fever in Brazil, 2000: implications of climatic alterations in disease spread. *Journal of Medical Virology*, Vol.65, No.3, (November 2001) 598–604, ISSN 0146-6615

Von, A., Zaragoza, E., Jones, D., Rodríguez-Morales, A.J., Franco-Paredes, C. (2007). New insights into Chagas Disease: a neglected disease in Latin America. *Journal of Infection in Developing Countries* Vol.1, No.2, (Oct 2007), 99-111.

WHO. (2003). *Climate change and human health – Risk and responses*, WHO, WMO, PNUMA.

WHO. (2009). *Facts on climate change and health*, PAHO, Washington, D.C.

Williams, R.J., Bryan, R.T., Mills, J.N., Palma, R.E., Vera, I. & de Velásquez, F. (1997). An outbreak of hantavirus pulmonary syndrome in western Paraguay. *American Journal of Tropical Medicine & Hygiene*, Vol.57, No.3, (September 1997) 274-282, ISSN 0002-9637

Woolhouse, M.E. & Gowtage-Sequeria, S. (2005). Host range and emerging and reemerging pathogens. *Emerging Infectious Diseases*, Vol.11, No.12, (December 2005) 1842-1847, ISSN 1080-6040

Zambrano, L.I., Sevilla, C., Reyes-García, S., Sierra, S., Kafati, R. & Rodriguez-Morales, A J. (2012). Potential impacts of climate variability on Dengue Hemorrhagic Fever in Honduras, 2010. *Tropical Biomedicine*, (accepted in press), ISSN 0127-5720.

Zeilhofer, P., Kummer, O.P., dos Santos, E.S., Ribeiro, A.L. & Missawa, N.A. (2008). Spatial modelling of *Lutzomyia (Nyssomyia) whitmani s.l.* (Antunes & Coutinho, 1939) (Diptera: Psychodidae: Phlebotominae) habitat suitability in the state of Mato Grosso, Brazil. *Memorias do Instituto Oswaldo Cruz*, Vol.103, No.7, (November 2008) 653-660, ISSN 0074-0276

Zell R, Krumbholz A, Wutzler P. Impact of global warming on viral diseases: what is the evidence? Curr Opin Biotechnol. 2008 Dec;19(6):652-60.

Chapter 13

Impact of Climate Change on the Geographic Scope of Diseases

Dziedzom K. de Souza, Priscilla N. Owusu and Michael D. Wilson

Additional information is available at the end of the chapter

1. Introduction

The Intergovernmental Panel on Climate Change [1] has shown that the climate will change, and provides different scenarios on what may happen and when, while the Millennium Ecosystem Assessment [2] has shown how the functioning of ecosystem services contribute to the maintenance or otherwise of human health. It is recognized that the main brunt of the effects of anthropogenic climate change will be borne by communities in developing countries, particularly in Africa. For example, the IPCC forecasts that some parts of Africa will become warmer and wetter, whereas others will become drier, and there will be higher frequencies of storms and floods. Changes in the distributions and amounts of rainfall will ensue, as water is a fundamental human requirement that also influences human wellbeing and health, and understanding how water systems will change is a pre-requisite to understanding potential future changes in disease epidemiology.

A possible scenario that could involve the spread of diseases into non-endemic regions is with the migration of infected people in places where vectors are already present but, as yet, there is no disease. For instance in South Africa, there are three newly described forms of S. *damnosum* (vectors of onchocerciasis) present, with one (the Pienaars form) [3] identified from sites where man-biting is known. This habit appears to be spreading in South Africa in the Johannesburg area and two separate sections of the Orange river [4]. Furthermore, it is known that there is extensive migration into South Africa (both legal and illegal) of people from countries where onchocerciasis is endemic. Such migration is likely to expand with increasing environmental degradation and poverty.

This chapter discusses how future climate change scenarios could influence disease transmission and distribution, for better or for worse; by delving into the past, present and future effects of climate on diseases using malaria as an example.It will also elaborate on the environmental factors affecting disease agents, the effects of climate on the physiological processes of pathogens and the host-pathogen interactions, vector diversity and vector borne diseases, as well as transmission zones related to disease distribution. Finally, the chapter examines the effects of climate change on migration and how this may impact disease distribution.

2. Changing patterns of disease distribution with climate: Past, present and future situation

The geographic distribution of diseases has revealed changes over history [5]. Several factors can be linked to the changes in disease distribution, but of interest to this review on climate change and disease distribution, are global warming, agricultural colonization, deforestation and reforestation. Changes in climate, either at the micro (country) or macro-geographical (continent) level influence the survival, reproduction, transmission of disease agents and vectors, and their interaction with the geophysical factors associated with climate: primarily precipitation, humidity, ambient and water temperature. A considerable range of diseases, including cholera [6], lymphatic filariasis [7] and tick-borne encephalitis [8] are affected by changing environmental conditions. However, alongside schistosomiasis and dengue, malaria is one of the diseases that has mostly been influenced by climate change events [9-11] and provides us with adequate information required to understand the roles that climate plays in driving disease prevalence and distribution.

Plasmodium falciparum malaria has had a long evolutionary history with humans [12-14], dating back to 10,000 years ago [15, 16], coinciding with human population growth and the change from hunter-gather's behavior to that of agriculture [17, 18]. The plausible distribution of malaria prior to intervention programs [19, 20], was believed to reach latitudinal extremes of 64° north and 32° south [21], at the turn of the 20th century. The 20th century, however, was undoubtedly a period of global climate change [1], and, using this era as a backdrop, many empirical and biological climate-malaria models [8, 10, 22-24] have sought to predict the future impact of climate change on the distribution of this disease. Climate variables such as temperature, humidity and rainfall affect the incidence and distribution of malaria, through changes in the vector and parasite life cycles and behavior [25, 26]. Studies in the Amazon Basin suggest that precipitation drives malaria incidence. However, this relationship varies in the uplands where more precipitation corresponds with more malaria, and is negative in areas dominated by wetlands and large rivers [27]. In determining the climate effects on malaria transmission, studies on *Anopheles stephensi* have shown that temperature affects sporogonic development of *P. falciparum* by altering the kinetics of ookinete maturation [28]. Ookinete development and blood meal digestion are lengthened as temperatures decrease from 27-21°C. Nevertheless, low temperatures (21–27°C) do not appear to significantly influence infection rates or densities of either ookinetes or oocysts. On the other hand, high temperatures (30 and 32°C) appeared to significantly impact parasite densities and infection rates by interfering with developmental processes occurring between parasite fertilization and ookinete formation, especially during zygote and early ookinete maturation [27]. Paaijmans and colleagues [29] further showed that the influence of climate on malaria transmission depends on daily temperature variations, with temperature fluctuations around low mean temperatures acting to speed up rate processes, whereas fluctuations around high mean temperatures act to slow down the processes. Despite all the evidence of climatic effects on malaria transmission, evidence-based maps of contemporary malaria endemicity [20] permit a re-evaluation of the changes in global epidemiology of the disease, when compared to the pre-intervention era [19] and enable an assessment of the observed changes in range and endemicity to those proposed to occur in response to climate change and observed under existing public health interventions [31].

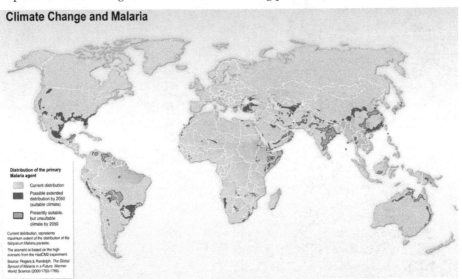

Figure 1. Climate change and malaria, scenario for 2050 (UNEP/GRID 2005). With climate conditions changing in the future, due to increased concentrations of carbon dioxide in the atmosphere, conditions for pests also change. The primary malaria agent, the *falciparum* malaria parasite, will be able to spread into new areas, as displayed in this map, by 2050 using the Hadley CM2 high scenario. Other areas, not displayed in the map, will be uninhabitable by the parasite.

Based on the biology of the vectors and their absence in regions that at present are too cool for their survival [10, 32], climate-malaria models depict an increase in the geographic range of the disease [33, 34], as temperature and global environmental conditions worsen. Figure 1, reproduced from UNEP/GRID 2005, represents the global distribution predictions of malaria by 2050. The changing climate conditions, due to increased concentrations of carbon dioxide in the atmosphere, will result in heat being trapped herein and thus modify the conditions for malaria vectors and *P. falciparum*, and malaria will be able to spread into new areas. However, despite the influence of climate on the distribution of the vectors, the predicted climate-malaria models hold true only if no actions are taken. Figure 2 reproduced from Gething *et al.* [31], represents the changing global malaria endemicity since the turn of the 20th century. Despite the global temperature increases in the 20th century [1, 35] the

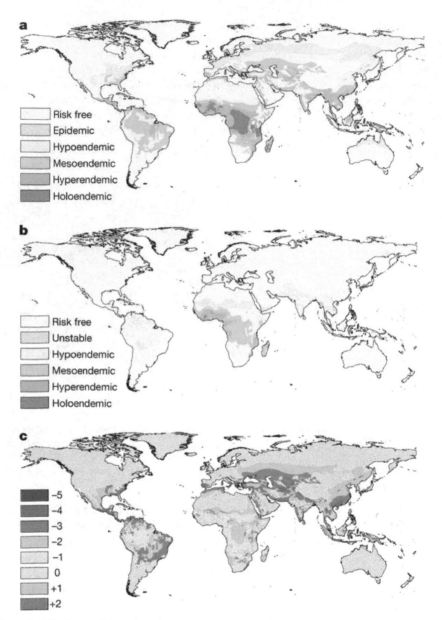

Figure 2. a) Pre-intervention endemicity (approximately 1900) as defined in Lysenko and Semashko [19]. b) Contemporary endemicity for 2007 based on a recent global project to define the limits and intensity of current *P. falciparum* transmission [30] and c) Change in endemicity class between 1900 and 2007. Negative and positive values denote reduction and increase respectively in endemicity. (Source: [31]).

figure depicts a marked, global decrease in the range and intensity of malaria transmission. The study also revealed that comparison of global-scale climate patterns with the historical and contemporary patterns of malaria endemicity, pointed out a disconnection in climate–malaria relationship over the twentieth century, indicating that non-climatic factors act as confounders of this relationship over time. Most malaria-climate models assume that: 1. all other factors affecting the distribution of malaria (such as disease and vector control efforts) either remain constant or have a relatively negligible effect; and 2. the link between climate and the global distribution and intensity of malaria is effectively immutable: an empirical climate–malaria relationship observed at one time period will be preserved even under changing climate and disease scenarios [31]. As such, contemporary endemicity maps provide a poor baseline for empirically-based predictions of future climate effects and for the climate-malaria predictions to hold true, the climate change effects must exceed the counteracting socio-economic development [36] and aggressive malaria control efforts [37-39].

3. Environmental factors affecting disease agents

Disease agents are affected by environmental factors, notably rainfall, humidity, and temperature. These, in turn, influence the incidence, distribution and spread of water-borne diseases. Water-borne and sanitation-related diseases are major contributors to global disease burden and mortality [40], especially in developing countries and in children under the age of 5 years. The main agents of these diseases are mostly viruses, bacteria, and parasitic worms [41]. Being mostly linked with water and sanitation, these diseases such as Cholera, typhoid, schistosomiasis, guinea worm and toxoplasmosis are prone to weather fluctuation events that affect their survival and dispersal through this medium.

Various studies have shown that many disease outbreaks associated with water are preceded by heavy rainfall events [42-48]. An example of this phenomenon is with the cryptosporidiosis outbreak associated with borehole extracted groundwater, where heavy rainfall led to running water from, fecally polluted, cattle grazing fields leaking into the borehole water supply [44]. Other studies have shown correlations between rainfall and the likelihood of detecting *Giardia* or *Cryptosporidium* oocysts in river water [49] and pathogenic enteric viruses in water [50], as well as coliform re-growth in water distribution systems, presumably because of increased nutrients in water [51].

Flooding, following heavy rainfall events, also poses substantial risk to the spread of infectious diseases -especially in developing countries where cases of leptospirosis, Hepatitis E, malaria and diarrheal diseases have been reported [52-57]. Most often, rainfall- and flood-associated outbreaks often come about as a result of contaminated water seeping into groundwater or water treatment systems [58] or the provision of suitable breeding grounds for disease vectors, in the case of malaria [59, 60].

Of all the climate change events, temperature poses the greatest threat to the spread of infectious diseases, as the increase in temperature has been linked to the blooming of various planktonic species [61-63] and pathogens [64, 65]. Planktonic blooms occur mostly in the summer months and the accumulation of toxins either through contact with water or food relating to these, such as blue-green algae [61] and dinoflagellates [63], results in cases being reported around these periods. However, the most evidence of the effect of temperature on risk from waterborne disease is in relation to cholera [64]. Evidence suggests that *Vibrio cholerae* survives in marine waters in a viable but non-cultural form that seems to be associated with algae and plankton [5, 64]. This is supported by observations that revealed that increases in sea-surface temperature, as a result of El-Nino events, have resulted in cholera outbreaks in Asia and South America [65, 67, 68].

Temperature does not only affect the growth of enteric pathogens, but also affects the spread of many viral, rickettsial, bacterial and parasitic diseases that can be transmitted by vectors [69]. Gubler and colleagues [25] list a range of possible mechanisms that may affect the risk of transmission of vector-borne diseases as a result of temperature changes. These mechanisms include the increase or decrease in vector survival, changes in rate of vector population growth, changes in feeding behavior, changes in susceptibility of vector to pathogens, changes in incubation period of pathogen, changes in seasonality of vector activity and changes in seasonality of pathogen transmission. These changes will be discussed in subsequent sections.

4. Climate change effects on pathogen strains, host-pathogen interactions and their physiological processes

Pathogen strain may play a significant role in the host-pathogen or vector-pathogen interactions. This section will try to reveal the latitudinal and temperamental changes on the diversity in pathogens and their interaction with their hosts and vectors.

Seasonal disease variations may come about as the result of many factors, including seasonally heightened host susceptibility resulting from seasonal stressors [70], changes in contact rates resulting from school terms in the case of childhood diseases [71] and seasonal changes in pathogen transmission rates resulting from climate variation [72]. Temperature, which is the main climate change driver, can affect disease by altering the susceptibility of hosts, the virulence of pathogens and the growth rates of both hosts and pathogens, which can in turn influence host pathology and disease emergence [73, 74]. The direct influences of climate variables (e.g. geography, temperature and rainfall) on the geographical range, growth rates of pathogens, variations in pathogen transmission rates and disease incidence is an active area of research in disease ecology [23, 64, 75, 76].

Cholera is an infectious disease endemic to many developing countries. It is caused by the bacterium *Vibrio Cholerae*. In Bangladesh, strains within the 01 serogroup of cholera are designated as being either of the "Classical" or the "El Tor" biotype, and both biotypes are affected by differing seasonal signatures with pronounced seasonality in disease incidence [72, 77]. The Classical strain was believed to be the dominant strain until the 1970s, when it was replaced by the El Tor strain which is more resilient to fluctuations in water quantity and quality, as evidenced by the less pronounced seasonal variability of its reproductive rates [70]. As such, Koelle and colleagues [70] have shown that the cholera strain sensitivity to environmental fluctuations can be considered a phenotypic character trait subject to evolution. Unlike the Classical, El Tor expresses *vps* (Vibrio polysaccharide) genes responsible for producing an exopolysaccharide that allows it to form a biofilm on abiotic surfaces, thus facilitating its environmental persistence [78]. However, the expression of this exopolysaccharide also appears to reduce intestinal colonization, and leads to a reduced infectivity and virulence [79], giving the Classical the advantage of easier infection and virulence. Despite the apparent evolutionary advantages or disadvantages, cholera cases due to both strains decrease during summer, and this is explained as the result of a dilution effect by the monsoon rains and the concurrent reduction in water salinity levels [80].

Small organisms tend to have faster generation times, stronger growth responses to temperature and wider thermal windows [81]. While it is easy to find variations in intra or extracellular bacterial pathogen strains, diversity in extracellular disease agents is not uncommon. For example, in West Africa, two geographic strains of *Onchocerca volvulus* (the parasite causing river blindness) have been observed and described as the forest and savannah forms [82], and these play important roles in the pathogenicity of the disease, as they transmit the severe –blinding form and the less severe, non-blinding onchocerciasis. The existence of geographic strains of parasites, influenced by environmental factors, has also been reported in *Wuchereria bancrofti* [83, 84].

As evidence suggests, climate-driven changes in interspecific interactions may lead to important consequences for host–pathogen relationships and disease emergence [85]. Because temperature patterns control growth and reproduction in a variety of organisms [86], changes in temperature are likely to influence the rate and timing of development of some species more strongly than others [87]. Thus, climate change will likely lead to both direct (i.e., physiological) and indirect (i.e., interspecific interactions) effects on parasite transmission, some of which may increase disease while others will reduce infection or pathology [87].

5. Vector diversity and vector borne diseases

The term "vector-borne diseases" describes illnesses in which pathogenic microorganisms such as viruses, bacteria, parasites, and fungi, are transmitted to humans or other animals through the feeding activity of blood- sucking arthropods.

The underlying mechanisms through which climate change may influence the infectivity of vector-borne diseases may be attributed to four main factors, including (i) expanding the range of vectors and hosts into new human populations, (ii) changes in vector or host population density. Furthermore, such changes in host/vector populations, as well as extreme environmental conditions affecting host immuno-competence may impinge on (iii) the frequency of infections, and (iv) pathogen load [88].

Under the second mechanism, high records of rainfall create convenient breeding grounds and increase food availability for vectors such as mosquitoes, ticks, and snails [88]. In South America, for example, following the abundant rainfall that marked the El Niño Southern Oscillation (ENSO) there came about a corresponding increase in *Aedes* mosquito populations [90, 91] while in Senegal and Niger, a lower incidence of malaria was attributed to diminished mosquito populations following reduced amounts of rainfall [92, 93].

Climate change results in the distribution of vectors into non-endemic areas, as well as their proliferation in endemic areas. Many vector-borne diseases are transmitted between a lower end range of 14-18°C and an upper end range of 35-40°C [89]. As lower end temperatures increase, the transmission periods of infectious diseases such as malaria, encephalitis, and dengue fever are enhanced or extended [94]. With malaria, the enhancement of transmission becomes imminent as temperature increases cause adult female mosquitoes to digest blood and feed more frequently, and as warmer waters shorten the incubation period of vector larvae [89]. Between the temperature ranges of 30-32°C, vectorial capacity increases substantially due to a decrease in the extrinsic incubation period, despite a reduction in the vector's survival rate. Mosquito species such as the *Anopheles gambiae, Anopheles funestus, Anopheles darlingi, Culex quinquefasciatus* and *Aedes aegypti* are responsible for transmission of most vector-borne diseases, and are sensitive to temperature changes as immature stages in the aquatic environment and as adults. Increases in water temperatures lead to faster maturation period for larvae [95] and a greater reproduction rate during disease transmission periods. In warmer climates, adult female mosquitoes digest blood faster and feed more frequently [96], thus increasing transmission intensity. Similarly, malaria parasites and viruses complete extrinsic incubation within the female mosquito in a shorter time as temperature rises [97], thereby increasing the proportion of infective vectors. It is also noted that warmer temperatures in temperate regions result in the "overwinter survivorship" of the North American deer mice, which are carriers of hantaviruses, in southwest United States following the 1997 ENSO [98]. On the other hand, extreme temperatures above the upper end generally disfavor the survival rate of vectors and pathogens [95], as witnessed in Senegal, where over the last three decades the incidence of malaria has been reduced by more than 60% [99]. Warming above 34°C generally has a negative impact on the survival of vectors and parasites [95]. Thus, as global temperature conditions worsen, there will be a shift in the occurrence and distribution of vector-borne diseases, as areas at the lower temperature range of disease transmission will see an increase in disease incidence, while areas at the upper temperature will see a decrease.

Stress, attributed to extreme environmental conditions, challenges the tolerance level of the hosts' immune system and makes them more susceptible to pathogen infections and recrudescence [88]. Temperature variations affect pathogen carrying capacity in arthropod vectors, and consequently, the epidemiology of infectious disease.

Moreover, climate change influences the seasonality of infectious disease patterns. Anomalies in weather patterns bring about irregularities in pathogen abundance, survival within both vectors and hosts, or virulence, impairing the accuracy of surveillance systems in predicting pandemic outbreaks [100]. It is hypothesized that seasonal changes in photoperiod affect vitamin D metabolism [101], and thus immune cell antibacterial and antiviral activity, leading to host immuno-suppression during the winter season. It is also hypothesized that in the northern and southern hemispheres, seasonal changes in photoperiods may bring about changes in flight and feeding activity of arthropod vectors to result in increased transmissibility of pathogens to hosts.

In terms of vector diversity, the environmental factors affecting the distribution of the vectors have the potential to results in their diversification. West Africa is one region with various ecological zones ranging from an extreme of dense forests, through the savannahs, to the other extreme of the Sahara desert. In this region, the *Anopheles gambiae* is highly diversified, with diversity reported in terms of sibling species, chromosomal and molecular forms. Studies have shown that environmental factors may predict the distribution of the observed diversity within this group of mosquitoes [102-104]. An important question, however, is the reason why these differences in the *An. gambiae* are only reported in west Africa and not anywhere else on the continent. The answer to this question may possibly be found in the existence of several ecological zones within this region. Similarly other examples of vector diversity exist as well. Until the discovery of cytotypes, *Simulium damnosum* was believed to be a single species, exhibiting little morphological variation, although differences in behavior and ecology were observed within the same area [105] and also behaved differently in different areas [106]. Other examples of diversity in vectors of diseases can be found in the *Glossina palpalis* group, *Lutzomyia longipalpis* and ixodid ticks.

6. Transmission zones related to disease distribution

Climate and temperature zones play an important role in the establishment of disease transmission zones. From the discussions and evidence presented in the sections above, it is becoming clearer that the effects of climate change will lead to the establishment of disease transmission risk areas, which will expand or decrease as the environmental factors become favorable or unfavorable for disease vectors and pathogens.

The spatial distribution of lymphatic filariasis and onchocerciasis can be classified in terms of high and low transmission areas. de Souza and colleagues [104] demonstrated the presence of low, medium and high lymphatic filariasis transmission zones in Ghana, influenced by the presence/absence of efficient vectors and the effects of environmental factors, specifically temperature. Transmission could also be described in terms of hypo/meso/hyperendemicity for onchocerciasis, depending on the level of prevalence and the risks of morbidity [107]. The importance of these transmission zones lies in the disease control options and the possible re-infection of areas previously declared as non-endemic, from areas of high or medium transmission zones.

Perhaps the idea of transmission zones could best be described using the example of meningitis. Meningococcal meningitis (MCM) is an infection of the meninges, caused by the bacteria *Neisseria meningitidis*, that causes high death rates in African communities.

Although epidemics of MCM occur worldwide [108], the conditions associated with the onset of the epidemic in sub-Saharan Africa, and the occurrence in both space and time of MCM cases and climate variability within the Sahel-Sudan area, result in the establishment of a disease transmission zone or belt. The geographical distribution of disease cases is called the "Meningitis Belt" and is more or less circumscribed to the bio-geographical Sahel-Sudan band [109, 110]. The transmission period in West Africa is climate dependent [111] and usually start at the beginning of February -with the harmattan winds, and then disappear in late May -with the onset of rainfall. Thus, the environmental dependability of MCM epidemics enables the establishment of transmission risk models which can be useful for directing surveillance activities.

7. Climate change and migration

Historically, the migration patterns of the early modern hominids out of Africa (some 100, 000 years ago) was believed to have coincided with climate related events and wet phases in the Sahara/Sahel region of North Africa [112-114]. This early migration may be linked with the possible co-migration of diseases to other parts of the world. Thus the early migration and disease distribution notion can be supported by studies that established that lice (which transmit diseases such as typhus, trench and relapsing fevers) had accompanied their human hosts in the original peopling of the Americas, near the end of the Pleistocene [115], probably as early as 10,000-15,000 years ago. Millions of people are forced to migrate when climate change brings about environmental disasters [116]. It is estimated that by 2050, between 25 million to 1 billion people will have been displaced as a result of climate change [117]. The impact of climate change on the environment comes about as a result of a) climate processes, characterized by the rise in sea levels, salinization of agricultural land, desertification, and drought, and b) climate events, such as flooding, storms, hurricanes, and typhoons [118]. Global climate change affects all parts of the world, however, the world's poorest regions, particularly in sub-Saharan Africa, are the hardest hit, since they lack adequate social and economic structures to enable them to cope with climate change induced environmental disasters and its associated problems.

Drought threatens agricultural productivity, and thereby, food security and economic stability in these poor world regions, as their economies are heavily reliant on climate-dependent activities such as farming and fishing. Many people are thus forced to travel outside their local communities in search of better conditions. A common pattern of climate migration is for individuals or groups to travel where environmental conditions are favorable. In West Africa, for example, drought seasons compel young men and women to travel outside their communities to engage in paid labor so as to increase the family income, as well as send remittances to support the rest of the family. The distance travelled by these climate migrants is either internal or external, depending on the family's resources, whereby they are enabled to only move to neighboring cities, or as far as Europe [118]. Typically, aside financial considerations, other factors such as existing social networks, past colonial relationships, and language determine how far migrants would travel. After the 2005

Hurricane Katrina disaster evacuees from the Gulf regions did not "stream across" to neighboring Mexico, but rather sought refuge elsewhere within the United States [118].

Many climate migration effects on disease distribution have been suggested. Over the short term, it has been suggested that climate change forced-migration will make the achievement of the Millennium Development Goals (MDGs) harder [118]. The provision of uninterrupted health services that underlie goals 4 and 5 of the MDGs (reducing child and maternal mortality and combating HIV/Aids, malaria and other diseases), is also likely to be affected as a result of large-scale climate migration [118]. The displacement of populations as a result of climate effects undermines the provision of medical care and vaccination programs; making infectious diseases harder to deal with and more deadly. It is well documented that refugee populations suffer worse health outcomes than settled populations [118]. Another effect of climate change migration is the spread of diseases as a result of overcrowding. One billion people, live in urban slums: in poor quality housing with limited clean water, sanitation and health services [118]. It is estimated that by 2030 this number will rise to 1.7 billion people [119]. This figure may even be higher, as it is estimated that 78 million people may be displaced by climate change by 2030 [120], and 250 million by 2050 [121]. The high population densities and high contact rates could help to spread disease, while health and education services are often inadequate. As an example, the spread of dengue fever in India has been associated with unplanned urbanization [122].

The most imminent impact of climate migration is the introduction of pathogens into new regions, thereby expanding the geographic range of infectious diseases. Northeastern Brazil, for instance, faces the challenge of outbreaks of visceral leishmaniasis as a consequence of migration during drought seasons [123]. Another example can be given of South Africa, where there are three newly described forms of *S. damnosum* present, with one identified from sites where man-biting is known [3]. This biting habit appears to be spreading in South Africa in the Johannesburg area and two separate sections of the Orange river [4]. Furthermore, it is known that there is extensive migration into South Africa (both legal and illegal) of people from countries where onchocerciasis is endemic. For instance, officially, 24,627 Nigerians entered South Africa in 2004, together with many from other onchocerciasis zones such as Cameroon and the Democratic Republic of the Congo (South Africa Statistics Statistical release P0351). Also, the spread of infectious diseases through migration becomes more serious and difficult to control as natives may have low level or non-immune resistance to newly introduced pathogens [124]. The converse may be true, where migrants into endemic regions may lack the resistance against pathogens. Rapid urban growth places heavy burdens on medical facilities, leads to poor sanitation, crowded housing situations, while the increase in human-human contact facilitates the spread of pathogens within urban communities. Nonetheless, climate migration is occurring at a time of unprecedented pressure on natural resources, especially water and arable land, as well as on social and health services [125].

The argument for disease distribution as a result of climate migration may not be attributed to humans alone, but also to animals and their roles in the transmission of zoonotic diseases.

In West Africa, the migration of the pastoral Fulanis has been a result many factors, including climate change [126, 127]. Cattle and livestock play an important role in the transmission of zoonoses including; anthrax, brucellosis, cryptosporidiosis, giardiasis and E. coli. Other animals such as wild birds also share with humans the capacity for moving over large distances, travelling across national and intercontinental boundaries. During migratory movements, birds have the potential of carrying and dispersing pathogens that can be dangerous for both humans and of course other animals and birds [128-130]. This spread of pathogens occurs at multiple geographic scales, consequently influencing disease dynamics [131-134]. Many disease pathogens can be spread in this manner and these include; viruses (e.g., West Nile, Sindbis, Newcastle), bacteria (e.g., borrelia, mycobacteria, salmonellae), and protozoa (e.g., cryptosporidia) [128]. However, the highly pathogenic avian influenza (H5N1) pandemic that started in China in 2005 [135] may be considered as the number one disease that opened up the world of migratory birds and the spread of diseases. In discussing the role of migratory birds in disease distribution, it is important to note that as the climate changes, so does the probability of zoonotic diseases increase. For example, studies have revealed that increasing temperatures alter bird migration patterns [136], and this may affect their duration of stay (influenced by food availability and suitable breeding conditions), thereby increasing the chances of transmitting disease pathogens.

8. Conclusion

In conclusion, most studies predict an increase in temperature - as ascribed to global warming, carbon dioxide concentration, rainfall, drought, and humidity. These factors influence the complex interactions within the public health triad comprising the environment, human host and disease vectors or pathogens. Although the effects may not be universal, the suitability of new environments will lead to the spread of disease vectors and pathogens into new areas, whereas the unsuitability (due to drought, for instance) may lead people to migrate into suitable areas, thereby introducing diseases in those areas.

Author details

Dziedzom K. de Souza*, Priscilla N. Owusu and Michael D. Wilson
Parasitology Department, Noguchi Memorial Institute for Medical Research, University of Ghana, Legon-Accra, Ghana

9. References

[1] Intergovernmental Panel on Climate Change (2007) Climate Change: The Physical Science Basis. Contribution of Working Group I to the Fourth Assessment Report of the Intergovernmental Panel on Climate Change (eds Solomon S. et al.) (Cambridge Univ. Press, 2007).

* Corresponding Author

[2] Millennium Ecosystem Assessment (2005) Ecosystems and Human Well-being: General Synthesis. Washington, DC: Island Press and World Resources Institute.

[3] Krüger A, Car M., Maegga BTA (2005) Descriptions of members of the *Simulium damnosum* complex (Diptera: Simuliidae) from southern Africa, Ethiopia and Tanzania. *Annals of Tropical Medicine and Parasitology*. 99: 293-306.

[4] Palmer RW, De Moor FC (1999) *Simulium damnosum* s.l. complex widespread in Southern Africa. *British Simulium Group Bulletin* 14: 10-14.

[5] Cliff A, Haggett P (1988) Atlas of disease distributions, analytical approaches to epidemiological data. Oxford: Blackwell.

[6] Pascual M, Bouma M, Dobson AP (2002) Cholera and climate: revisiting the quantitative evidence. Microbes Infect. 4(2): 237–245.

[7] Sattenspiel L (2000) Tropical environments, human activities, and the transmission of infectious diseases. Am J Phys Anthropol. Suppl 31:3–31.

[8] Randolph SE, Rogers DJ (2000) Fragile transmission cycles of tick-borne encephalitis virus may be disrupted by predicted climate change. Proc R Soc Lond B. 267: 1741–1744.

[9] Martens P (1998) Health and Climate Change: Modelling the Impacts of Global Warming and Ozone Depletion. London: Earthscan Publications Ltd.

[10] Martens P, Kovats RS, Nijhof S, de Vries P, Livermore MTJ, Bradley DJ, Cox J, McMichael AJ (1999) Climate change and future populations at risk of malaria. Global Environ Change. 9 (Suppl 1): S89–S107.

[11] Rogers DJ, Randolph SE (2000) The global spread of malaria in a future, warmer world. Science. 289: 1763–1766.

[12] Wiesenfeld SL (1967) Sickle-cell trait in human biological and cultural evolution. Science. 157: 1134–40.

[13] Coluzzi M (1999) The clay feet of the malaria giant and its African roots: hypotheses and inferences about origin, spread and control of *Plasmodium falciparum*. Parassitologia. 41: 277–83.

[14] Joy DA, Feng XR, Mu JB, Furuya T, Chotivanich K, Krettli AU, Ho M, Wang A, White NJ, Suh E, Beerli P, Su XZ (2003) Early origin and recent expansion of *Plasmodium falciparum*. Science. 300: 318–21.

[15] Hume JCC, Lyons EJ, Day KP (2003) Human migration, mosquitoes and the evolution of *Plasmodium falciparum*. Trends Parasitol. 19: 144–49.

[16] Hart DL (2004) The origin of malaria: mixed messages from genetic diversity. Nature Reviews Microbiol. 2: 15–22.

[17] De Zulueta J (1987) Changes in the geographical distribution of malaria throughout history. Parassitologia. 29: 193–205.

[18] De Zulueta J (1994) Malaria and ecosystems: from prehistory to posteradication. Parassitologia. 36: 7–15.

[19] Lysenko AJ, Semashko IN (1968) Geography of malaria. A medico-geographic profile of an ancient disease. In: Lebedew AW, editor. Itogi Nauki: Medicinskaja Geografija. Academy of Sciences, USSR; Moscow. pp. 25-146.

[20] Russell PF (1956) World-wide malaria distribution, prevalence and control. Am J Trop Med Hyg. 5: 937–56.

[21] Snow RW, Gilles HM (2002) The epidemiology of malaria. In: Warrell DA, Gilles HM, editors. Essential malariology. 4th Edn. Arnold; London. pp. 85-106.

[22] Parham PE, Michael E (2010) Modeling the effects of weather and climate change on malaria transmission. Environ. Health Perspect. 118: 620–626.

[23] Pascual M, Ahumada JA, Chaves LF, Rodo X, Bouma M (2006) Malaria resurgence in the East African highlands: temperature trends revisited. Proc. Natl. Acad. Sci. U.S.A. 103(15): 5829–5834.

[24] van Lieshout M, Kovats RS, Livermore MTJ, Martens P (2004) Climate change and malaria: analysis of the SRES climate and socio-economic scenarios. Global Environ. Change-Hum. Policy Dimensions 14: 87–99.

[25] Gubler DJ, Reiter P, Ebi KL, Yap W, Nasci R, Patz JA (2001) Climate variability and change in the United States: potential impacts on vector- and rodent-borne diseases. Environmental Health Perpectives. 109(suppl 2): 223–233.

[26] Koenraadt CJM, Githeko AK, Takken W (2004) The effects of rainfall and evapotranspiration on the temporal dynamics of *Anopheles gambiae* s.s. and Anopheles arabiensis in a Kenyan village. Acta Tropica. 90(2): 141–153.

[27] Olson SH, Gangnon R, Elguero E, Durieux L, Guégan JF, Foley JA, Patz JA (2009). Links between climate, malaria, and wetlands in the Amazon Basin. Emerging Infectious Diseases. 15(4): 659-662.

[28] Noden BH, Kent MD, Beier JC (1995) The impact of variations in temperature on early *Plasmodium falciparum* development in *Anopheles stephensi. Parasitology, 111(5): 539-545.*

[29] Paaijmans KP, Blanford S, Bell AS, Blanford JI, Read AF, Thomas MB (2010) Influence of climate on malaria transmission depends on daily temperature variation. PNAS 107(34): 15135–15139.

[30] Hay SI, Guerra CA, Gething PW, Patil AP, Tatem AJ, Noor AM, Kabaria CW, Manh BH, Elyazar IR, Brooker S, Smith DL, Moyeed RA, Snow RW (2009) A world malaria map: *Plasmodium falciparum* endemicity in 2007. PLoS Med. 6(3): e1000048.

[31] Gething PW, Smith DL, Patil AP, Tatem AJ, Snow RW, Hay SI (2010) Climate change and the global malaria recession. Nature. 465: 342–345.

[32] Martens WJM, Niessen LW, Rotmans J, Jetten TH, McMichael AJ (1995) Potential impact of global climate change on malaria risk. *Environ. Health Perspect.* 103(5), 458-464.

[33] Martin MH, Lefebvre GM (1995) Malaria and climate: sensitivity of malaria potential transmission to climate. *Ambio* 24: 200-207.

[34] Lindsay SW, Martens WJM (1998) Malaria in the African highlands: past, present and future. Bulletin of the World Health Organization. 76(1): 33-45.

[35] Small J, Goetz SJ, Hay SI (2003) Climatic suitability for malaria transmission in Africa, 1911–1995. Proc. Natl Acad. Sci. USA 100: 15341–15345.

[36] Beguin A, Hales S, Rocklov J, Astrom C, Louis VR, Sauerborn R (2011) The opposing effects of climate change and socio-economic development on the global distribution of malaria. *Global Environmental Change.* 21: 1209–1214.

[37] Fillinger U, Ndenga B, Githeko A, Lindsay SW (2009) Integrated malaria vector control with microbial larvicides and insecticide-treated nets in western Kenya: a controlled trial. Bull. World Health Organ. 87: 655–665.

[38] Kleinschmidt I, Sharp B, Benavente LE, Schwabe C, Torrez M, Kuklinski J, Morris N, Raman J, Carter J (2006) Reduction in infection with Plasmodium falciparum one year after the introduction of malaria control interventions on Bioko Island, Equatorial Guinea. Am. J. Trop. Med. Hyg. 74(6): 972–978.

[39] Bhattarai A, Ali AS, Kachur SP, Mårtensson A, Abbas AK, Khatib R, Al-Mafazy AW, Ramsan M, Rotllant G, Gerstenmaier JF, Molteni F, Abdulla S, Montgomery SM, Kaneko A, Björkman A (2007) Impact of artemisinin-based combination therapy and insecticide-treated nets on malaria burden in Zanzibar. PLoS Med. 4(11): e309.

[40] Pruss A, Havelaar A (2001) The global burden of disease study and applications in water, sanitation and hygiene. In: Fewtrell L, Bartram J, editors. Water Quality: Guidelines, Standards and Health. London: IWA Publishing. pp. 43–59.

[41] Hunter PR (1997) Waterborne Disease: Epidemiology and Ecology. Chichester: Wiley.

[42] Smith HV, Patterson WJ, Hardie R, Greene LA, Benton C, Tulloch W, Gilmour RA, Girdwood RWA, Sharp JCM, Forbes GI (1989) An outbreak of cryptosporidiosis caused by post-treatment contamination. Epidemiology and Infection. 103: 703–715.

[43] Joseph C, Hamilton G, O'Connor M, Nicholas S, Marshall R, Stanwell-Smith R, Sims R, Ndawula E, Casemore D, Gallagher P, Harnett P (1991) Cryptosporidiosis in the Isle of Thanet: an outbreak associated with local drinking water. Epidemiology and Infection. 107: 509–519.

[44] Bridgman S, Robertson RMP, Syed Q, Speed N, Andrews N, Hunter PR (1995) Outbreak of Cryptosporidiosis associated with a disinfected groundwater supply. Epidemiology and Infection. 115: 555–566.

[45] Willocks L, Crampin A, Milne L, Seng C, Susman M, Gair R, Moulsdale M, Shafi S, Wall R, Wiggins R, Lightfoot N (1998) A large outbreak of cryptosporidiosis associated with a public water supply from a deep chalk borehole. Communicable Disease and Public Health. 1: 239–243.

[46] Anon (2000) Waterborne outbreak of gastroenteritis associated with a contaminated municipal water supply, Walkerton, Ontario, May– June 2000. Canada Communicable Disease Report. 26: 170–173.

[47] Curriero FC, Patz JA, Rose JB, Lele S (2001) The association between extreme precipitation and waterborne disease outbreaks in the United States, 1948–1994. American Journal of Public Health. 91: 1194–1199.

[48] Miettinen IT, Zacheus O, von Bonsdorff CH, Vartiainen T (2001) Waterborne epidemics in Finland in 1998-1999. Water Science and Technology. 43(12): 67–71.

[49] Atherbolt TB, LeChevallier MW, Norton WD, Rosen JS (1998) Effect of rainfall on *Giardia* and *Cryptosporidium*. Journal of the American Water Works Association. 90: 66–80.

[50] Miossec L, Le Guyader F, Haugarreau L, Pommepuy M (2000) Magnitude of rainfall on viral contamination of the marine environment during gastroenteritis epidemics in human coastal population. Revue d Epidemiologie et de Sante Publique. 48(suppl 2): 2S62–2S71.

[51] LeChevallier MW, Schulz W, Lee RG (1991) Bacterial nutrients in drinking water. Applied and Environmental Microbiology, 57: 857–862.

[52] Barcellos C, Sabroza PC (2001). The place behind the case: leptospirosis risks and associated environmental conditions in a flood-related outbreak in Rio de Janeiro. Cadernos de Saude Publica. 17(suppl): 59–67.

[53] Easton A (1999) Leptospirosis in Philippine floods. British Medical Journal. 319: 212.

[54] Homeida M, Ismail AA, El Tom I, Mahmoud B, Ali HM (1988) Resistant malaria and the Sudan floods. Lancet. 2(8616): 912.

[55] Novelli V, El Tohami TA, Osundwa VM, Ashong F (1988) Floods and resistant malaria. Lancet 2(8624): 1367.

[56] Shears P (1988) The Khartoum floods and diarrhoeal diseases. Lancet. 2: 517.

[57] McCarthy MC, He J, Hyams KC, el-Tigani A, Khalid IO, Carl M (1994) Acute hepatitis E infection during the 1988 floods in Khartoum, Sudan. Transactions of the Royal Society of Tropical Medicine & Hygiene. 88: 177.

[58] Hunter PR (2003) Climate change and waterborne and vectorborne disease. J. Appl. Microbiol. 94(S), 37S–46S.

[59] Bouma MJ, Dye C, Van der Kaay HJ (1996) Falciparum malaria and climate change in the North West Frontier Province of Pakistan. Am J Trop Med Hyg. 55: 131–137.

[60] Bouma MJ, Proveda G, Rojas W, Chavasse D, Quinones M, Cox J, Patz J (1997) Predicting high-risk years for malaria in Columbia using parameters of El-Nino southern oscillation. Trop Med Int Health. 2: 1122–1127.

[61] Hunter PR (1998) Cyanobacterial toxins and human health. Journal of Applied Bacteriology. 84(suppl): 35S–40S.

[62] Morris Jr JG (1999) Pfiesteria, "the cell from hell", and other toxic nightmares. Clinical Infectious Diseases. 28: 1191–1196.

[63] Hungerford JM (2001) Seafood toxins. In: Labbe RG, Garcia S, editors. Guide to Foodborne Pathogens. New York: John Wiley and Sons. pp. 267–283.

[64] Colwell R (1996) Global climate and infectious disease: the cholera paradigm. Science 274: 2025–2031.

[65] Lobitz B, Beck L, Huq A, Wood B, Fuchs G, Faruque ASG, Colwell R (2000) Climate and infectious disease: use of remote sensing for detection of *Vibrio cholerae* by indirect measurement. Proceedings of the National Academy of Science. 97: 1438–1443.

[66] Islam MS, Draser BS, Bradley DJ (1990) Long-term persistence of toxigenic *Vibrio cholerae* O1 in the mucilaginous sheath of a blue-green alga, *Anabaena variabilis*. Journal of Tropical Medicine and Hygiene. 93: 133–139.

[67] Pascual M, Rodo X, Ellner SP, Colwell R, Bouma MJ (2000) Cholera dynamics and El-Nino-Southern Oscillation. Science. 289: 1766–1769.

[68] Speelmon EC, Checkley W, Gilman RH, Patz J, Caleron M, Manga S (2000) Cholera incidence and El-Nino-related higher ambient temperature. Journal of the American Medical Association. 283: 3072–3074.

[69] Cook GC (1996) Manson's Tropical Diseases, 20th edn. London: WB Saunders

[70] Koelle K, Pascual M, Yunus M (2005) Pathogen adaptation to seasonal forcing and climate change. *Proc. R. Soc. B.* 2: 971-977.

[71] Fine PE, Clarkson JA (1982) Measles in England and Wales—I: an analysis of factors underlying seasonal patterns. Int. J. Epidemiol. 11: 5–14.

[72] Spira WM (1981) Environmental factors in diarrhea transmission: the ecology of Vibrio cholerae O1 and cholera. In: Holme T, Holmgren J, Merson MH, Mollby R, editors. Acute enteric infections in children: new prospects for treatment and prevention Amsterdam: Elsevier.

[73] Cairns MA, Ebersole JL, Baker JP, Wigington PJ, Lavigne HR, Davis SM (2005) Influence of summer stream temperatures on black spot infestation of juvenile coho salmon in the Oregon Coast Range. Transactions of the American Fisheries Society. 134: 1471–1479.

[74] Raffel TR, Rohr JR, Kiesecker JM, Hudson PJ (2006) Negative effects of changing temperature on amphibian immunity under field conditions. Functional Ecology. 20: 819–828.

[75] Kutz SJ, Hoberg EP, Polley L, Jenkins EJ (2005) Global warming is changing the dynamics of Arctic host–parasite systems. Proceedings of the Royal Society B-Biological Sciences. 272: 2571–2576.

[76] Bradbury J (2003) Beyond the fire-hazard mentality of medicine: the ecology of infectious diseases. PLoS Biol. 1: 148–151.

[77] Glass RI, Becker S, Huq SI, Stoll BJ, Khan MU, Merson MH, Lee JV, Black RE (1982) Endemic cholera in rural Bangladesh, 1966–1980. Am. J. Epidemiol. 116: 959–970.

[78] Reidl J, Klose KE (2002) Vibrio cholerae and cholera: out of the water and into the host. FEMS Microbiol. Rev. 26: 125–139.

[79] Watnick PI, Lauriano CM, Klose KE, Croal L, Kolter R (2001) The absence of a flagellum leads to altered colony morphology, biofilm development, and virulence in Vibrio cholerae 0139. Mol. Microbiol. 39: 223–235.

[80] Miller CJ, Drasar BS, Feachem RG (1984) Responses of toxigenic Vibrio cholerae 01 to physicochemical stresses in aquatic environments. J. Hyg. Camb. 93: 475–495.

[81] Portner HO (2002) Climate variations and the physiological basis of temperature dependent biogeography: systemic to molecular hierarchy of thermal tolerance in animals. Comp Biochem Physiol A 132: 739-761.

[82] Zimmerman PA, Dadzie KY, De Sole G, Remme E, Alley ES, Unnasch TR (1992). *Onchocerca volvulus* DNA Probe Classification Correlates with Epidemiologic Patterns of Blindness. J Infect Dis 165: 964-968.

[83] Kumar NP, Patra KP, Hoti Sl, Das PK (2002). Genetic variability of the human filarial parasite, *Wuchereria bancrofti* in South India. Acta Tropica 82: 67-76.

[84] Thangadurai R, Hoti SL, Kumar NP, Das PK (2006). Phylogeography of human lymphatic filarial parasite, Wuchereria bancrofti in India. Acta Tropica 98: 297-304.

[85] Gilman SE, Urban MC, Tewksbury J, Gilchrist GW, Holt RD (2010) A framework for community interactions under climate change. Trends in Ecology and Evolution. 25: 325–331.

[86] Stenseth NC, Mysterud A (2002) Climate, changing phenology, and other life history and traits: nonlinearity and match-mismatch to the environment. Proceedings of the National Academy of Sciences of the United States of America. 99: 13379–13381.

[87] Paull SH, Johnson PTJ (2011). High temperature enhances host pathology in a snail–trematode system: possible consequences of climate change for the emergence of disease. Freshwater Biology. 56: 767–778.

[88] Mills JN, Gage KL, Khan AS (2010) Potential influence of climate change on vector-borne and zoonotic diseases: a review and proposed research plan. *Environ Health Perspect.* 118:1507–1514.

[89] Githeko AK, Lindsay SW, Confalonieri UE, Patz JA (2000) Climate change and vector-borne diseases: a regional analysis. *Bulletin of the World Health Organization.* 78:9

[90] Anyamba A, Linthicum KJ, Tucker CJ (2001) Climate-disease connections: Rift Valley fever in Kenya. *Cad Saude Publica.* 17(suppl):133–140.

[91] Linthicum KJ, Anyamba A, Tucker CJ, Kelley PW, Myers MF, Peters CJ (1999) Climate and satellite indicators to forecast Rift Valley fever epidemics in Kenya. *Science.* 285:397–400.

[92] Julvez J, Mouchet J, Michault A, Fouta A, Hamidine M (1997) The progress of malaria in Sahelian eastern Niger. An ecological disaster zone. *Bull Soc Pathol Exot.* 90:101–104.

[93] Mouchet J, Faye O, Juivez J, Manguin S (1996) Drought and malaria retreat in the Sahel, West Africa. *Lancet.* 348:1735–1736.

[94] Patz JA, Riesen WK (2001) Immunology, climate change and vector-borne diseases. *TRENDS in Immun.* 22:4.

[95] Rueda LM, Patel KJ, Axtell RC, Stinner RR (1990) Temperature-dependent development and survival rates of *Culex quinquefasciatus* and *Aedes aegypti* (Diptera: Culicidae). *Journal of Medical Entomology.* 27: 892-898.

[96] Gillies MT (1953) The duration of the gonotrophic cycle in *Anopheles gambiae* and An. *funestus* with a note on the efficiency of hand catching. *East African Medical Journal.* 30: 129-135.

[97] Turell MJ (1989) Effects of environmental temperature on the vector competence of *Aedes fowleri* for Rift Valley fever virus. *Research in Virology.* 140: 147-154.

[98] Yates TL, Mills JN, Parmenter CA, Ksiazek TG, Parmenter RR, Vande Castle JR, Calisher CH, Nichol ST, Abbott KD, Young JC, Morrison ML, Beaty B, Dunnum JL, Baker RJ, Salazar-Bravo J, Peters CJ (2002) The ecology and evolutionary history of an emergent disease: hantavirus pulmonary syndrome. *Bioscience* 52:989–998.

[99] Faye O, Gaye O, Fontenille D, Hébrard G, Konate L, Sy N, Herve JP, Toure YT, Diallo S, Molez JF (1995) Malaria decrease and drought in the Niayes area of Senegal. *Sante´* 5: 299–305.

[100] Fisman DN (2007) Seasonality of infectious diseases. *Annu. Rev. Public Health.* 28:127–43.

[101] Cannell JJ, Vieth R, Umhau JC, Holick MF, Grant WB, Madronich S, Garland CF, Giovannucci E (2006) Epidemic influenza and vitamin D. *Epidemiol. Infect.* 134:1129–40.

[102] Matthews SD, Meehan LJ, Onyabe DY, Vineis J, Nock I, Ndams I, Conn JE (2007) Evidence for late Pleistocene population expansion of the malarial mosquitoes, *Anopheles arabiensis* and *Anopheles gambiae* in Nigeria. *Medical and Veterinary. Entomology.* 21: 358-469.

[103] Yawson AE, Weetman D, Wilson MD, Donnelly J (2007) Ecological zones rather than molecular forms predict genetic differentiation in the malaria vector *Anopheles gambiae* s.s. in Ghana. *Genetics.* 175: 751-761.

[104] de Souza D, Kelly-Hope L, Lawson B, Wilson M, Boakye D (2010) Environmental factors associated with the distribution of *Anopheles gambiae* s.s in Ghana; an important vector of lymphatic filariasis and malaria in West Africa. *PLoS ONE.* 5(3):e9927.

[105] Service MW (1982) Importance of vector ecology in vector disease control in Africa. *Bull. Soc. Vector Ecol.* 7: 1-13.

[106] Freeman P, de Meillon B (1953) Simuliidae of the Ethiopian Region. *British Museum of Natural History, London,* 224 pp.

[107] Katabarwa MN, Eyamba A, Chouaibou M, Enyong P, Kuete T, Yaya S, Yougouda A, Baldiagai J, Madi K, Andze GO, Richards F (2010) Does onchocerciasis transmission take place in hypoendemic areas? A study from the North Region of Cameroon. *Trop. Med. Int. Health.* 15: 645–652.

[108] Molesworth AM, Cuevas LE, Connor SJ, Morse AP, Thomson MC (2003) Environmental Risk and Meningitis Epidemics in Africa. *Emerging Infectious Diseases*. 9(10): 1287-1293.

[109] Lapeyssonnie L (1963) [The meningococcal meningitis in Africa] (in French). Bull World Health Organ. 28-3: 114 p.

[110] Cheesbrough JS, Morse AP, Green SDR (1995) Meningococcal meningitis and carriage in western Zaire: A hypoendemic zone related to climate? Epidemiol Infect. 114: 75–92.

[111] Sultan B, Labadi K, Guegan JF, Janicot S (2005) Climate drives the meningitis epidemics onset in West Africa. PLoS Med 2(1): e6.

[112] Castaneda IS, Mulitza S, Schefuß E, dos Santos RAL, Damste' JSS, Schouten S (2009) Wet phases in the Sahara/Sahel region and human migration patterns in North Africa. Available: www.pnas.org/cgi/doi/10.1073/pnas.0905771106.

[113] Carto SL, Weaver AJ, Hetherington R, Lam Y, Wiebe E (2009) Out of Africa and into an ice age: On the role of global climate change in the late Pleistocene migration of early modern humans out of Africa. J Hum Evol 56:139–161.

[114] Osborne AH, Vance D, Rohling EJ, Barton N, Rogerson M, Fello N (2008) A humid corridor across the Sahara for the migration of early modern humans out of Africa 120,000 years ago. Proc Natl Acad Sci USA 105:16444 –16447.

[115] Raoult D, Reed DL, Dittmar K, Kirchman JJ, Rolain JM, Guillen S, Light JE (2008) Molecular Identification of Lice from Pre-Columbian Mummies. Journal of Infectious Diseases 197 (4): 535-543

[116] Lonergan S (1998) The role of environmental degradation in population displacement. *Environmental Change and Security Project Report*. 4: 6.

[117] Stern N (2006) The economics of climate change: the Stern review. *Cambridge University Press*, Cambridge, 56.

[118] Brown O (2008) Migration and climate change. *International Organization for Migration*. Available: http://www.iisd.org/pdf/2008/migration_climate.pdf. Accessed 2012 Feb 29.

[119] Sclar ED, Garau P, Carolini G (2005) The 21st century health challenge of slums and cities. The Lancet 365: 901-903.

[120] Global Humanitarian Forum (2009): The Anatomy of a Silent Crisis. Geneva: Global Humanitarian Forum

[121] Christian Aid (2007): Human tide: The real migration crisis. Christian Aid Report. London. http://www.christianaid.org.uk/Images/human-tide.pdf. Accessed: 6/11/2012).

[122] Shah I, Deshpande GC, Tardeja PN (2004) Outbreak of dengue in Mumbai and predictive markers for dengue shock syndrome. J. Trop. Pediatrics, 50: 301-305.

[123] Franke CR, Ziller M, Staubach C, Latif M (2002) Impact of the El Niño Oscillation on Visceral Leishmaniasis, Brazil. *Emerging Infectious Diseases*. 8(9): 914-7

[124] Population Action International (2011) Why population matters to infectious diseases and HIV/AIDS. Available: http://populationaction.org/wpcontent/uploads/2012/02/PAI-1293-DISEASE_compressed.pdf. Accessed 2012 Feb 29.

[125] International Centre for Migration, Health and Development (2010) Climate change, migration, and health. A changing universe. Available: http://icmhd.wordpress.com/2010/08/19/climate-change-migration-and-health/. Accessed 2012 Feb 29.

[126] Adebayo AG (1991) Of Man and Cattle: A Reconsideration of the Traditions of Origin of Pastoral Fulani of Nigeria. History in Africa, Vol. 18, pp. 1-21

[127] Morrissey J (2009) Environmental Change and Forced Migration: A State of the Art Review, Refugee Studies Centre, Oxford Department of International Development, Queen Elizabeth House, University of Oxford, Oxford.

[128] Jourdain E., Gauthier-Clerc M., Bicout D., Sabatier P (2007) Bird Migration Routes and Risk for Pathogen Dispersion into Western Mediterranean Wetlands. Emerging Infectious Diseases 13(3): 365-372.

[129] Reed KD, Meece JK, Henkel JS, Shukla SK (2003) Birds, migration and emerging zoonoses: West Nile virus, Lyme disease, influenza A and enteropathogens.Clin Med Res. 1:5–12.

[130] Hubalek Z (2004) An annoted checklist of pathogenic microorganisms associated with migratory birds. J Wildl Dis. 40:639–59.

[131] Kilpatrick AM, Chmura AA, Gibbons DW, Fleischer RC, Marra PP, Daszak P. (2006) Predicting the global spread of H5N1 avian influenza. Proc Natl Acad Sci U S A 103: 19368–19373.

[132] Marra PP, Griffing S, Caffrey C, Kilpatrick AM, McLean R, Band C, Saito E, Dupuis AP, Kramer L, Novak R (2004) West Nile virus and wildlife. Bioscience 54: 393–402.

[133] Altizer S, Bartel R, Han BA (2011) Animal migration and infectious disease risk. Science 331: 296–302.

[134] Gunnarsson G, Latorre-Margalef N, Hobson KA, Van Wilgenburg SL, Elmberg J, Olsen B, Fouchier RA, Waldenström J (2012) Disease Dynamics and Bird Migration — Linking Mallards Anas platyrhynchos and Subtype Diversity of the Influenza A Virus in Time and Space. PLoS ONE 7(4): e35679.

[135] Normile D (2006). Avian influenza. Evidence points to migratory birds in H5N1 spread. Science 311:1225.

[136] Hurlbert AH, Liang Z (2012) Spatiotemporal Variation in Avian Migration Phenology: Citizen Science Reveals Effects of Climate Change. PLoS ONE 7 (2): e31662

Permissions

The contributors of this book come from diverse backgrounds, making this book a truly international effort. This book will bring forth new frontiers with its revolutionizing research information and detailed analysis of the nascent developments around the world.

We would like to thank Prof. Netra Chhetri, for lending his expertise to make the book truly unique. He has played a crucial role in the development of this book. Without his invaluable contribution this book wouldn't have been possible. He has made vital efforts to compile up to date information on the varied aspects of this subject to make this book a valuable addition to the collection of many professionals and students.

This book was conceptualized with the vision of imparting up-to-date information and advanced data in this field. To ensure the same, a matchless editorial board was set up. Every individual on the board went through rigorous rounds of assessment to prove their worth. After which they invested a large part of their time researching and compiling the most relevant data for our readers. Conferences and sessions were held from time to time between the editorial board and the contributing authors to present the data in the most comprehensible form. The editorial team has worked tirelessly to provide valuable and valid information to help people across the globe.

Every chapter published in this book has been scrutinized by our experts. Their significance has been extensively debated. The topics covered herein carry significant findings which will fuel the growth of the discipline. They may even be implemented as practical applications or may be referred to as a beginning point for another development. Chapters in this book were first published by InTech; hereby published with permission under the Creative Commons Attribution License or equivalent.

The editorial board has been involved in producing this book since its inception. They have spent rigorous hours researching and exploring the diverse topics which have resulted in the successful publishing of this book. They have passed on their knowledge of decades through this book. To expedite this challenging task, the publisher supported the team at every step. A small team of assistant editors was also appointed to further simplify the editing procedure and attain best results for the readers.

Our editorial team has been hand-picked from every corner of the world. Their multi-ethnicity adds dynamic inputs to the discussions which result in innovative outcomes. These outcomes are then further discussed with the researchers and contributors who give their valuable feedback and opinion regarding the same. The feedback is then collaborated with the researches and they are edited in a comprehensive manner to aid the understanding of the subject.

Apart from the editorial board, the designing team has also invested a significant amount of their time in understanding the subject and creating the most relevant covers. They scrutinized every image to scout for the most suitable representation of the subject and create an appropriate cover for the book.

The publishing team has been involved in this book since its early stages. They were actively engaged in every process, be it collecting the data, connecting with the contributors or procuring relevant information. The team has been an ardent support to the editorial, designing and production team. Their endless efforts to recruit the best for this project, has resulted in the accomplishment of this book. They are a veteran in the field of academics and their pool of knowledge is as vast as their experience in printing. Their expertise and guidance has proved useful at every step. Their uncompromising quality standards have made this book an exceptional effort. Their encouragement from time to time has been an inspiration for everyone.

The publisher and the editorial board hope that this book will prove to be a valuable piece of knowledge for researchers, students, practitioners and scholars across the globe.

List of Contributors

Pashupati Chaudhary, Keshab Thapa, Krishna Lamsal and Puspa Raj Tiwari
Local Initiatives for Biodiversity, Research and Development (LI-BIRD), Pokhara, Kaski, Nepal

Netra Chhetri
School of Geographical Sciences and Urban Planning the Consortium for Science, Policy and Outcomes, Arizona State University, USA

Ann Marie Raymondi
School of Sustainability, Arizona State University, Tempe, USA

Sabrina Delgado Arias
Consortium for Science Policy and Outcomes, Arizona State University, Washington DC, USA

Renée C. Elder
School of Geographical Sciences and Urban Planning, Arizona State University, Tempe, USA

Molly E. Brown
NASA Goddard Space Flight Center, Greenbelt MD, USA

Vanessa M. Escobar
Sigma Space/NASA Goddard Space Flight Center, Greenbelt, MD, USA

Heather Lovell
School of GeoSciences, The University of Edinburgh, UK

Attila Fur and Flora Ijjas
Budapest University of Technology and Economics, Department of Environmental Economics, Hungary

Wilhelm Kuttler
Applied Climatology and Landscape Ecology, University of Duisburg-Essen, Germany

Tayfun Kindap, Alper Unal, Deniz Bozkurt and Mehmet Karaca
Istanbul Technical University, Eurasia Institute of Earth Sciences, Turkey

Huseyin Ozdemir and Goksel Demir
Bahcesehir University, Environmental Engineering Department, Turkey

Ufuk Utku Turuncoglu
Istanbul Technical University, Informatics Institute, Turkey

Huseyin Ozdemir and Mete Tayanc
Marmara University, Environmental Engineering Department, Turkey

Andrew Chen
Cox School of Business, Southern Methodist University, Dallas, Texas, USA

Jennifer Warren
Dallas Committee on Foreign Relations, Dallas, Texas, USA, Cox School of Business, Southern Methodist University, Dallas, Texas, USA, Concept Elemental, Dallas, Texas, USA

Kihwan Seo
School of Geographical Sciences and Urban Planning, Arizona State University, USA

Natalia Rodriguez
School of Sustainability, Arizona State University, USA

Nedka Ivanova and Plamen Mishev
University of National and World Economy, Sofia, Bulgaria

Subana Shanmuganathan, Ajit Narayanan and Philip Sallis
Auckland University of Technology, Auckland, New Zealand

Majid Mathlouthi
Research Laboratory in Science and Technology of Water in INAT, Tunis, Tunisia

Fethi Lebdi
National Agronomic Institute of Tunisia (INAT), University of Carthage, Tunis, Tunisia

Alfonso J. Rodríguez-Morales
Research Group Infection and Immunity and Department of Community Medicine, Faculty of Health Sciences, Universidad Tecnológica de Pereira (UTP), Office of Scientific Research, Cooperativa de Entidades de Salud de Risaralda (COODESURIS), Working Group on Zoonoses, International Society for Chemotherapy (ISC) and Committee on Zoonoses and Haemorrhagic Fevers, Asociación Colombiana de Infectología (ACIN), Pereira, Risaralda, Colombia
Instituto Experimental José Witremundo Torrealba, Universidad de Los Andes, Trujillo, Venezuela

Carlos A. Delgado-López
Faculty of Sciences for Health, Universidad de Caldas, Manizales, Caldas, Colombia

Dziedzom K. de Souza, Priscilla N. Owusu and Michael D. Wilson
Parasitology Department, Noguchi Memorial Institute for Medical Research, University of Ghana, Legon-Accra, Ghana